P9-EMO-643

Also by Judith Miller

God Has Ninety-nine Names: Reporting from a Militant Middle East

Saddam Hussein and the Crisis in the Gulf (with Laurie Mylroie)

One, by One, by One: Facing the Holocaust

Also by William Broad

The Universe Below: Discovering the Secrets of the Deep Sea

Teller's War: The Top-Secret Story Behind the Star Wars Deception

Star Warriors: A Penetrating Look into the Lives of the Young Scientists Behind Our Space Age Weaponry

Betrayers of the Truth: Fraud and Deceit in the Halls of Science (with Nicholas Wade)

Judith Miller
Stephen Engelberg
William Broad

GERMS

Biological Weapons and America's Secret War

SIMON & SCHUSTER

New York   London   Toronto   Sydney   Singapore

SIMON & SCHUSTER
Rockefeller Center
1230 Avenue of the Americas
New York, NY 10020

Book design by Susan Hood

Manufactured in the United States of America

9 10 8

Library of Congress Cataloging-in-Publication Data
Miller, Judith.
Germs: Biological weapons and America's secret war /
Judith Miller, Stephen Engelberg, William Broad.
p. cm.
Includes bibliographical references and index.
1. Biological warfare—Safety measures—United States.
2. Biological warfare—Research—United States.
I. Engelberg, Stephen. II. Broad, William J. III. Title.
UG447.8.M54 2001
358'.38'0973—dc21 2001042690

ISBN 0-684-87158-0

For Tanya, Gabrielle, and Jason

Contents

. . . untune that string,
And, hark, what discord follows!

William Shakespeare,
Troilus and Cressida

Preface

In December 1997, six years after the Persian Gulf War, the Pentagon announced that it had decided to vaccinate its 2.4 million soldiers and reservists against anthrax. It seemed to be a curious move. Saddam Hussein's biological weapons program had been exposed more than two years earlier. So what had changed? Was the decision driven by a new, even more frightening danger from germ weapons? Was the Clinton administration looking for an international issue on which it could appear tough?

We set out to explore for the *New York Times* what had motivated the decision. From the beginning, we worked as a team: a science writer with a knowledge of weapons, a veteran foreign correspondent who had tracked international terrorism, and an editor who had investigated the intelligence agencies and the Pentagon. We were skeptical, well aware of how federal agencies often inflate such dangers to justify their existence and budgets.

We quickly learned that the anthrax decision was part of a much larger government effort to combat what officials believed was a growing danger from germ weapons. Over the next three years, we followed the story from Washington to Kazakhstan to Japan to Russia, eventually deciding to write this book. The issues were as complex and intellectually challenging as any we have ever examined, cutting across science, intelligence, and foreign affairs. We came to see the debates of the 1990s over what to do about germ weapons in a much larger context—a half century of largely secret history.

That history is framed by science. The biological advances of the latter twentieth century had vanquished diseases and extended the human life span. But these discoveries also gave unprecedented power to those who would use germs to destroy. For more than twenty-five years, the United States and the Soviet Union competed in a race to develop biological arms. The threat seemed to vanish in 1972 when the world's nations signed a treaty banning them. But germ weapons did not disappear. No sooner was the treaty signed than the Soviets secretly decided to expand their program on a vast, industrial scale. Inevitably, as scientific knowledge spread, other nations, terrorist groups, cults, and even individuals dreamed of acquiring their own biological bombs.

We learned that a small group of experts had prodded the Clinton administration to take the threat of germ weapons more seriously. Their campaign gained momentum from a series of catastrophic events and some startling intelligence reports from Iraq, Russia, and elsewhere.

Our story begins with a terrifying incident that at the time went largely unnoticed.

1

The Attack

IT was noon on Sunday, September 9, 1984. Parishioners from Saint Mary's were drifting into Shakey's Pizza in The Dalles, Oregon. Dave Lutgens and his wife, Sandy, who had bought the simple eatery with its wooden tables and linoleum floors in 1977, were sharing a pizza with Dan Ericksen, a friend who served on Wasco County's land-use planning commission. They were discussing the topic that had come to obsess nearly everyone in this stable community of ten thousand in the spectacular Columbia River Gorge, not far from snow-capped Mount Hood: the growing tensions between the county and its controversial newcomers, a religious cult known as the Rajneeshees.

In 1981, followers of the Bhagwan Shree (meaning "Sir God" in Sanskrit) Rajneesh had paid $5.75 million for a remote sixty-four-thousand-acre ranch in Wasco County, a two-hour drive from The Dalles, the county seat. Their plan was to build a "Buddhafield," an agricultural commune in which they could celebrate their "enlightened master's" credo of beauty, love, and guiltless sex. Detractors said it was the sex, not the meditation, that had attracted thousands of followers, many of them wealthy Westerners, first to Poona, India, the commune's original home, and then to Oregon. The group had left Poona under growing political pressure stemming from reports that its leaders made money not only from followers but also from drugs and other illicit activities, charges the cult denied. But the guru, a balding, bearded Indian with a religious charlatan's permanent smile, preached that it was blessed to be rich. He owned a collection of diamond-studded watches and ninety Rolls-Royces.

His devoted followers, the "sannyasins," dressed only in shades of red, sang while working ten to twelve hours a day, every day, at the ranch. In only three years, they had built a small city on this once barren land, which now supported dozens of modular buildings and mobile homes; a 2.2-acre meeting hall; a 160-room hotel; a two-block-long shopping mall; a casino and disco; a dam and a lake; networks of new roads; sophisticated water, sewage, and transportation systems; an airstrip on the valley floor for the sect's five jet planes and helicopter; and a seemingly thriving community of four thousand. When the group sold tickets to its annual summer festivals, the numbers swelled to fourteen thousand. Some locals had tried to stop what they called the cult's "idolatrous orgies," but the state's tradition of tolerance prevailed.

Like most county officials, Dan Ericksen didn't like the group. There were plenty of reasons to be suspicious of them. The Rajneeshees had built their community on land zoned primarily for agriculture. They had harassed local residents who opposed their expansion plans and had threatened neighbors who had initially welcomed them but had become alarmed by the group's aggressiveness.

In 1982 the sannyasins had moved into the neighboring town of Antelope, whose population had stood at seventy-five before the influx. After winning electoral control of the town council, they had ruled that all of its meetings must begin and end with a joke, and they had infuriated residents by insisting on taking over the local school. They had renamed the town Rajneesh and turned Antelope's sole business, a combination store, restaurant, and gas station, into a vegetarian health-food café called Zorba the Buddha. Locals had responded with bumper stickers that proclaimed BETTER DEAD THAN RED and MONEY CAN'T BUY ANTELOPE'S HERITAGE.

The Bhagwan's followers also created a separate city within the ranch's borders, which they called Rajneeshpuram, and controlled zoning there too. They created their own city police department—the 60-member "Peace Force," which they equipped with many weapons and military gadgetry. Oregonians who drove down county public roads on Rajneesh property complained of being stopped and mistreated by the Bhagwan's police.

Incorporation gave the Rajneeshee police access to state-run law-enforcement training programs and Oregon's crime-data networks. But the FBI, which was investigating civil-rights complaints against the group from several county residents, denied the sect access to the sensitive information available on its National Crime Information Center database.

Now the commune was trying to expand further, again in violation of zoning and legal restrictions, by inviting onto the ranch some three thousand homeless people from New York and other cities throughout the country under its Share-a-Home program. Exploiting Oregon's liberal voter-registration laws, the Rajneeshees wanted to win electoral control of the county commission and other posts by registering the homeless to vote in the November 1984 elections. Once the Rajneeshees controlled Wasco County, Ericksen and other locals feared, nothing would stop them.

The Rajneeshees had deluged Ericksen's planning commission and the county commission with requests, petitions, and lawsuits. When the county commission challenged their demands, they responded with harsh attacks and dire, unspecified threats.

Though he traced his own family's roots back to Oregon's fur-trapping founders, Lutgens felt that the area needed new blood. The Rajneeshees worked hard and attracted educated, talented followers—lawyers, doctors, engineers. And they spent money. They had invested more than $35 million in the ranch since their arrival. This was a blessing in The Dalles, whose main industry, an aluminum smelter, had shut down two years earlier. Lutgens also knew that a lot of beer had been delivered out to the ranch. Local farmers, who grew sweet cherries and produce, and other businessmen had also sold the commune equipment and supplies.

They may have been weird, he thought, but the Rajneeshees seemed determined to make their patch of desert bloom. Set in a steep-sided valley forged by two streams and surrounded by rocky cliffs and high rolling hills, the property was parched by blistering heat in summer and sometimes flooded by winter downpours that turned its volcanic soil instantly into gooey clay—which was why locals called the ranch the Muddy. Yet the sect had filled the overgrazed land with vegetable gardens and fruit orchards.

Sannyasins occasionally ate at Dave Lutgens's restaurant. They were vegetarians and liked his salad bar, which was always well stocked with fresh vegetables, mixed salads, several varieties of lettuce, garnishes, and dressings. Lutgens liked it too, and went to the bar to get a large helping of macaroni for himself and his wife and some salad for Dan.

THE stomach cramps began later that day. They were mild at first. Lutgens felt them along with nausea as he sat at his cash register, counting the afternoon's take. By early evening, the symptoms were too severe to ignore. Dizzy and disoriented, he could barely make it across his bedroom to the toilet. Chills, fever, and intense diarrhea and vomiting left him weak and dehydrated.

Two days later, his wife started getting ill, and Dan Ericksen was even sicker than they were. He had been taken to Mid-Columbia Medical Center, the town's only hospital. By week's end, thirteen of Lutgens's twenty-eight employees were sick. And dozens of his customers had also called to complain that they had gotten violently ill after eating at his restaurant. Some were threatening to sue him.

ON September 17, the Wasco-Sherman Public Health Department received a call from someone who complained of gastroenteritis after eating at another restaurant in The Dalles. In the next few days, the department received at least twenty more complaints, involving two more restaurants. Less than forty-eight hours after the outbreak began, a pathologist at Mid-Columbia Medical Center had determined from a patient's stool that the bacteria making people sick was salmonella, one of nature's hardiest germs, though infection usually is not fatal. On September 21, within four days of the first report, scientists at the Oregon State Public Health Laboratory in Portland analyzed the stool samples further and identified the bacteria as *Salmonella typhimurium,* a common agent in food poisoning. In this case, they found they were looking at a very unusual strain that was treatable with most antibiotics. Since there were roughly twenty-five hundred known strains of salmonella, this was speedy scientific sleuthing. But by late September, reports of new cases were diminishing.

Carla Chamberlain, a no-nonsense nurse who headed the county public health office, knew that between 1980 and 1983, the department had reported only sixteen isolates of salmonella, eight of which were *Salmonella typhimurium*. None resembled this strain. When the wave began to crest, Chamberlain thought the outbreak was over.

In fact, the citizens of The Dalles were under biological siege. Once the rod-shaped bacteria entered a victim's body, the invaders multiplied wildly over hours and days, damaging tissues and overwhelming rival bacteria. Their main weapons were toxins and sticky hairs on their cell walls that let them seize hold of the mucous membranes of the colon and small intestine and then force their way inside. The toxins caused the intestines to exude waves of watery fluid. Typically, abdominal pain began anywhere from twelve to forty-eight hours after infection, followed by diarrhea, chills, fever, and sometimes vomiting. Lasting up to four days, salmonellosis could be severe in the young, the elderly, and patients with weakened resistance. The diarrhea could be life-threatening if dehydration was not treated aggressively with fluids.

On September 21, the day after the Oregon state lab identified the salmonella strain that had caused the initial outbreak, Chamberlain's office began receiving a second wave of reports involving people who had fallen ill at ten different restaurants in The Dalles. Because of its location on Interstate 84, a major east-west highway, the town had a disproportionately large number of restaurants—some thirty-five in all.

Arthur Van Eaton, the pathologist at Mid-Columbia, and his small staff were overwhelmed with patients and work. Their new laboratory, only a year old, was stacked high with specimens destined for the state's lab in Portland, ninety miles away. Normally, the lab used up one shipment of media, the broth in which specimens are cultured to see what kind of bacteria will grow, every two or three weeks. But during the outbreak's second wave, the lab went through three shipments a week. The twenty petri dishes of tests in a normal week mushroomed to two hundred every other day. At the peak of reports, the lab ran out of media altogether.

For the first time ever, all of Mid-Columbia's 125 beds were filled; some patients had to be kept in corridors. Many were angry and hostile, and very frightened; doctors had difficulty treating them. Violent patients and their families demanded their test results; some even threw stool and urine samples at the hospital's doctors and technicians.

By the end of the outbreak, almost a thousand people had reported symptoms to their doctors or the hospital; 751 were confirmed to have salmonella, making it the largest outbreak in Oregon's history. No one had died, miraculously. But a pregnant woman had given birth prematurely, and her baby was suffering from the poison's effects.

Meanwhile, Chamberlain and her small staff were struggling. A colleague was inspecting restaurants and talking to their owners and employees, while Chamberlain interviewed patients to identify their common denominators.

On September 25, in the midst of the second wave, the state sought help from the U.S. Centers for Disease Control and Prevention, based in Atlanta, and its Epidemic Intelligence Service, seventy mostly young doctors who learned epidemiology firsthand by investigating suspicious disease outbreaks throughout the country. Founded during the early days of the Cold War, the EIS's original mission was to help detect a germ attack against the United States.

By the time Thomas Török, a new EIS officer who had done his medical internship in Eugene, reached Oregon, sixty cases of *Salmonella typhimurium* had already been confirmed. Chamberlain's office had found a connection among the patients: most of those who had gotten sick had recently eaten at a salad bar. That very day, the local health sanitarian recommended that restaurants voluntarily discontinue salad-bar service. All did so.

Assisted by Chamberlain's office, more than twenty local public-health workers, including local sanitarians, state health officials, Török, and three other EIS officers, began an exhaustive epidemiological investigation of the outbreak. There was no mainframe computer in the county health office. Until an EIS doctor rigged up a primitive portable computer, interview material and collected data were all recorded and collated by hand.

The investigators interviewed hundreds of patients, as well as their families and friends. They tracked down out-of-state visitors who had paid for meals with credit cards to see how they felt and ask what they had eaten. They talked to all of the 325 food handlers who worked at the ten restaurants; about 100 of them had been infected, many of them falling ill before their patrons. They measured salad-bar temperatures and inspected food-handling practices. They visited an uncertified dairy in neighboring Washington State and tested the cows, cow feces, raw milk, and even the farm's pond water for salmonella. They didn't find it.

They checked the two local water systems that supplied The Dalles for contamination, as well as water at the restaurants. They visited a farm that had sold cucumbers and tomatoes to one restaurant and discovered that a nearby trailer court had suffered from septic-tank malfunctions in early September. Tests, showed however, that the adjacent vegetable patch had not been contaminated.

Another farm had provided cantaloupes to a restaurant whose patrons had gotten sick. But inspectors found that all the melons had been harvested and sold. There were none left to test.

Several suspect food items that had been served at affected restaurants were sampled and tested. Again, the tests were negative. Interviews with 120 people who had ordered home delivery showed that none of them had fallen ill. Nor had people who had eaten food served at banquets. It was the people who had eaten from salad bars—or who had ordered side dishes of mixed salads like macaroni or potato salad—who had become sick.

There was no common source for the food. The lettuce had come from different suppliers; so had the other vegetables. The salad dressings were from different wholesalers. Each item was traced back to its source. The investigators even checked for contamination in the kale that one restaurant had used as decoration on the bar. They found nothing. They did find salmonella in the milk in the coffee creamers in one of the restaurants and in the blue-cheese dressing of another, though not in the dry mix that had been used to prepare the dressing. This suggested that the dressing had been contaminated during or after its preparation. But how, why, and by whom?

ONE man in The Dalles was sure that the outbreak was not natural. Judge William Hulse, the head of the three-member county commission that ruled on contested land-use issues, among other things, feared that the Rajneeshees had poisoned his fellow citizens with salmonella, because he and another commissioner had come down with it a day after they had visited the commune the previous year.

Hospital records showed that Hulse had nearly died, though doctors did not know from what. They had not tested him for salmonella. The trip to the ranch had been unpleasant—but he had expected that. The commissioners were doing a mandatory inspection of the ranch before its annual summer festival. Ma Anand Sheela, the Bhagwan's personal secretary and the commune's de facto leader, had instructed him and another commissioner to get into the back of the van that had taken them around the ashram.

When the men had returned to their car, one of the tires was flat. While the Rajneeshees changed it, they offered the commissioners paper cups of water. Eight hours later, they became violently ill with symptoms that were virtually identical to Dave Lutgens's. Hulse and the other commissioner suspected that the Rajneeshees must have put something in their water, but they had no proof. So neither filed a complaint, requested an investigation, nor took further action until a year later, when their neighbors and friends started getting sick.

Hulse shared his suspicions with his colleague Carla Chamberlain. She had visited the ranch to discuss the county's health-reporting requirements and knew that the Rajneesh medical lab was better equipped than the county's. Many people in town were suspicious because they, too, had experienced run-ins with the cultists. But Hulse and Chamberlain agreed that they would have difficulty convincing Tom Török, his team of federal disease investigators, or even state health officials from Portland or Salem who had not dealt directly with the cult that the outbreak was deliberate. Among the most skeptical was Török's mentor, Laurence R. Foster, the most senior state epidemiologist and a widely respected figure in the medical community. Foster, a civil libertarian who came from the region, ardently believed that the Rajneeshees were being unfairly harassed because of their strange religious beliefs.

Relations between the cult and the county had deteriorated dramatically by the fall of 1984. Rajneeshee officials were now openly insulting their critics, calling them idiots, bigots, rednecks, and liars. Sheela and other commune leaders had been shown on television carrying weapons and threatening revenge against the "United States of Aggression" if sannyasins were hurt or their commune's plans thwarted.

County officials had received threats in the mail and had been threatened at public meetings. Sannyasins had watched their homes and offices, jotting down names and license-plate numbers of visitors. And the cult, which had retained many of the state's most prominent lawyers, had deluged them with lawsuits. Karen LeBreton, the deputy county clerk, estimated that more than 60 percent of her work involved responding to the cult's suits and petitions. The local people's terror was laced with the frustration of knowing that many outsiders believed that they were intolerant hicks.

The perception that the Rajhneeshees were not to blame for the outbreak was reinforced by state public-health officials and the CDC in the fall of 1984 and early 1985, when the medical investigators circulated initial reports on the outbreak to other health experts. In a preliminary report written in November 1984, Foster, the deputy state epidemiologist, concluded there was "no evidence" to support the hypothesis that the outbreak was the result of deliberate contamination. The "weight of the evidence," he wrote, supported the idea that food handlers at the various restaurants were responsible. While acknowledging that he could find no common source of infection of the food handlers or the customers, Foster concluded that the contamination "could have occurred where food handlers failed to wash their hands adequately after bowel movements and then touched raw foods."

Foster's findings were echoed two months later by Tom Török. In a preliminary report issued in January 1985, Török's team of federal scientists said that it, too, was unable to find the source of the outbreak and that food handlers were probably to blame. Because workers preparing food at the affected restaurants had fallen ill before most patrons had, the report reasoned, and because some minor violations of sanitary practices at a few restaurants had been detected, food handlers "may have contaminated" the salad bars, the scientists concluded. Török's federal team asserted that there was "no epidemiologic evidence" to suggest that the contamination had been deliberate. The report infuriated locals, who were convinced that the cult was responsible for the outbreak, as well as law-enforcement officials, who now lacked the "probable cause" needed to open a criminal investigation of the group.

MORE than a year would pass after the outbreak before Oregonians learned that Rajneeshees had poisoned the town.

On September 16, 1985, a year after the outbreaks, the Bhagwan, who had recently emerged from a self-imposed, four-year vow of public silence, held a press conference at the ranch. He accused Sheela and her allies, who had resigned their posts and flown to Europe two days earlier, of having betrayed his faith. Creating a "fascist regime" at the ranch, he announced, Sheela and her "fascist gang" had not only tried to kill fellow sannyasins who challenged her authority but had stolen money, mismanaged the commune's affairs, and left it $55 million in debt. He said that Sheela had also poisoned his doctor and dentist and the district attorney of neighboring Jefferson County, and had tried to contaminate a water system in The Dalles; furthermore, she had conducted experiments on white mice at a secret lab to test poisons that could kill people slowly without detection. He called for a government investigation.

AFTER the guru's charges, federal and state police formed a joint task force under Dave Frohnmayer, Oregon's attorney general, that included FBI agents, local and state police, the sheriff's office, the Immigration and Naturalization Service, and the National Guard.

Law-enforcement officials and police set up headquarters at the ranch, but at the Bhagwan's invitation and, hence, his sufferance. As a result, they had to negotiate access to each and every cubbyhole in every room of any building they sought to enter. In the ensuing weeks, the Bhagwan's blessing was granted and revoked so many times that investigators concluded that they would need greater authority to gather enough evidence to build cases that would stand up in court. The police had cause for concern: the sannyasins were bugging their makeshift offices at the ranch and secretly monitoring their phone conversations with colleagues in Washington, D.C., and Salem, the state capital. The cultists were also destroying evidence, including medical evidence, of their crimes.

On October 2, some fifty investigators armed with search warrants entered the ranch with one hundred subpoenas for sannyasins and the authority to search seven locations, among them the Pythagoras Medical Clinic, the Rajneesh Medical Corporation, and a secret room underneath Sheela's house that had a tunnel that opened near a road. In the clinic's lab, Michael Skeels, the director of the state public-health lab, found glass vials containing salmonella "bactrol disks" that Ma Anand Puja, a nurse, had ordered from VWR Scientific, a medical supply company in Seattle. Such disks held germs normally used in diagnostic testing. Carefully removing them as a germ-warfare expert from Washington looked on, Skeels shipped the disks and other substances found in the lab to the CDC in Atlanta. Analysis revealed that the salmonella in the disks was identical to that which had sickened people in The Dalles the previous year. Skeels had found the biological equivalent of a smoking gun. The salmonella mystery was apparently over.

The failure to recognize the illnesses in The Dalles as germ terrorism was not surprising. Robert V. Tauxe, a CDC scientist who had investigated the epidemic, said the germ detectives had been swayed by Dr. Foster's concerns about bigotry and had begun the inquiry determined not to leap to conclusions. The sicknesses, he said, could easily have been caused by a natural eruption of salmonella, an amazingly complex bacteria that had caused far larger, more virulent outbreaks. A 1986 case at a Chicago milk processing plant involved seventeen thousand confirmed cases. Scientists had come to respect the microorganism's ability to "get into things." Salmonella continued to teach researchers lessons. "It's rare that we have had to invoke evil intentions when bad luck and stupidity are usually to blame," Tauxe said. Moreover, the investigators' initial prescription for stemming the illnesses—closing the salad bars—had been right, albeit for the wrong reasons.

Experienced scientists and physicians are trained to analyze disease by thinking through the most probable causes of a mysterious illness. "If it looks like a horse, don't think about zebras," they are taught. For American scientists in 1984, bioterrorism was, in effect, a zebra. Why would anyone want to harm innocent people in a remote small town in Oregon?

"Call us naïve," Lutgens said years later, "but we never imagined people could have done such a thing. You don't expect bioterrorism in paradise."

As the criminal investigation unfolded, FBI agent Lynn Enyart, who supervised such investigations in Oregon, and Robert Hamilton, the state attorney who coordinated the task force, were stunned by the scope of the cultists' crimes.

Investigators uncovered what remains the largest, most sophisticated illegal electronic eavesdropping system in American history. Beginning in 1983, a year before the salmonella poisonings, a Rajneeshee security team had bugged entire floors of their hotel, many of their disciples' homes, the ranch's public pay phones, the Zorba the Buddha café, and even the Bhagwan's bedroom.

They also discovered a series of other plots in 1984 and 1985 to kill or sicken people on an eleven-person enemies list, among them Charles Turner, the U.S. attorney; several county officials; a former disciple who had won a lawsuit against the cult; and a journalist from the *Oregonian* newspaper.

Two key officials at the ranch began cooperating with the police in exchange for pledges of reduced prison time, telling investigators how Sheela and her allies had targeted and poisoned rivals and dissidents within the cult.

THE investigation eventually established that the cult had experimented in 1984 and 1985 with poisons, chemicals, and bacteria. The commune's germ-warfare chief was a thirty-eight-year-old American nurse of Philippine origin who had been a close ally of Sheela's since their days in Poona, India. Ma Anand Puja, whose real name was Diane Ivonne Onang, supervised medical care at the commune. One of the "Big Moms," as the commune's three women leaders were known, Puja wielded enormous power.

A stocky woman with narrow eyes, a fixed sneer, and jet-black hair, Puja was known to some sannyasins as "Nurse Mengele." She was obsessed with poisons, germs, and disease.

David Berry Knapp, known as Krishna Deva, or K.D., the mayor of Rajneeshpuram, later testified for the prosecution after being taken into a federal witness-protection program. He said that Puja had given Haldol, a powerful tranquilizer, to many of the violent, mentally disturbed homeless people whom the cult had brought to the ranch to help win the county elections. Acting under orders from Puja and Sheela, sannyasins had injected hypodermic syringes filled with the prescription drug into beer kegs and had poured the drug into tea consumed by the homeless. They had also stirred it into their mashed potatoes.

Another key witness told prosecutors that when state medical authorities asked Puja to account for the large Haldol purchases, she ordered her assistants to fabricate records to disguise its actual use. Several had left the commune rather than break the law.

Puja was responsible for buying the prescription and over-the-counter drugs that the Rajneesh Medical Corporation kept in the Pythagoras pharmacy, as well as its medical supplies. Because she headed a medical corporation, she was entitled to buy such products from commercial medical supply companies like VWR Scientific and even obtain dangerous pathogens from the American Type Culture Collection, the giant private germ bank located first in Maryland and later in Virginia from which doctors, clinics, and hospitals order germs for research and standard diagnostic tests.

An invoice from the ATCC shows that the cult ordered and received a variety of such pathogens, among them *Salmonella typhi,* which causes typhoid fever, an often fatal disease that prompts at least two weeks of high fever. Had this bacterium been used in the salad bars instead of *Salmonella typhimurium,* said Skeels, people would almost certainly have died in the outbreak.

The invoice also listed *Salmonella paratyphi,* which causes a similar illness, though not as severe, and, most startling of all, *Francisella tularensis,* which causes tularemia, a debilitating and sometimes fatal disease. U.S. Army scientists in the 1950s had turned *F. tularensis* into a weapon, and it still remains on the nation's list of germs a foe might use in a biological-warfare attack.

Finally, Puja had obtained orders of *Enterobacter cloacae, Neisseria gonorrhoeae,* and *Shigella dysenteriae.* Fewer than one hundred organisms of shigella are needed to cause very severe dysentery—profuse diarrhea, bloody mucoid stools, and cramping—and death in as many as 10 to 20 percent of cases, even in previously healthy persons.

The ATCC invoice was dated September 25, 1984, indicating that the agents had been delivered between the two waves of the salmonella outbreak.

No pathogens ordered from the germ bank were ever found at the ranch, and the order's implications got little attention. The investigation had begun more than a year after the poisonings, giving the cult time to destroy evidence before the authorities obtained search warrants. When the invoices were seized in the search, they were not shown to public health officials who would have understood their significance. But those officials, who learned of the orders only years later, considered both the agents and the timing of their arrival at the ranch ominous. Skeels said that *Francisella tularensis, Salmonella typhi, Salmonella paratyphi,* and *Shigella dysenteriae* were all unnecessary in a clinical lab the size of Puja's. And all these bacteria, he said, could have been used for bioterrorism.

Puja was also particularly fascinated by the AIDS virus, about which relatively little was known at the time. The Bhagwan had predicted that the virus would destroy two-thirds of the world's population. For Puja, it was a means of control and intimidation. She repeatedly tried to culture it for use as a germ weapon against the cult's ever-growing enemies. Her apparent failure was not for lack of effort. After she was told by a technician at the ranch, for instance, that her lab lacked the necessary equipment to stabilize and dry the virus, the corporation promptly bought a "quick-freeze dryer" in September 1984.

Experts doubted that Puja had the skills, expertise, or supplies needed to culture the AIDS virus or the other dangerous pathogens she had ordered. But several sannyasins told police that in at least one instance she had injected blood drawn from a homeless man who had tested positive for the AIDS antibody into the veins of a cult rival. The fate of the man is unknown.

STATE and federal investigators eventually concluded that the plot to poison people in The Dalles with a biological agent, which involved about a dozen people, had grown out of the cult's legal war with the county and its determination to win control of the county government in the November elections.

Sometime during the spring of 1984, according to sworn affidavits and court testimony in 1985 and 1986, the commune's inner circle began brainstorming about how the commune's four thousand or so members could defeat the roughly twenty thousand residents of Wasco County. At one meeting, Sheela fastened on the idea of making non-Rajneeshees too sick to vote. Together, she and Puja began reading books like *How to Kill: Volumes 1–4,* and *The Handbook of Poisons,* trying to locate various bacteria that would sicken people without killing them. They also asked a urologist at Puja's clinic about poisons and bacteria that would be difficult to trace. The urologist, apparently unalarmed since he was told that the women were trying to defend the commune against germ attacks by the Rajneeshees' numerous enemies, mentioned salmonella as a possibility.

Rajneeshpuram's mayor, K.D., told another cult member that Puja did some experiments with a hepatitis virus and had initially considered using it to sicken local residents. Puja also proposed using *Salmonella typhi,* the bacteria that causes typhoid fever, which she wound up purchasing for the lab. But a Rajneesh Medical Corporation lab technician warned her that orders of typhoid cultures would be easy to trace if an outbreak occurred. As far as K.D. knew, the plan was never implemented.

K.D. also testified that Puja had considered sickening people by putting dead rodents—rats and mice—into the water system. Puja believed that dead beavers would be especially effective because they harbored a natural pathogen—*Giardia lamblia,* which causes diarrhea. When other plotters complained that the county's water reservoirs were covered by screens, K.D. recalled, someone "jokingly" suggested that the beavers be put in a blender and liquified.

Sheela and Puja finally settled on *Salmonella typhimurium* as their germ weapon of choice, known by the American Type Culture Collection's designation 14028. Before they began plotting in the spring of 1984, the cult ordered the bactrol disks from VWR Scientific, the Seattle-based company. Puja then used the bacteria in the disks to culture and produce large amounts of the bacteria in a part of her lab known as the "Chinese Laundry." Ava Kay Avalos, known as Ma Ava, another star prosecution witness who was granted partial immunity, said that the lab was later moved, at another technician's insistence, to a complex closer to where people with AIDS and other infectious diseases were kept—a more isolated part of the ranch.

Two rooms in the A-frame housed the production unit, with its gloves, masks, white robes, pills, syringes, containers, a large freeze-dryer, and what Ava described as a small, green "apartment-type refrigerator" in which Puja kept the petri dishes filled with colonies of salmonella.

K.D. and Ava confessed in sworn statements to having been among the cult's first germ warriors. Puja and Sheela needed to test their product in the summer and fall of 1984 to ensure that the bacteria would incapacitate voters later that year.

Of the two commandos, Ava proved the more gifted. The previous year, Puja and Sheela gave her a vial of light brown "murky water" that had a putrid smell and sent her to the Zorba the Buddha café in Antelope to poison Judge Hulse, who was eating there before his scheduled visit to the ranch. Though she initially feared that Hulse might be alarmed by the tint or foul smell of the water she served him, Hulse suspected nothing, and thanked her for the glass. He was given a second salmonella cocktail after his tire was flattened during his visit to the ranch.

Though prosecutors were able to document poisonings in the fall of 1984 at only four of the ten contaminated restaurants and at a supermarket on the west side of The Dalles, they believed that the sannyasins had carried out similar attacks in Salem, Portland, and other Oregon cities. The plotters had boasted of having attacked a nursing home, and even the salad bar at Mid-Columbia Medical Center, which two Rajneeshees were said to have infiltrated. Such attempts were never proven in court.

In the fall of 1985, K.D. and Ava gave grand juries vivid accounts of the group's outings the year before. The group had entered an Albertson's grocery store and sprinkled what they called their "salsa" on the lettuce in the produce department. Puja had wanted to inject the germs into milk cartons, but K.D. said he had dissuaded her because shoppers would see that the cartons had been tampered with.

At about the same time, according to K.D. and Ava, other sannyasins, many of whom had donned wigs and dressed in neutral-toned street clothes for their raids, had put salmonella in restaurant coffee creamers and in blue-cheese dressing; on several occasions they had scattered it over fruits and vegetables at several salad bars, foods that are not usually convivial hosts for the otherwise hardy germs.

To reassure those who doubted that their guru would bless the poisonings, Sheela had played for the disciples at a meeting that fall segments of a taped conversation with the Bhagwan. They heard Sheela ask their "enlightened master" what should be done with people who opposed his vision. K.D., who attended the meeting, testified that he heard the Bhagwan respond that Hitler, too, had been misunderstood when he wanted to create a new man. Hitler, the Bhagwan had said, "was a genius" whose only mistake was "to attack the Russians and open two fronts."

"If it was necessary to do things to preserve [the Bhagwan's] vision, then do it," K.D. reported the guru as saying on the muffled tape. Sheela had interpreted this to mean that killing people in the name of the guru was fine. If a few people had to die so that the Bhagwan's message could prevail, the disciples were "not to worry," Sheela told doubters at the meeting. If the commune had to choose between saving one thousand unenlightened people or one enlightened master, she added, "you should always choose the enlightened master." If necessary, Ava remembered hearing Sheela say, "one thousand people would have to be killed."

As it turned out, Sheela and Puja decided that mass murder was not needed to swing the election. Having successfully sickened at least 751 people in the fall of 1984—a test for the planned election forays—the cultists expanded their campaign in October by deciding to import homeless people and registering them to vote.

But Wasco County officials outsmarted them. Shortly before the registration period closed, county officials, citing the appearance of a suspiciously large number of registration cards, insisted that all prospective new voters be questioned by a special panel at the armory. Realizing that the homeless would never be able to pass such scrutiny, the cult abandoned both the poisonings and the registration schemes. Most of the cult members never even registered.

Wasco County residents, by contrast, sensing that their town was in jeopardy, registered and voted in record numbers. The turnout for the November 1984 election was proportionately the largest in Oregon's history. The Rajneeshee-backed candidates were defeated.

ON Sunday, October 27, 1985, Dave Frohnmayer received a call at home from Charles Turner, the U.S. attorney in Portland whom Sheela and her gang had targeted and nearly killed that year. "The Bhagwan's gone!" he shouted. "He's fled."

Knowing that the guru was about to be indicted for conspiring to evade immigration laws, the Bhagwan's lawyers had been discussing with federal officials the terms for their client's surrender. Meanwhile, the head of the local FBI office had drafted a contingency plan to "snatch" the Bhagwan from the ranch should those talks fail, which seemed likely.

Frohnmayer and others felt that bloodshed would have been inevitable had the FBI proceeded with its plan. In addition to Rajneeshpuram's sixty-member, heavily armed "Peace Force," the guru was protected by an Uzi and Kalashnikov machine-gun-wielding personal security force whose members were prepared to die protecting him.

The guru, apparently unaware that the police were tracking his plane, was arrested as soon as his jet landed in Charlotte, North Carolina. With him were a small group of disciples and twenty-one suitcases containing a revolver, $58,522 in several currencies, thirty-five jeweled wristwatches, and seventeen pairs of designer eyeglasses; his disciples had also managed to load his throne onto the plane.

On the day of the Bhagwan's arrest, the West German police detained Sheela, Puja, and another cult leader in a posh resort hotel. Extradition proceedings began. In the following months, a flurry of indictments were issued by state and federal grand juries. Since there was no antiterrorism law at the time, federal investigators charged the cultists with violating immigration laws and a consumer-product tampering statute that Congress had passed in 1982 after the poisoning of Tylenol capsules.

In the summer of 1986, Sheela and Puja, who had been extradited from Germany, pleaded no contest to charges of attempted murder, illegal wiretapping, the poisoning of Judge Hulse, causing the salmonella outbreak in The Dalles, and other crimes. Both women received maximum twenty-year sentences. Sheela was fined nearly $400,000 and ordered to pay Wasco County $69,353.31 in restitution. But thanks to shrewd bargaining, both women wound up serving less than four years in a federal prison in Pleasanton, California, a jail for nonviolent white-collar offenders. While Oregon had intended to seek additional remedies against them when their federal sentences were finished, Sheela and Puja were released early for good behavior and fled to Europe before the Justice Department notified the state.

The Bhagwan received a ten-year suspended prison sentence, paid $400,000 in fines, and left the United States, forever.

Most of the contaminated restaurants never recovered from the poisonings. Dave Lutgens, having lost nearly all of his business, changed the name of his restaurant. Dave's Hometown Pizza was the only restaurant to survive long after the attacks at its original location and under its original ownership.

THE Rajneeshees' attack did not attract much attention. It occurred before the days of competing twenty-four-hour cable news shows. There was no "Outbreak in Oregon" to rivet viewers, no nonstop footage of victims staggering into hospitals. The prosecutions of cult members unfolded in the Pacific Northwest, far from the nation's media centers. Dave Frohnmayer, Oregon attorney general, had been unable to interest high-level friends at the Justice Department and White House in the case. When public health officials figured out how easily the Rajneeshees had spread the disease, they decided not to publish a study of the incident. No one wanted to encourage copycats.

But the attack was nonetheless significant. It was the first large-scale use of germs by terrorists on American soil, the union of a modern phenomenon and an age-old means of destruction. The unusual case exposed numerous shortcomings. One was the ease with which pathogens could be ordered from a germ bank. The Rajneeshee lab, because of its small size, never had to register with the state. Its status as a medical corporation was enough to let it buy germs legally. And the cult's poisonings, officials said, prompted no changes in state or federal rules or, as far as could be determined, in germ-bank procedures.

The case revealed another problem. The partnership between law-enforcement officials and their scientific colleagues was rocky, a clash of cultures. Information was not shared; opportunities were missed. It proved difficult to establish that a crime had been committed. Even to trained eyes, a natural outbreak and a germ assault look much the same: large numbers of people become violently ill. The inquiry underscored the importance of intelligence from insiders. Investigators cracked the case only after members of the cult came forward and confessed their crimes.

Quietly, the small cadre of experts and federal officials who understood the power of germ weapons began to wonder if the attack in Oregon was an anomaly or a harbinger.

2

Warrior

DECADES before the Rajneeshees started their experimentation, the United States and the Soviet Union were already veterans at turning germs into weapons, having produced arsenals to cripple and kill millions of people. There had been challenges. On the American side, a main problem solver was Bill Patrick. For two decades, he had done biological research at Fort Detrick, Maryland, the U.S. Army's sprawling base for the design of germ weapons. He had been awarded five patents and risen to become chief of what was known as the Product Development Division. In 1969, when President Nixon suddenly ended the program, he had stayed on at Detrick doing defensive work.

So when the Federal Bureau of Investigation needed a top expert to study the attack of the Rajneeshees and examine the cult's compound, it turned to Patrick. Flown out from Detrick to Oregon in late 1985, he sensed that the cult was to blame for the outbreak. For one thing, he found a germ incubator—an unusual piece of equipment for a health clinic. Patrick knew salmonella well. Scientists at Detrick had investigated the bug as a weapon, with at least one worker falling ill.

As the evidence grew, Patrick could see that the people in The Dalles had been victims of crude bioterrorism. The Rajneesh crowd had chosen one of the mildest pathogens and had ignored the airborne delivery of biological agents—the traditional and most effective way of mobilizing microbes for war. Growing germs was easy; the process was like brewing beer. It was the subtleties of biological engineering—drying germs, encapsulating them in special coatings, making them hardy and stable enough for wide dissemination by aerosol sprayers, learning how to extend their shelf life—that turned pathogens into deadly weapons.

Beginning in the 1950s, Patrick and his men had devised little biological time bombs that would float invisibly in the air for hours, like dust, or sail far on the wind. The small size of the particles meant that once they encountered a person, they could penetrate deep into the lungs. The natural defenses of the human respiratory tract, ranging from hairs in the nose to cilia along the windpipe, easily block large particles. But small ones zip right by. Inside the lung, multiplying in moist tissues, a single invader could produce millions of offspring. Patrick and his coworkers had also learned how to concentrate the tiny bomblets, ensuring wide destruction. In time, they had managed to increase the potency of anthrax—a usually fatal disease of coughing, high fevers, hard breathing, chest pain, heavy perspiration, and a bluish discoloration of the skin caused by lack of oxygen—so that a single gallon held up to eight billion lethal doses, enough in theory to kill every man, woman, and child on the planet. A bigger challenge was distributing the poison, most of which would dissipate in the wind. American scientists had spent decades investigating how to make lethal dissemination as efficient as possible.

Such progress had a price, of course. Painstakingly, the germ-development program at Fort Detrick had tested prospective germ weapons on nearly a thousand American soldiers, in sealed chambers and the wilds of the Utah desert. Reaching beyond the military, it had exposed prisoners at the Ohio State Penitentiary, where volunteers were carefully monitored. Clandestinely, it also sprayed American cities with mild germs to investigate the likely impact of deadly pathogens.

The price of progress also included a number of accidents in which the experimenters, including Patrick, were inadvertently turned into human guinea pigs. Women were largely banned from the work after two gave birth to children with severe birth defects. Both babies died. In all, the scientists made enough mistakes to become victims of their own pathogens 456 times. All but three survived. Two men fell to anthrax. One was consumed by Bolivian hemorrhagic fever, an exotic ailment that eats away at the body's internal organs and causes profuse bleeding from the nose, mouth, anus, and other mucous membranes. Patrick himself came down with Q fever. This relatively mild disease was meant to cripple foes with chills, coughing, headaches, hallucinations, and fevers of up to 104 degrees.

Patrick had no regrets. After decades of living with burdensome secrecy rules, he spoke freely in his early seventies about the experiments, his own illness, and the deaths. It had all been done in the line of duty, driven by fear of the Communists, he said. What Moscow's germ program was in fact doing turned out to be far worse than anything Patrick and his colleagues ever imagined.

In speaking out, Patrick wanted to rebut critics who argued that germ weapons were impractical and too dangerous to use. He was eager to show not only the seriousness of the foreign menace but also how far the black art had come in American hands. In addition, he wanted to warn that the Rajneeshees' attack was but a foretaste of things to come. Terrorism was on the rise, and increasingly its targets were American. Late in life, he worked hard with many federal agencies to encourage better defenses, worried about the threat not only to Americans but also his own family.

Nothing in his past had suggested that Bill Patrick would become a germ warrior. Like a surprising number of his peers who developed weapons for biological warfare, he had begun his career eager to save lives and aid the revolution that was turning medicine from an art into a science. To all appearances, he was just a regular guy.

WILLIAM Capers Patrick III was born July 24, 1926, the only child of a southern couple whose families were of Scotch-Irish descent. His middle name was taken from a relation who was a Methodist bishop. His hometown of Furman, South Carolina, was a tiny speck of civilization that lay in the Low Country near the Savannah River. While provincial, with a population of about one hundred people, the town and the surrounding area were a boy's dream.

Patrick was serving in the army during the Second World War when he first encountered penicillin, which was just coming into wide use as a way to battle microbes in the human body. Fascinated by the development, he went back to school ready to cure diseases and aid the new field of antibiotic medicine.

He received a bachelor's degree in biology from the University of

South Carolina in 1948 and, a year later, a master's degree in microbiology from the University of Tennessee in Knoxville. Upon graduation in 1949, he began working for a company in Indiana that made the new wonder drugs of antibiotic medicine. A research microbiologist, he helped pioneer ways to mass-produce penicillin and bacitracin, another antibiotic. His work centered on learning how best to grow microbes and extract their antibiotics, which are natural compounds that germs use to fight one another in microscopic wars. Most fundamentally, he helped find ways to keep the bacterial colonies alive and reproducing, a process that required the constant renewal of food supplies and the removal of metabolic wastes.

In his antibiotic work, Patrick was aiding one of the greatest and most inconspicuous of human revolutions. People in antiquity had lived an average of twenty or thirty years. By Patrick's day, the figure had risen to more than sixty years. By the century's end, in industrial nations, it stood at roughly eighty years. The lengthening of the human life span was due largely to the decline of infectious disease. One by one, history's great killers and cripplers—plague, cholera, tuberculosis, smallpox, typhus, leprosy, diphtheria, polio, influenza, dysentery, pneumonia, whooping cough, and a dozen other scourges—had been vanquished or tamed.

One day, Patrick got a call from a former teacher in Tennessee who had gone to work for the military on secret projects. His work, the mentor confided, centered on how highly pathogenic organisms might be deployed to help deter and win wars. He could give no details; such talk was forbidden. But the research was fascinating. He urged the biologist to sign up, appealing to his love of country and his sense of scientific adventure.

The young scientist was eager for challenges and flattered by the attentions of a former teacher. In 1951 after a background check that took more than half a year, Patrick won a top-secret clearance and permission to work at what was then called Camp Detrick, the heart of the federal government's research on germ warfare.

GERMS and warfare are old allies. More than two millennia ago, Scythian archers dipped arrowheads in manure and rotting corpses to increase the deadliness of their weapons. Tatars in the fourteenth century hurled dead bodies foul with plague over the walls of enemy cities. British soldiers during the French and Indian War gave unfriendly tribes blankets sown with smallpox. The Germans in World War I spread glanders, a disease of horses, among the mounts of rival cavalries. The Japanese in World War II dropped fleas infected with plague on Chinese cities, killing hundreds and perhaps thousands of people.

Despite occasional grim successes, germ weapons have never played decisive roles in warfare or terrorism. Unintended infection is another matter. European conquests around the globe were often made possible because the indigenous peoples lacked immunity to the invaders' endemic diseases, including smallpox, measles, influenza, typhus, and plague. But intentional warfare with germ weapons has been relatively rare, especially in modern times, and has been widely condemned as unethical and inhumane. Even so, in the early twentieth century, Canada, France, Germany, Japan, the Soviet Union, and the United Kingdom were among the many countries that investigated how to wage biological war.

All understood that the weapons they were developing were fundamentally different from bombs and bullets, grenades and missiles. These munitions were alive. They could multiply exponentially and, if highly contagious, spread like wildfire. Strangest of all, given war's din, they worked silently.

In the days before atom bombs, germ weapons were seen as an ideal means of mass destruction, one that left property intact. Their main drawback was their unpredictability. In the close confines of a battlefield, the weapons followed the dictates of nature, not military commanders. They might kill an adversary, or they might bounce back and devastate the ranks of the attacker and his allies. Their best use seemed to be against a distant enemy, reducing the chance that the disease would boomerang.

With intelligence agencies warning that Tokyo and Berlin had biological weapons, Washington began to mobilize against germ attacks in 1942. President Franklin D. Roosevelt publicly denounced the exotic arms of America's foes as "terrible and inhumane," even while preparing to retaliate in kind. The man chosen to lead the secret U.S. program was George W. Merck, the president of a drug company. Merck was a household name, and generations of physicians had come to rely on *The Merck Manual* as a trusted guide for diagnosing and treating disease. But the new effort was designed to be nearly invisible, its degree of secrecy matched only by America's project to build the atom bomb.

This germ initiative had its headquarters at Camp Detrick, an old army base in rural Maryland that was close enough to Washington for quick responsiveness, but far enough away to ensure a margin of isolating safety. The work got under way in 1943 and expanded quickly. From a rural outpost in farm country, the base grew overnight into a dense metropolis of 250 buildings and living quarters for five thousand people.

The post was ringed by fences, towers, and floodlights. Guards, under orders to shoot first and ask questions later, kept their machine guns loaded. The scientists were issued pistols, which they kept at their sides or nearby on workbenches. The headquarters building at the heart of the compound had its own set of armed guards on alert around the clock. All personnel had identity passes with employee photos—a security precaution that would become widespread in future decades. Persons leaving the post surrendered their photo passes to the guards; accidentally keeping one could lead to arrest and interrogation.

The scientists toiled on anthrax for killing enemy troops and agricultural blights for destroying Japanese rice and German potatoes. It wasn't easy work. For instance, they had to coax the anthrax bug into its best form. Toward the end of its growth cycle, they used heat or chemical shock to force the rod-shaped bacteria to convert into spores, a dormant state. When the process worked properly, the spores were very hardy, resisting heat, disinfectants, sunlight, and other environmental factors. Anthrax spores had been known to remain viable for decades. The scientists harvested the spores and put them into weapons. Upon being inhaled, the spores would convert back to rods and establish an infection.

The Detrick scientists also learned how to reap the poisons that some bacteria excrete—a tactic that sidestepped the necessity of infection and instead yielded deadly toxins that could be sprayed directly on foes. One was botulinum toxin, the most poisonous compound known to science. It paralyzes muscles, including the diaphragm, without which the lungs cannot function, and its victims quickly die. In time, the scientists learned how to make botulinum toxin so concentrated that a pound of it, if properly dispersed, could in theory kill a billion people.

None of the biological arms developed by the United States were used on the battlefield during the war, and afterward the effort slowed down markedly and shrank in size. But it endured. One reason was that the Americans obtained thousands of records from Japan documenting the Imperial Army's germ-warfare program during World War II. Japan had killed thousands of Chinese in widespread attacks with anthrax, typhoid, and plague on Manchurian towns and cities, Western scholars say. Doctors in the army's infamous Unit 731 had also conducted gruesome experiments on Chinese and other prisoners of war, including Americans. Doctors had infected healthy prisoners with pathogens to learn how diseases spread. Many victims, or "logs" as the Japanese called them, were deliberately starved and frozen to death. Some were dissected alive. While nine Japanese doctors and nurses were convicted after the war of having vivisected eight captured American fliers, no senior Japanese official was tried for having waged biological warfare. And American officials granted Unit 731's chief, Shiro Ishii, and several of his associates immunity from prosecution in exchange for the voluminous records of Japan's germ program and their help in deciphering them. The scientific data were considered a windfall and carefully studied.

The American military was fascinated by a weapon of mass destruction whose costs were so low compared with those for chemical arms and the atom bomb, recently invented. The federal government worried that not only the Soviets but other adversaries making similar comparisons would be tempted to develop pathogens for warfare. Pound for pound, germ weapons were seen as potentially rivaling nuclear blasts in their power to maim and kill, and some were considered even more destructive.

In a secret report of July 1949, a panel of more than a dozen senior federal and private experts told the secretary of defense that germ weapons deserved more attention in planning and development. Such warfare was "in its infancy," the panel said, and foreseeable advances would raise weapon effectiveness "by a very large factor." Germ weapons, silent but deadly, were ideal for covert attacks. "The resemblance of the results of such sabotage to natural occurrences," the panel said, would greatly aid clandestine use. It warned that germ attacks on the United States "might be disastrous" and urgently recommended crash programs of "home defense, involving collaborate efforts of federal, state and private agencies."

Some veterans of the secretive work disagreed with the government's reasoning. Theodor Rosebury, a microbiologist at Detrick during the war, assailed germ weapons in his 1949 book *Peace or Pestilence.* He warned that the field's promises were illusory and that its munitions had no real military value, since the outcome of germ attacks would always be impossible to predict or control. The expertise, he argued, should instead be turned to attacking infectious disease. His plea had no immediate impact.

IN April 1951, Patrick arrived at the Detrick army base on the outskirts of Frederick, Maryland. He was twenty-five.

Barbed wire ran atop its fences. CAMERAS ARE UNAUTHORIZED read the sign at the front gate.

Guards, armed and alert, stood at the entrance.

Like all new employees, Patrick signed a waiver that granted the United States government rights to his body if he died from an illness acquired at Detrick. Having done that, he received a series of vaccinations, which were required before new employees could go into "hot zones" teeming with disease germs.

He quickly learned the other survival rituals—the eating of antibiotics, the washing of hands, the bathing of people and labs in ultraviolet light, the wavelength best suited for killing germs. Caution also called for protective hoods and masks, rubber gloves and boots. The men often donned protective suits that made them sweat and itch. They breathed purified air. They stood for hours at "hot boxes"—glass housings with attached rubber gloves so the men could reach inside to handle glassware swarming with microbes or to assemble the guts of biological bombs. Despite the dangers, Patrick moved his family onto the post in 1952. It had its own housing, theater, restaurants, and child care. The social life revolved around the officers' club.

Patrick joined Detrick just as it was beginning to stir. The outbreak of the Cold War and the Korean War led Washington to put new emphasis on planning for germ battles. The testing of prototype nuclear arms at sites in the Soviet Union and the United States was already shaking the globe.

At Detrick, construction crews built a hollow metal sphere four stories high. Employees called it the eight ball. Inside, germ weapons were to be exploded, creating mists of infectious aerosols for testing on animals and people. Workers also erected Building 470, a windowless prototype factory for making anthrax. It rose eight stories, a skyscraper among the low buildings.

Under military orders, often clandestinely, Detrick experts fanned out to probe the nation's vulnerability to saboteurs. The scientists sprayed mild germs on San Francisco and shattered lightbulbs filled with bacteria in the New York City subway, all to assess the ability of pathogens to spread through urban centers. The germs were meant to be harmless. But years later critics charged that some had produced hidden epidemics, especially among the old and infirm. After the army sprayed the San Francisco area with *Serratia marcescens,* eleven patients at the Stanford University hospital came down with that type of infection. One patient died there. The doctors were so mystified by the outbreak that they wrote it up in a medical journal. The government later denied any responsibility for the death or the other infections, producing evidence in court that its germs were not to blame. The scientific dispute was never resolved.

The army also studied the threat of enemies wielding a speculative class of munitions known as ethnic weapons—germs that selectively target particular races. One military worry centered on *Coccidioides immitis,* a fungus that causes fever, cough, and chills and, if left untreated, kills blacks far more often than whites. The military feared that it would be used against bases, where blacks tended to do the manual labor. In 1951, at navy supply depots in Mechanicsburg, Pennsylvania, and Norfolk, Virginia, the Detrick scientists staged mock attacks with a nonlethal variant of the deadly fungus. The depots, said a report on the action, employed "many Negroes, whose incapacitation would seriously affect the operation of the supply system."

American scientists also did outdoor experiments to assess how Soviet cities could be attacked with anthrax germs. Dry runs were made against Saint Louis, Minneapolis, and Winnipeg, cities whose climates and sizes were judged similar to the Soviet targets. The effort was code-named Project Saint Jo. The clandestine tests, involving 173 releases of noninfectious aerosols, were meant to determine how much agent would have to rain down on Kiev, Leningrad, and Moscow to kill its residents. Each cluster bomb in the planned attacks held 536 bomblets. Upon hitting the ground, each bomblet would emit a little more than an ounce of anthrax mist. The disease, if untreated, kills nearly every infected person—a very high rate of mortality, even compared with plague and most other pathogens.

The snow was deep and the sky clear when experimenters in a special car drove into a Minneapolis suburb of homes, light industry, trees, and pine foliage to release the test mist. There was very little wind, and the winter night was marked by a strong temperature inversion. Overhead, a dome of warm air trapped cool air below. Air samplers showed that the release traveled nearly a mile. The "dosage area," experimenters wrote, was "unusually large."

UNTIL Patrick's arrival, America's hunt for living weapons had focused mainly on bacterial diseases, including anthrax, plague, and tularemia, a disease which kills one out of twenty people and leaves the rest very sick. Tularemia produced not only the usual chills, fever, and coughing of infectious disease but also skin lesions larger than those of smallpox—ulcers up to an inch wide, their centers raw, their edges turned up in reddish mounds.

But the shortcomings of bacteria as weapons were becoming obvious. Infections acquired in attacks on cities or battlefields could be successfully treated by large doses of antibiotics—the wonder drugs that Patrick as an industry researcher had been pioneering. That emerging fact of medical life diminished the role of bacteria as killers and cripplers for war.

Viruses were a beguiling alternative. Compared with bacteria, they were less complex and often more deadly. To Detrick scientists, their microscopic size offered a range of potential military advantages.

A single human egg is just visible to the naked eye and has a width of about one hundred microns, or millionths of a meter. Human hairs are seventy-five to one hundred microns wide and easier to see because they are long. An ordinary human cell is about ten microns wide and by definition invisible. Most bacteria are one or two microns wide. They and their cousins, such as the mycobacteria, are considered the smallest of the microscopic world's fully living things.

By contrast, viruses are hundreds of times smaller, and occasionally a thousand times. If bacteria were the size of cars and minivans, viruses would be the size of cell phones. One of the tinier ones, the yellow fever virus, is only two one-hundredths of a micron wide. The foot-and-mouth virus is smaller. Viruses are small because they lack most of life's usual parts and processes, such as metabolism and respiration. Scientists consider them barely alive, seeing them more as robots than organisms. To thrive and reproduce, they invade a cell and take over its biochemical gear, often at the expense of the host.

Over the ages, this biological intimacy has made viruses one of the most dangerous of all humanity's foes. They include the causative agents of influenza, smallpox, and Ebola, the scourge from Africa that bleeds its victims dry.

People can be powerless against them. Viruses are small enough to slip into cells, where they are safe from the assaults of the human immune system. By contrast, anthrax bacteria, lumbering giants at up to four microns wide, must battle their way into the body, with many thousands of them often needed to start an infestation.

Moreover, viruses are largely invulnerable to attack by antibiotics or other weapons of science because they are nearly indistinguishable from their human hosts. As an army reference book on germ warfare put it, viruses "may be particularly attractive" because so few treatments are available against them.

As Detrick scientists investigated such issues, they did know of one treatment that worked against viruses—immunization. Most vaccines are made of viruses that are dead, weakened, or harmless yet biologically akin to noxious ones. When injected—or, in some cases, swallowed—the vaccine sends a false alarm of pending attack to the body's immune system, which then forms antibodies to fight a particular type of invader. The defensive buildup is slow. So, to ward off invaders effectively, vaccines must often be given weeks to months in advance. They seldom work right away.

From the start, the army knew that the protective action of vaccines could be turned on its head to make viruses more suitable for war. An aggressor could use immunization to protect his troops, while an unvaccinated enemy would be vulnerable.

As Patrick settled into Detrick, the genocidal power of viruses was driven home by two dramatic episodes of pest control. The target was rabbits. The creatures had overrun Australia, their numbers competing so vigorously for sheep and cattle pasture that livestock production began to fall. In 1950, scientists responded with the virus that causes myxomatosis, a disease that often kills rabbits after leaving them blind and twitching. It spread fast and killed more than 99 percent of the infected animals. Europe in the postwar period had suffered a similar explosion of rabbits, which ate farmers' crops. In 1952, French experts released a few animals infected with the virus at Eure-et-Loire, not far from the palace at Versailles. By the next year, the disease had swept not only through France but as far as Belgium, the Netherlands, Switzerland, and Germany, killing up to 90 percent of the rabbits. Farmers were elated. In time, the exterminations were seen as vital to the postwar revival of European agriculture.

The rabbit killing was of special interest to the American germ warriors because the myxomatosis virus is part of the pox family, whose most famous member causes smallpox. So the rabbit drama was considered useful in studying how the smallpox virus might spread through populations of unvaccinated humans.

PATRICK'S credentials won him a leading role in Detrick's expanding operations. He evaluated viral agents that the scientists were developing and did experiments to see if the microbes could be produced easily in bulk and still maintain their virulence. He was a production engineer, though at this time, in the early 1950s, he was also working toward a doctorate in microbiology at the University of Maryland.

The army needed large amounts of viruses to test for deadliness and, if it came to that, for making bombs. Patrick knew a lot about microbial mass production and quality control from his experience in industry. He advanced quickly, soon becoming manager of Detrick's effort to make prototype virus factories.

His foremost tool was the chicken egg. Cheap and simple to procure, eggs are rich in the proteins and nutrients needed for viral growth and reproduction and are surrounded by a hard shell that makes them easy to handle. Drug companies often incubate viruses for vaccines in eggs. At Detrick, in a similar way, fertilized chicken eggs were found ideal for viral mass production.

Wielding syringes, Patrick's workers injected viruses into eggs and sealed them for incubation in warm ovens. The germs would infect the growing embryo and multiply, producing trillions of offspring. After a few days, the germs and their dying hosts would be harvested, the resulting compounds often pink in color.

The viral agents would then be tested on mice, guinea pigs, rabbits, rhesus monkeys, and—in time—American soldiers. One insight drawn from the preliminary tests and analysis was surprising and counterintuitive. Early on, the military concluded that killing enemy soldiers was unnecessary. In fact, viral crippling came to be seen as preferable to death, since an ailing soldier tied up more enemy transports, doctors, nurses, hospitals, drugs, and bureaucrats. Moreover, compared with the alternative, the incapacitation of troops could be billed as humane—no small thing in war, especially in a democracy, where public disapproval could quickly end a military plan or operation.

Viruses turned out to lend themselves to this subtle range of debilitation. Most affecting humans were not lethal. In fact, like the common cold and the flu, most tended to be merely bothersome, if generally debilitating and only occasionally deadly. Over evolutionary time, the microbes had often found it in their best interest to keep their human hosts alive.

Among the viruses that Patrick and his colleagues developed as weapons were those that give rise to encephalitis, a brain disease of fevers, seizures, comas, and in some cases death. Another was the yellow fever virus, which causes chills, stomach bleeding, and yellow skin due to liver failure and bile accumulation. The scientists also investigated rickettsiae, which range in size between viruses and bacteria. Like viruses, most burrow into cells to reproduce. Unlike viruses, some are slowed by antibiotics. One rickettsia that Patrick studied was the Q-fever microbe, an extremely hardy germ that causes fevers, chills, and a throbbing headache, usually behind the eyes.

In time, Patrick and his successors identified about fifty different viruses and rickettsiae that were good candidates for germ warfare. That was nearly three times the number of suitable bacteria. The advances meant that antibiotics alone would be insufficient to protect people from germ attacks.

THE army, eager to assess the effectiveness of the Q-fever microbe, *Coxiella burnetii,* approved its being tested on people, which Patrick hailed as an experimental breakthrough. Tests on surrogates, such as guinea pigs and rhesus monkeys, left scientists unsure about the effect on humans. The test subjects were Seventh-Day Adventists, conscientious objectors who, following the Old Testament's injunction not to kill, refused to bear arms for their country but were persuaded to volunteer their bodies for germ-warfare studies.

At Detrick, Patrick and his team made up batches of *Coxiella burnetii* while working in a brand-new structure of yellow bricks known as the virus pilot plant, Building 434. Its thirty-four employees could process about one thousand eggs a day, making quarts of test agent that swarmed with trillions of germs.

Patrick had the slurries of Q-fever germs carefully transported to the test sites. The first was Detrick's own eight ball. There, starting in early 1955, the Adventists gathered around the ball's periphery to don face masks and breathe deeply, inhaling mists of germs through rubber hoses connected to the ball's interior. Army experimenters administered a range of doses and droplet sizes to the men.

Once infected, the Adventists were carefully monitored. If symptoms broke out, they would be given antibiotics to cut the illness short and avoid serious complications. Though often no worse than the flu, if left untreated, Q fever can progress to severe chills, trembling, blinding headaches, muscle and joint pain, diarrhea, weight loss, visual and auditory hallucinations, facial pain, speech impairments, heart inflammation, and congestive heart failure. About one in a hundred infected people die. The Adventists were carefully supervised. All survived.

The tests produced a surprise. It turned out that just one microbe of *Coxiella burnetii*—a single invader—was enough to bring on the disease. That discovery was a medical first. Such powers of human infection had been suspected but never demonstrated.

Over the years, Patrick was often present at the sites where field testing took place, edgy over how his germs would do. At Dugway Proving Grounds in the Utah desert, thirty Adventists were assembled in the summer of 1955 for the first American field trial of germ warfare agents on human beings. The test measured how well the germs spread. At the center of the circular test grid were five sprayers, each holding five ounces of Patrick's Q-fever slurry. On the night of July 12, after almost a week of false starts, a fine mist emerged from the sprayers and sailed on a gentle breeze. The test subjects were more than a half mile away, waiting nervously, increasingly cold in the early evening, trying to follow instructions to breathe normally.

It worked. The pattern of infection among the Adventists showed that *Coxiella burnetii* was ideal for aerosol dissemination.

The next hurdle was figuring out how to spread the germ in war. A key test centered on the nation's newest jet fighter, the F-100 Super Sabre, the first production aircraft to break the sound barrier in level flight. The jet was built to carry nuclear and conventional bombs. Now a test model was rigged up for scattering germs.

Mechanics strapped a tank onto the plane's belly and connected it to special nozzles that expelled the germs into the wind. The question was whether the turbulence would break enough of the spray into particles one to five microns wide, the ideal range for penetrating human lungs. The jet thundered over the Utah desert, and the method turned out to be remarkably effective.

Only animals were exposed intentionally in this test series. But one F-100 pilot fell ill when he got out of his plane before it was decontaminated. The man had been on his honeymoon, Patrick recalled, and his wife looked like Marilyn Monroe. Three soldiers were also stricken as they manned distant barricades on a road at the proving ground's edge. Their job was to stop traffic if the winds shifted suddenly. The germs had sailed fifty miles. "We were overjoyed," Patrick recalled. The incident showed that the scientists' mathematical models for germ dispersion were correct. More important, he said, it demonstrated that his team was beginning to produce "a liquid product that was very, very good." The finding was celebrated, even as news of the accident was suppressed.

AMERICAN officials suspected that Moscow was engaged in similar experiments. This belief was strengthened in 1956, when the Soviet defense minister, Georgi Zhukov, told a Communist Party Congress that any modern war would certainly include the use of biological weapons.

Soon Washington uncovered hard evidence. An American U-2 spy plane, while flying high over a desolate island in the Aral Sea, snapped photo after photo of dense clusters of buildings and odd geometric grids. Analysts at the Central Intelligence Agency saw a link to biological weapons.

One day, Patrick and dozens of other managers at Detrick were summoned to a special meeting. Security was tight. Guards at the conference room door checked the security badges of arriving personnel against names on a list. After brief introductions, officials from the Central Intelligence Agency passed around the spy photos.

The structures on the Soviet island were unmistakably similar to the bull's-eye pattern of rings in the Utah desert, where roads, sensors, electrical poles, and test subjects were arrayed at increasing distances from germ sprayers.

The confirmation of what had long been suspected reinforced the conviction among the scientists that their labors were not only justified but also crucial, if for no other reason than to threaten the Soviets with retaliation. What they saw in examining the photos turned out to be a glimpse of a large enterprise that had long predated the American effort. It is now known that the Soviet program for germ warfare began in the 1920s and 1930s and grew steadily into one of the earliest and largest of the modern era, developing arms to infect people with anthrax, typhus, and other diseases. At that time, the United States had no germ weapons. Stalin's totalitarian rule poured vast resources into the endeavor, though in time his purges killed or imprisoned many leading microbiologists.

The turning point for Moscow was the Second World War. The capture of a Japanese germ unit, the study of its techniques, and the resolve to strengthen the Soviet military after the wartime slaughter of millions of Russians helped renew interest in germ weapons. In 1946 at Sverdlovsk, military engineers set up a factory that specialized in anthrax. In 1947 outside Zagorsk, the Soviet military built a complex for making viral weapons, including smallpox. By 1956, when the U-2 planes started flying over the Soviet Union, Moscow had built many secret bases across the land for developing and producing germ weapons.

WORRIED about the Soviet program and impressed with the powers of viruses and rickettsiae to cripple and kill, the United States in the late 1950s prepared to build factories capable of producing enough pathogens and biological toxins to fight wars. Officially, American policy at that time was no first use. Biological weapons were to be fired only in response to an enemy's germ attacks.

In 1956, at the age of thirty, Patrick won a promotion and soon became responsible for designing a distant plant where the production methods perfected at Detrick would be reproduced on a large scale so that viruses could be made not by the ounce but also by the gallon and the drum. The site was the Pine Bluff Arsenal, an army base that had been carved out of the woods of central Arkansas. The state was a leading producer of eggs.

The army was already using Pine Bluff to make weapons from bacteria, including those that cause tularemia. Under Patrick, it expanded into viruses. He worked with contractors and engineers on the design of the Pine Bluff virus plant, which was known as X1002. Its main function was to infect and harvest many thousands of eggs quickly. Workers at Detrick pushed around heavy metal trays each holding sixty eggs. At Pine Bluff, Patrick mechanized the process with conveyor belts.

By trial and error, he found that if egg trays moved past work stations slowly, at fifteen inches per minute, the operators would lose their concentration. "But if you moved it at twenty-two or twenty-three inches per minute, emotionally they could not accept that speed and would throw their hands up and quit," he recalled. "We found that twenty inches per minute was just right."

In time, Patrick's production lines began to hum and produce the germ responsible for Q fever. In a week, workers at X1002 could infect and process about 120,000 eggs, which in turn produced 120 gallons of agent—enough to cripple millions of people.

The pace of production was even higher for the virus that caused Venezuelan equine encephalitis, or VEE. In a week, workers on the mechanized line could infect and harvest about 300,000 eggs, which in turn produced nearly 500 gallons of noxious agent. Patrick knew the VEE symptoms well because an accidental outbreak at Detrick had once sickened fifteen members of his crew. "It's not lethal," he said of the disease. "It just makes you think you want to die. Your eyes want to pop out of your head."

Public health workers also knew the disease well because VEE is indigenous to Florida, Trinidad, Mexico, and Central and South America—places where mosquitoes carry the virus. The disease strikes suddenly to cause malaise, severe headache, high fevers, painful sensitivity to light, nausea, vomiting, cough, and diarrhea. The fatality rate, less than 1 percent in adults, is higher among the old, the young, and the infirm. In epidemics, one child in twenty-five develops signs of central nervous system infection, with convulsions, coma, and paralysis. In children with severe encephalitis, the fatality rate may rise to as high as 20 percent. Survivors can have permanent neurologic damage.

PRESIDENT Dwight D. Eisenhower was briefed on Fort Detrick's advances just before he left the White House. The full meeting of the National Security Council took place on February 18, 1960. Absent was Richard M. Nixon, the vice president, who was preparing his own run for president.

Eisenhower was trying to leave a legacy of growing trust between the superpowers, with disarmament under way and defense budgets declining. But the Democrats were calling for steep rises in defense spending, and Eisenhower deeply resented their charge that America had fallen behind in missile forces. He saw the attacks as partisan.

At the meeting, Herbert F. York, a nuclear arms designer newly appointed as the Pentagon's chief scientist and its number three official, told Eisenhower that the field of controlled incapacitation promised to "open up a new dimension of warfare." The weapons, he said, were nonlethal. Instead of killing, they caused lethargy, irritation, blackout, paralysis, illness, and the lack of will to fight. The effects would be temporary, York said, though minor repercussions "might persist permanently." Agents such as Q fever and VEE were becoming more concentrated, he said, and scientists were finding ways to extend their storage lives from one to three years. In the future, he said, army scientists would produce a new agent—an African virus that caused Rift Valley fever, a hemorrhagic disease of chills, bleeding, and stupor in most victims rather than death—as well as "tailored variants" of other agents. The new arms promised a military advantage. By contrast, York noted, most Soviet biological agents were lethal.

General Lyman Lemnitzer, the army chief of staff, described how the weapons might be used. They could blanket large areas and inflict heavy casualties without destroying buildings or property. Their potential was especially great, he said, "where friendly civilians may be present in an area occupied by enemy forces," a situation that might arise in the Philippines or Indochina. Lemnitzer gave a hypothetical example: if Communist forces seized an important region, the United States could respond with an attack of the encephalitis virus. Planes could drop bomblets or spray agent on the contested area. Two medium bombers, he said, might be enough. In three days, after the agent had taken effect, planes would fly in American or allied parachutists to reclaim the territory.

York said such weapons were potentially revolutionary. The diseases could be as mildly disabling as influenza or as deadly as atomic bombs. Over the next five years, he added, the Pentagon wanted to invest more money in research and weapons production. George B.

Kistiakowsky, the president's science adviser, said the field's "prospects were definitely bright" and backed a budget increase.

Eisenhower concurred.

Thomas S. Gates Jr., the secretary of defense, asked what the world would think about the use of such weapons, and wondered if they should be thought of as tantamount to nuclear arms. Gordon Gray, the president's national security adviser, noted that under current U.S. policy, the use of either chemical and biological weapons required the president's approval.

Eisenhower, who had more combat experience than anyone else in the room, called incapacitating agents "a splendid idea" that nonetheless posed "one great difficulty." An enemy might consider their use as full-fledged germ warfare and retaliate with lethal agents.

General Nathan F. Twining, chairman of the Joint Chiefs, agreed. He said that if the United States were ever to use such incapacitating agents, "we should publicize their nonlethal effects to the greatest extent possible."

In an interview years later, York said that in time his own enthusiasm for nonlethal weapons gave way to deep skepticism because of the problem that Eisenhower had identified. An assault with incapacitating agents, though billed by the attacker as humane, he said, would surely invite terrible revenge.

SPENDING on biological weapons rose dramatically after John F. Kennedy took office in January 1961. Robert S. McNamara, the new secretary of defense, did a sweeping review of military programs, including biological weapons. The Joint Chiefs of Staff found that the arms had unique advantages, especially their ability to incapacitate rather than kill. They strongly endorsed a crash program to prepare the germs for war.

The virus work, already a high priority, was redoubled, and such companies as General Electric, Booz-Allen, Lockheed, Rand, Monsanto, Goodyear, General Dynamics, Aerojet General, North American Aviation, Litton Systems, and even General Mills, maker of Cheerios and Wheaties, joined the germ program. Quietly, the purveyor of dry cereals built a "line-source disseminator," a device meant to spray germs continuously from airplanes to ensure wide swaths of infective mists.

Eager for flexibility and greater range, Pentagon planners readied a half dozen missiles for biological warheads, including the Pershing, the Regulus, and the Sergeant. The payload for the Sergeant consisted of 720 bomblets, each three inches wide, holding seven ounces of disease agent. Released ten miles up, at the fringes of the atmosphere, the bomblets would scatter during their downward plunge to cover more than sixty square miles, spraying agent into the air as a fine spray.

Military leaders also directed that field trials fan out from the Utah desert to sites more typical of the target zones. To determine the usefulness in Southeast Asia, testing was undertaken in Okinawa, Panama, and the central Pacific. Much of Russia was cold, so germs were tested in Alaska.

The expansion produced a new military bureaucracy. Founded in May 1962, the unit had its headquarters at Fort Douglas, near Salt Lake City. Its hundreds of personnel oversaw the movement of ships, jets, people, gear, microbes.

AFTER Fidel Castro seized power in Cuba in 1959, American military officials drew up contingency plans to invade the island nation and topple the dictator. One of the most secretive was referred to by Fort Detrick as the Marshall Plan, an ironic reference to the American effort to rebuild Europe after the Second World War. The plan, never before revealed publicly, called for a huge assault by American troops that would begin with a biological strike against Cuba's soldiers and civilians.

War plans are written to cover every imaginable scenario. But the Marshall Plan was a serious option, according to several American officials involved in the preparations. Over the years, the work included not only close collaborations with Fort Detrick's germ experts but agent selection, casualty forecasts, and study of the weather patterns over Cuba. There were even drills to see how fast the required germs could be produced.

The plan's specifics evolved as new agents became available and as the Kennedy administration's increasing investment in military biology began to pay off. Initially, Detrick could provide mainly lethal agents, such as anthrax. But in the early 1960s, as production increased at the new Pine Bluff factories, Detrick was able to offer an increasingly wide selection of incapacitants meant to leave most of the target population alive. Military biologists and planners argued for using such pathogens and biological toxins if the president ordered an invasion. Incapacitants could immobilize Cuban defenses, significantly reducing projected American combat casualties.

As Detrick refined its plans for a biological strike, the Pentagon also investigated how it might produce fake incidents that could create popular outrage and backing for a Cuban invasion. Castro was to be falsely blamed for such covert American acts as hijacking planes, sinking boats carrying Cuban refugees, terrorizing Miami or Washington, D.C., and even blowing up an American ship in Cuban waters. The architect of the pretext strategy was General Lemnitzer, the same man who told President Eisenhower about the military uses of incapacitating germ weapons. Under Kennedy, he was chairman of the Joint Chiefs. Acts of subterfuge, Lemnitzer wrote McNamara in March 1962, "would provide justification for US military intervention in Cuba." Kennedy ultimately rejected Lemnitzer's ideas, and he was later shifted out of his post.

The Cuban missile crisis erupted in October 1962, bringing Washington and Moscow to the brink of nuclear war. President Kennedy ordered a large force to be assembled for a possible invasion. The Pentagon set more than a million people in motion. It estimated that in a conventional assault, the number of Americans wounded or killed in the first ten days of battle could easily reach 18,500, or roughly 10 percent of the 180,000-man invasion force.

The Kennedy administration considered a wide range of military options against Cuba during the crisis. But there is no indication that attacks with germs ever figured in the high-level debate over how to dislodge the Soviet Union's missile bases. McNamara, Kennedy's defense secretary, said in an interview that he had never heard of the Marshall Plan. He added that neither he nor President Kennedy ever had "any intention" of using pathogens or biological toxins against Cuba during the crisis. Philip D. Zelikow, a scholar who received security clearances to review the secret records of the confrontation, said he knew of no tape recording or document sent to the president and his senior advisers that made any mention of germ weapons.

But some in the American military, especially those serving at Fort Detrick, saw the Cuban missile crisis as a lesson suggesting that germ weapons might some day save thousands of American lives. And with the completion of new factories at Pine Bluff, it became possible to attack all of Cuba with incapacitating agents.

A Pentagon official who had access to military archives said the options for a possible invasion of Cuba eventually included biological attack. He said that the director of the staff officers working for the Joint Chiefs had approved it. This official said the plan specified a combination of germs that could be used, and he estimated that a biological strike would affect millions of Cubans. It remains unclear how seriously this option was considered. The staff officers were less powerful than they became decades later, and their approval of plans advanced by the army and other services and commands were often routine.

Riley D. Housewright, Fort Detrick's scientific director at the time, recalled that the planning was directed by Pentagon officials who encouraged the germ scientists to refine how, exactly, such an attack would work. "I'd get maps half the size of my desk" that indicated the position of Russian troops and weapons in Cuba, Housewright said. The military officials, he added, "were very good at keeping us informed," while he and his colleagues, in turn, helped the officers judge "the potential for success." In the early 1960s, Housewright said, his men prepared agents that could incapacitate or kill large numbers of Cubans. One lethal alternative that was considered, he said, involved spraying enemy troops with botulinum toxin. Housewright considered the Marshall Plan "a good thing" because it could have saved American lives in the event of an invasion of Cuba.

Federal officials and germ experts, most speaking on the condition of anonymity, described the Marshall Plan's details and its range of destructive goals. Patrick would say nothing, even though other officials said the plan was modified over time to draw heavily on his work at Detrick with incapacitants. But in a talk to military officers in 1999, Patrick did say that a Cuban attack plan had been drawn up which relied on two incapacitating germs that worked sequentially to lengthen the time of disability. "The concept was that if we got into a shooting war that we would spray these organisms concomitantly," Patrick told the group. "We would incapacitate the Cuban population from three days to a little over two weeks." No infrastructure was to be destroyed. "We'd just make a lot of people sick. Very few of them were going to die." The germs would strike down "old folks like me," he added, killing less than 2 percent of the population. "We could move our forces in and take over the country and that would be it."

In interviews, Patrick eventually did describe what other experts said was a Marshall Plan drill. He said it was done in the early 1960s, under secret orders from Detrick and the army hierarchy in the Pentagon. The purpose was to test whether the United States could mobilize quickly for a germ conflict.

American germ-warfare scientists had developed a special cocktail of two germs and one biological toxin designed to work sequentially so that victims would come down with uncommonly long periods of sickness and debilitation. Animal testing had suggested that the brew was potent. Now Patrick helped prepare amounts that dwarfed all previous efforts. In all, teams at Pine Bluff made thousands of gallons of the cocktail, enough to fill a swimming pool.

The toxin of the cocktail was staphylococcal enterotoxin B, known as SEB, a poison excreted by the *Staphylococcus aureus* bacterium and a main cause of food poisoning. The germ warriors made it into a weapon by cultivating up trillions of the bacteria and then concentrating the poison. Whoever breathed the vapor would fall ill three to twelve hours later. The symptoms included chills, headache, muscle pain, coughing, sudden fever up to 106 degrees (close to what produces coma, seizures, and death), and, less frequently, nausea, vomiting, and diarrhea. The fever lasted days and the cough weeks.

The virus in the mix caused Venezuelan equine encephalitis. Its incubation period varied from one to five days, followed by the sudden onset of the nausea and diarrhea often associated with serious infection, as well as spiking fever up to 105 degrees. The acute phase lasted from one to three days, followed by weeks of weakness and lethargy.

The final element was the bug that caused Q fever. Its incubation period was ten to twenty days, after which it generally produced up to two weeks of debilitating symptoms, including headaches, chills, hallucinations, facial pain, and fevers of up to 104 degrees. Chronic Q fever, the U.S. Army found, was rare, but if the disease progressed to that stage it was frequently fatal.

In the Marshall Plan drill at Pine Bluff, Patrick's work was considered essential because he had helped perfect the means of mass producing the cocktail's three agents, SEB, VEE, and Q fever. "We tried to have a lot of different agents available," Housewright recalled. The incapacitants, he added, "showed that there was a humane aspect to the whole situation. It was not the same as putting an atomic bomb down their throat, which would have been just as easy or easier to deliver. It was a humane act."

Exposure to the agents would debilitate Cubans within a few

hours, and the incapacitation would last up to three weeks. Jets would tank up at Pine Bluff's airfield, fly to Cuba, and spray the concoction over key towns, ports, and military bases, moving east to west with the trade winds. Havana, in the west, was to get special attention. The isolated Caribbean island of Cuba was seen as an ideal target for biological warfare. And the germs would pose no threat to American troops. None of the diseases was contagious, and according to forecasts, all the cocktail's mists would have dissipated before an invasion.

Cuba's population at the time was about 7 million. The plan's estimates were that about 1 percent of the population, or 70,000 people, might die if all individuals exposed to the cocktail became sick.

Though the Marshall Plan was never implemented, the emerging ability to conduct a new kind of warfare had potential repercussions far beyond Cuba. For instance, military planners also saw a possible use for incapacitating agents in Laos. Washington feared that the Chinese Communists might try to take the country in a southward drive. According to a once secret report that reviewed American and allied capabilities in 1962 for limited military operations, using biological weapons in Laos had obvious drawbacks. "The spreading of sickness among the people," it said, "would be regarded as a friendly act by few of them." But if a biological attack in Laos produced "a successful pacification at small human cost," such action might "come to be regarded in a different light."

IN 1963, the Kennedy administration invited a number of academic specialists to Washington for the summer to offer advice on arms control. Among them was Matthew S. Meselson, a Harvard biologist who had recently done pioneering research on how DNA copies itself. At the Arms Control and Disarmament Agency, Meselson asked if he could look into biological issues. The agency's senior officials, busy negotiating a treaty with Moscow meant to ban most nuclear tests, told him to go ahead. Armed with secret clearances, Meselson set off.

At Detrick, he recalled, officials showed him the munitions and explained how they could cause many casualties. What was the rationale, Meselson asked, since the government already had nuclear arms? Germ weapons were cheaper, came the reply. Meselson then asked the CIA if any other nations were developing biological arms. No, came the answer, there were suspicions about the Soviet Union but no hard evidence. Meselson thought about this for a few days. The best thing for the wealthiest nation on earth to do, he reasoned, was to keep war expensive. Making it cheap enough for any dictator or undeveloped nation was a bad idea. Moreover, with biology every day becoming more powerful, germ weapons were increasingly dangerous.

Meselson went to see McGeorge Bundy, who a few years earlier, as Harvard's dean of faculty, had originally hired him. Now he served in Washington as the president's national security adviser. At the White House, Bundy told the biologist not to worry.

"We'll keep it out of the war plans," he told Meselson. "But we can't get rid of it because we have so many other things to do." In fact, the germs stayed in the war plans, former officials said. The budgets for biological warfare by this time were quite large, and there was no way the enterprise was about to declare itself useless.

As the Vietnam War intensified, scientists at Fort Detrick redoubled their work on smallpox. This virus was no mere incapacitant. Smallpox was ancient, highly contagious, and killed about a third of its victims, mainly from blood loss, cardiovascular collapse, and secondary infections as pustules spread over the body. Many survivors were scarred and blinded. Health authorities estimated that it had killed more people over the ages than any other infectious disease. In the twentieth century alone, it was estimated to have killed a half billion people—more than all the wars and epidemics combined, including the great flu pandemic of 1918–19.

As a rule, the American military generally avoided contagious diseases because of their unpredictability. But the rule had exceptions. Washington kept the smallpox virus on hand as a weapon for special military actions and the clandestine wars of the Central Intelligence Agency, which maintained its own supply. The secretive work on smallpox took place in the 1960s as world and American health authorities began a global effort to eradicate the dread disease.

The highly contagious virus had been studied by Patrick and his peers at Detrick but, unlike SEB, VEE, and *Coxiella burnetii,* it was never listed publicly as part of the official arsenal of weapons the government acknowledged years and decades after the biological program ended. Instead, officials tended to say that smallpox had been found unsuitable as a weapon and implied that it had been abandoned.

The smallpox virus, known as variola major, is large, measuring four-tenths of a micron—big enough to be seen easily in a light microscope and near the magical size for ease of human inhalation. It is also robust. Variola can survive outside the human host for days. Its size and hardiness explain why it spreads so easily. Typically, an infected person passes the virus to three or four others in close contact, often by coughing. But the virus can also be spread by contaminated bedsheets, clothes, blankets, and handkerchiefs. Its contagiousness and high fatality rate are what make it so destructive and feared.

Detrick experts learned how to multiply the germ in chicken eggs and human tissue cultures and tested it extensively on rhesus monkeys, which came down with high fevers, pustules, and symptoms of metabolic crisis. "Exposed monkey 3912," a report noted, "developed a facial twitch and paralysis of the right arm, and died on the sixth day after exposure."

The biologists at Detrick also found a way to alter the variola virus to make it even more durable. The trick was suspended animation.

In the battle between man and microorganisms, refrigeration slows down germ reproduction and metabolism. Typically, cooling also lengthens the lives of microbes. When microbes are frozen in exactly the right way (surrounded by sugar and protein, cooled quickly, and put under high vacuum to remove ice in a process known as sublimation), they can enter a dormant state in which they behave like vegetative spores. This process is known as lyophilization, or freeze-drying. Once dormant, the germ remains asleep even when returned to room temperature, staying that way for years or even decades. Industry uses the trick to make dry yeast for baking. It is only when such hibernating organisms are doused with water that they revive and multiply, in effect rising from the dead.

American weaponeers in the 1950s applied the new technique to microbial agents for war. They found that in some cases the process was so efficient it could make agents too strong. One solution was to dilute the dried agent with inert material.

At Detrick, investigators found that the drying process killed some microbes. But not smallpox. The virus easily withstood freeze-drying. The drying produced "no significant loss in virus viability," one smallpox study noted, adding that the tested strain showed "high mortality for man."

Virulence could thus be preserved for months and years without refrigeration, making the old scourge ideal for modern war. After three years, dried smallpox retained a quarter of its strength, one study found. Moreover, methods were developed that turned the dried virus into a fine powder that was found to be "easily disseminated."

The scientists devised means of clandestinely spreading the disease. Tiny aerosol generators, or atomizers, that could be hidden in everyday objects were developed. In May 1965, experts from Detrick's special-operations unit took the system on the road, using men's briefcases to spray mock smallpox germs in Washington National Airport, just outside the nation's capital. The lengthy report on the secret test, which was done to gauge America's vulnerability to enemy smallpox attacks, concluded that one in every twelve travelers would have become infected, quickly producing an epidemic as the disease spread across the country.

Smallpox would be an excellent choice for terrorism, the report said, because, among other things, "a long incubation period of relatively constant duration permits the operatives responsible for the attacks to leave the country before the first case is diagnosed." The incubation period of smallpox averages twelve days, at which point the disease shows its first symptoms, including malaise, fever, headache, and vomiting.

The public had no idea at the time about the secret tests. But popular culture reflected a growing fear of germ weapons. A 1965 movie, *The Satan Bug,* starring Richard Basehart and Anne Francis, was the story of a madman who stole a lethal virus from a secret government lab to take over the world. To demonstrate his seriousness, he first killed the inhabitants of a small Florida town and then threatened to destroy Los Angeles.

In the mid 1960s, as American forces began to bomb North Vietnam, the military came to Fort Detrick to ask for help with one of its most vexing problems: the Ho Chi Minh Trail. The North Vietnamese and the Viet Cong were using the jungle road as the lifeline for their war against the South. Nothing seemed to stop the movement of arms—not ambushes, booby traps, high-tech sensors, heavy artillery, or B-52 strikes. American military commanders asked their colleagues at Fort Detrick if biological weapons might stanch the flow.

The planning turned to assessments of smallpox, a potential move that was logical, if desperate. A boomerang effect seemed unlikely, since American troops were routinely vaccinated against the contagious disease. And North Vietnamese troops appeared to be vulnerable. In some ways, the setting was ideal. Though Vietnam had experienced no smallpox outbreaks since 1959, the disease still lurked in neighboring countries, allowing an epidemic to be attributed to natural causes. In the argot of covert operations, the strike would be plausibly deniable, a key requirement.

But attacking with such a devastating disease had enormous potential drawbacks. The trail wove through three countries. Sickness and death could spread uncontrollably, possibly infecting allies and civilians. And North Vietnam could retaliate with other germ agents. The Communist government had powerful allies in China and the Soviet Union, both of which had biological arsenals.

The idea was shelved. In the years that followed, the American government took a leading role in the global effort to eradicate smallpox, which was still killing millions of people each year. With a determined program of vaccinations, doctors and public health workers wiped out history's worst killer in just a few years.

Meanwhile, the American military's deliberations about a smallpox attack remained secret.

WITH public opposition to the Vietnam War growing, protestors took aim at the federal government's program for biological weapons. Its outlines were known from news articles and such books as Theodor Rosebury's *Peace or Pestilence.* In 1967, thousands of scientists signed a petition questioning the military's efforts. The opponents included Meselson, the Harvard biologist. His voice carried weight because he was a prestigious scientist with secret clearances. Another critic was Seymour M. Hersh, the investigative journalist who wrote an exposé in 1968 that was subtitled *America's Hidden Arsenal.* Fiction echoed the rising criticism. In his debut novel, *The Andromeda Strain,* Michael Crichton imagined the military flying into space to gather pathogens for germ warfare. In the story, a satellite crashes to Earth and begins a deadly outbreak that threatens human extinction. The contrast to *The Satan Bug* was telling: the military had now become the villain.

At Fort Detrick, crowds of antiwar protesters marched outside the gates, past armed guards and barbed wire. The American public, while knowing no details of what went on inside, understood that Fort Detrick was the hub of germ warfare. "They'd turn their back on you," Patrick said of the protesters. "But we thought we were doing the patriotic thing because we were risking life and limb."

The risk was real. In late 1968, a high-speed centrifuge that was spinning test tubes suddenly broke down and shattered the glass vials, exposing Patrick's men to a fog of psittacosis bacteria. The debilitating germ causes high fever and a severe pneumonia with hacking coughs. Just before Christmas, the illness sent five workers into the base hospital.

Opposition grew and found new outlets. At an airport, Meselson bumped into his old Harvard colleague Henry Kissinger, a history professor whom Richard M. Nixon, the new president, had recently named his national security adviser. The two men had worked next door to each other at Harvard and had talked on more than one occasion. Kissinger knew of the biologist's unease about germ weapons. "What should we do?" Kissinger asked, Meselson recalled. "I said, 'Let me think about it. I'll write you some papers.' "

One of Meselson's studies, dated September 1969, argued that the weapons, while extremely destructive, were unnecessary. A light aircraft could deliver enough to kill populations over several thousand square miles, he wrote, but the disease could spread far beyond the target area or create a long-term epidemic hazard. The same fickleness was true of incapacitating germs, Meselson argued. In an attack they might actually cause a large number of deaths among both enemy personnel and intermingled civilians, or might cause too little incapacitation to be militarily effective.

Meselson asserted that the nation's strategic nuclear forces were enough to deter attack. "We have no need to rely on lethal germ weapons and would lose nothing by giving up the option," he wrote. "Our major interest is to keep other nations from acquiring them," he noted, since the living munitions constituted cheap atom bombs. "Germ weapons that could threaten a large city are much simpler and cheaper to acquire than the corresponding nuclear weapons."

Meselson's arguments fell on fertile ground. For all the nation's work on germ weapons, the military chiefs and defense secretary found them of little use for the projection of American military power. And President Nixon needed a dramatic gesture amid growing discontent over the Vietnam War.

On November 25, 1969, Nixon ended it all—the tests, the accidents, the war preparations. "The U.S.," he said, "shall renounce the use of lethal biological agents and weapons, and all other methods of biological warfare. The U.S. will confine its biological research to defensive measures." The human race, he added, "already carries in its hands too many of the seeds of its own destruction." Nixon had been in office ten months. He made his announcement at Fort Detrick, smiling and clearly relishing his role as a peacemaker.

Patrick was there in the crowd, just feet away from the president. He later admitted that it was one of the darkest days of his life. He felt that the threat of responding in kind to foreign germ attacks was the best possible deterrent, and that the president was foolish to throw it away.

Biological toxins that were extracted from bacteria, such as staphylococcal enterotoxin B, were spared from the ban and reclassified as chemical weapons to evade the new policy. In January 1970 Meselson wrote another paper for the White House, which he said President Nixon read. It argued that classing the toxins produced by germs as chemical weapons was a technical distinction that undermined the administration's policy goals, as well as the president's credibility. "The arms control benefits of our newly decided policy," Meselson wrote, "will be reduced if we maintain biological laboratories where secret work is done." An active U.S. toxin weapons program, he argued, "would prevent us from demilitarizing and declassifying our biological research laboratories at Fort Detrick and our germ weapons production facility at Pine Bluff."

The next month, Nixon extended his renunciation of germ weapons to cover biological toxins. Detrick was out of the offensive business, after more than a third of a century. The White House directed the military to concentrate on germ defense, the mirror image of its previous work. Now the main goal was to foil a foe who might attack the United States with viruses, bacteria, or toxins. The Nixon administration became the world's leading advocate for a treaty banning germ warfare, which Washington now condemned as immoral and repugnant. Starting in 1972, the United States, the Soviet Union, and more than one hundred other nations signed the Biological and Toxin Weapons Convention. The accord prohibited the possession of deadly biological agents except for research into such defensive measures as vaccines, detectors, and protective gear. It was the world's first treaty to ban an entire class of weapons.

But it was only a pledge, and the treaty was filled with loopholes. No limits were set on the quantities of germs that could be used for research. No standards were stated for distinguishing between defensive and offensive work. No forms of consultation were specified. No mechanisms of enforcement were established.

Analysts inside and outside the government nonetheless saw the international ban as entirely in America's self-interest. Their logic was simple: when war was costly and military action required tons of advanced equipment and nuclear arms, only the wealthiest countries could play the game. Cheap arms of devastating force leveled the playing field. And that, experts agreed, was something that Washington wanted to avoid. It was a point that Meselson, among others, had raised early and had argued in public and private for years.

But Patrick and his colleagues felt that many countries were unlikely to honor their pledges to renounce germ weapons, which were so inexpensive and effective that they were destined to proliferate clandestinely. Moreover, he and his colleagues worried that such weapons would eventually be used against America. "Most of the people I worked with—the chief of the pilot plant division, the chief of munitions—all these people thought, 'Jeez, it's going to come around to bite us,' " he recalled. "This stuff is too damn good to go away."

Despite his misgivings, Patrick helped Fort Detrick go from "black to white," as he put it. He aided the three-year effort of destroying the offensive arsenal and then began working to develop germ defenses at the newly established United States Army Medical Research Institute of Infectious Diseases. This was the center of the government's new defensive effort. It was based at Fort Detrick, in the shadow of its predecessor. The defensive work had far fewer experts and buildings, and a much smaller budget, than the offensive program did. In time, the army lab became but one of many tenants on the sprawling Fort Detrick base.

But Patrick was uneasy. He not only disliked the gamble he felt the nation was taking but found that the new work ran counter to sensibilities he had developed over the decades. "It's a different world," Patrick said of finding ways to blunt germ attacks. "Defense studies are so much more complicated. It takes eighteen months to develop a weapons-grade agent and ten years to develop a good vaccine against it."

Patrick was frank about relishing his days as a germ warrior, saying he was comfortable with memories of killing animals, infecting people, and finding new ways to produce death. It was all part of military readiness, of deterring foes, of keeping the nation strong. "At the time we were doing this, the objective was to solve the problem and not consider the philosophical ramifications of what we were doing," he said. "On Fridays, when we'd sit around and bullshit, we wouldn't say, 'We have a moral obligation to curtail this!' It would be, 'How do we increase the concentration?' You never connected it to people. Maybe that's bad. But there was danger involved here.

"We used to think about the Chinese and the Russians. And if we had known what they were really doing, we would have worked harder."

3

Revelations

As the world renounced germ weapons, science was shattering the barriers between man and nature. For the first time, biologists learned how to manipulate the instruction books of life that reside inside all living things. It was a breakthrough that gave biological warfare new possibilities.

Nature had written the rulebooks over the eons. Science first glimpsed them in 1953, with the discovery of how DNA, or deoxyribonucleic acid, governs our heredity. The genetic code turned out to be locked in a double spiral whose structure was shared by everything from microbes to manatees. Surprisingly, the diverse rulebooks shared not only the same alphabet but the same language.

In the early 1970s, scientists took advantage of this commonality to found the field of gene engineering, also known as bioengineering or gene splicing. Now scientists could edit, rewrite, and rearrange life's script, moving a line or a whole chapter from one creature to another. It was biological cut and paste. With growing ease, genes were recombined and rearranged.

An early test centered on the firefly. Its gene for bioluminescence was added to a carrot, and lo, a new variety of carrot was produced that glowed in the dark. In the mid-1970s, scientists of a more practical bent spliced the human gene for insulin into bacteria and grew the hybrids in vats. In time, most diabetics came to rely on this mass-produced insulin, which was the first product of what became a multibillion-dollar industry. Thousands of recombinant projects sought to cure diseases, save lives, and improve human health.

The consequences were also profound for the world's militaries. The new biology might produce diseases unknown to modern medicine, and for which there were no cures. Superbugs might be made that were more potent, dangerous, and adept. Germs, it seemed, might become the ultimate weapon.

Scientists at the forefront of the genetics revolution were among the first to understand its potential for evil. A generation earlier, the physicists who invented the atom bomb devoted themselves to nuclear arms control. Now, biologists with the most intimate knowledge of the new science began searching for ways to contain the destructive power of their discoveries.

JOSHUA Lederberg grew up in New York City, the first son of an Orthodox rabbi. In 1958, at age thirty-three, he won the Nobel Prize for work he began at twenty. His research demonstrated that bacteria transferred genes among one another in a sexlike activity. In time, his discovery became central to the new science of gene engineering. Lederberg became a public advocate for the developing field but also saw its potential for making germ warfare even more dangerous. "To be enhancing that technology," he said in an interview, "I thought was in the long run suicidal."

His discovery of genetic exchanges in bacteria led him to warn about the dangers of antibiotic resistance decades before the issue caught public attention. Lederberg was among the first scientists to talk about cloning as a practical matter. He helped found not only microbial genetics but the fields of artificial intelligence in computers and the study of the likelihood of extraterrestrial life.

His lifelong war against biological weapons was informed by his personal knowledge of the military and its struggles with infectious disease. In 1943, while still a teenager, he'd won entry into a navy undergraduate program at Columbia University for future medical officers. In 1944, he advanced to become a medical student at Columbia's College of Physicians and Surgeons. His studies alternated with spells of duty at the U.S. Naval Hospital at Saint Albans, on Long Island. In the clinical pathology laboratory there, Lederberg peered through a microscope to examine stool specimens for parasite ova and blood smears for malaria. The disease had sickened Marines returning from the Guadalcanal campaign.

After receiving his doctorate at Yale, he took up a teaching post at the University of Wisconsin, which had played a major role in Fort Detrick's establishment as the nation's center for making germ weapons. A Wisconsin microbiologist, Edwin B. Fred, had advised Roosevelt's secretary of war on the need, and Ira L. Baldwin, another microbiologist at the school, had selected the Detrick site and become its first scientific director.

While at Wisconsin, Lederberg met Leo Szilard, who had helped make the atom bomb in the Second World War and afterward had gone to the University of Chicago to study biology. Chicago was full of Manhattan Project veterans—Szilard, Enrico Fermi, Edward Teller—debating what to do with the destructive force they had unleashed, some even advocating world government. Szilard favored treaties to limit nuclear arms. He and Lederberg hit it off, meeting first in biology seminars and then informally. Szilard, Lederberg said, became "my first role model" of a public-policy scientist.

Building on his earlier research, Lederberg in 1967 began attempting to splice genes. The experiments did not succeed, but Lederberg saw their promise. It might soon be possible, he reasoned, to create new forms of life in the laboratory. He did the experiments after he had moved from the University of Wisconsin to Stanford, the California university that in a few years would become the center of the genetics revolution.

During the Vietnam era, as the national revolt against biological warfare was gaining momentum, Lederberg became a columnist for the *Washington Post* and wrote in an August 1968 article that the creation of synthetic genes threatened humanity with the "systematic construction" of new disease agents and "the most perilous genocidal experimentation." He called for curbs on germ warfare, saying "time is running out."

Lederberg also campaigned in private to end the government's program and informed its leaders of the dangers he foresaw. In March 1969, he wrote Riley Housewright, Fort Detrick's scientific director, to propose issues for an upcoming discussion. "Where will it all end?" Lederberg asked of biological weapons. "Can BW proliferation be controlled? Will BW be adopted as a guerrilla tactic, and what countermeasures are possible?" It was an early reference to what would eventually be known as germ terrorism. "I am *very* worried about this," he wrote, "but hardly dare mention it for fear of putting an evil idea in someone's head."

Later in 1969 he told the House Foreign Affairs Committee that allowing new strides in germ weaponry "is akin to our arranging to make hydrogen bombs available at the supermarket." After Nixon's ban took effect later that year and Washington began lobbying for a global treaty, Lederberg was asked by the administration to advise the United Nations. In August 1970 he delivered a speech in Geneva to the U.N.'s Committee on Disarmament that was seen decades later as a turning point. Recent advances in science promised an arms race, he warned, "whose aim could well become the most efficient means for removing man from the planet." Infestations had wiped out the American chestnut and the European grapevine, he said. Now mankind faced similar dangers—"of our own invention." The peril, he said, had provoked "my own moral preoccupation with whether my own career will have been labeled a blessing or a curse to the humanity from which I spring." It was a dilemma, he added, that "faces my entire generation of biological research scientists and our younger students at this very moment."

In 1972, the Biological and Toxin Weapons Convention was opened for signature in London, Moscow, and Washington. In time, more than one hundred nations embraced the treaty's goal of abolishing biological weapons and declared germ warfare "repugnant to the conscience of mankind."

At exactly the same moment, civilian scientists were starting the gene revolution in earnest.

IT began over a midnight snack. In late 1972, Stanley N. Cohen, a young colleague of Lederberg's at the Stanford School of Medicine, and Herbert W. Boyer, a biochemist at the University of California at San Francisco, were attending a conference in Hawaii. As the scientists talked in a deli late one night, they realized that by teaming up they might do something that had never been done before: move DNA from one organism into another, creating a new form of life. They wanted to merge the genes of organisms that were wholly unrelated—for example, those of dissimilar germs or maybe even plants and animals.

Scientists had already taken a first step, discovering how enzymes could cut loose individual genes from a strand of DNA. Moving a gene from one organism to another was a greater challenge. Cohen and Boyer saw promise in the early experiments that showed how bugs accomplish such transfers in nature. Genes were found to move on tiny rings of DNA that pass freely between bacteria. Lederberg named them plasmids. Researchers had also recently figured out how to glue together snippets of DNA, suggesting an intriguing possibility. Perhaps genes snipped from one organism could hitch a ride on a plasmid as it moved into another. In theory, Cohen and Boyer realized, genetic engineering might be simple. The two scientists could slice the DNA from one organism, paste it into a plasmid, and move it into an unrelated form of life.

Back in California, the two scientists and their colleagues set to work. In months, they succeeded in creating the first recombinant life. They did not invent the basic technology, but their novel way of applying it quickly created a thriving new field.

The serious work began in early 1973. Boyer and Cohen's first experiments centered on *Escherichia coli,* the normally benign microbe of the human intestinal tract. From an unrelated microbe, they cut out a gene that conferred resistance to penicillin and spliced it into *E. coli.* Subsequent tests showed it had worked. The new microbe was now immune to the drug. In July 1973, they took a bigger step, splicing genes from the South African clawed toad into *E. coli.* The new germ successfully reproduced. And its progeny had toad genes. To a limited degree, science had merged bacterium and amphibian. Experimentation showed that not only plasmids but also viruses could be modified for the purpose of injecting foreign genes, and that targets could include not only microbes but plant and animal cells.

Until the 1973 successes, nobody knew whether genes from one organism would function or even survive in another. Cohen and Boyer showed that they did. Moreover, they demonstrated that crossing the ancient barriers between species, and in time even between plants and animals, was relatively easy.

Among the many repercussions was a very practical one. As germs were customized and multiplied, scientists found that their creations could be programmed to make many different kinds of products, such as insulin or other biological compounds. That process is known as gene expression. A germ can divide every half hour. So in a day, one microscopic parent can produce billions of identical offspring. Germs could be turned into vast factories producing a range of products.

As the feat was made public, scientists began to worry that the new biology could accidentally release into the environment novel organisms that would wreak havoc. Cohen and Boyer, aware of the danger, had been careful to insert genes that already existed in some populations of natural bacteria or were harmless. But anxiety spread, and scientists called for a temporary global moratorium on such work.

At an old chapel overlooking the Pacific, top biologists gathered in February 1975 to hammer out guidelines for research in the nascent field. The youthful leaders were present at the Asilomar conference center, as were its elder statesmen, including Lederberg, who was wary of restrictions on peaceful work, fearing they would slow the advance of genetic research. Altogether there were ninety American and fifty foreign scientists, including some from the Soviet Union.

One group of scientists at the meeting warned that the Pentagon might use genetic engineering to make new germs for biological warfare and called for an international treaty prohibiting gene splicing aimed at making weapons. That suggestion went nowhere. The organizing committee did, however, after considerable debate, win agreement for the most stringent safety precautions. The sticking point was a proposal to prohibit the most dangerous class of experiments—those that involved transplanting genes that are harmful to humans, such as genes that make botulinum toxin. In the end, the motion for a complete ban on such work passed, though five scientists cast dissenting votes.

The next month, March 1975, the Biological and Toxin Weapons Convention went into force, the culmination of three years of diplomacy. But the ban was silent on the central issue raised at Asilomar—the military implications of the new biology. The treaty said any kind of research was permitted, as long as its purpose was protective. But defensive research could look a lot like offensive work. And the gene splicing pioneered by Cohen and Boyer added an even more troubling complexity—the power to create superbugs. If a foe could defeat old vaccines and antidotes with new bugs, military scientists had reason to try the most lethal experiments imaginable. How else could they assess or counter new threats to their nations? The research was akin to manufacturer of bulletproof vests acquiring a new kind of pistol to assess the old fabric.

In the months and years that followed, the moratorium on potentially dangerous experiments ended as governments set up review panels to evaluate whether proposed research was safe. Gene engineering took off, first on college campuses, then in industry. Lederberg and Cohen, the Stanford colleagues, signed up as consultants to Cetus Corporation of Berkeley, California, which sought to capitalize on the gene-splicing discoveries. Soon, Cetus and its sister companies were churning out designer germs that produced human insulin, interferon, and growth hormone. The new field, Lederberg wrote in a 1975 article, "promises some of the most pervasive benefits for the public health since the discovery and promulgation of antibiotics."

NIXON resigned the presidency in August 1974. Soon afterward, Democratic leaders, in a newly combative mood and in firm control of the Senate, began investigating the government's recently abandoned efforts to harness pathogens for war. Lengthy hearings were held in the autumn of 1975, led by Senator Frank Church of Idaho, who was intent on exposing wrongdoing in the shadowy world of American intelligence. Nothing was said of government planning for biological attacks on Cuba or Vietnam. But the hearings and reports disclosed a new chapter of American history that featured Fort Detrick and the Central Intelligence Agency planning assassinations and stockpiling germs to cripple or kill foreign leaders. The partnership, which extended from 1952 to 1970, had no known successes. The investigation revealed a series of amateurish plots, failed hits, and embarrassing errors.

For instance, the Eisenhower administration, fearing a tilt toward Moscow in the newly independent Congo, had planned to kill Patrice Lumumba, a former postal clerk who in 1960 was elected prime minister. The CIA considered a number of candidate germs, including the virus that causes smallpox. It decided on botulinum toxin, which acted faster and was more reliable, its paralysis of lungs and muscles quickly resulting in the victim's death. The toxin was shipped to Africa. But tight security frustrated CIA agents and the plan was rendered moot when rebels seized power and killed Lumumba.

Another revelation centered on the Kennedy administration, still eager to undo Castro after the Cuban missile crisis. It produced a bizarre scheme to paralyze the dictator, who was a diving enthusiast, by giving him diving gear dusted with two pathogens. The wet suit was to carry a toxic fungus, and the breathing apparatus was to be laced with a bacterium that causes tuberculosis, a potentially fatal illness of coughing and lung hemorrhaging. The plot went awry when an American lawyer gave Castro a wet suit untreated with the CIA's poison.

As congressional investigators dug further, they found that even after Nixon's ban the CIA had kept a small arsenal of pathogens, germ toxins, and biological poisons that were strong enough to sicken or kill millions of people. Most of the agents were stored at Detrick, the inventory surprisingly diverse. It included not only relatively large amounts of the anthrax bacillus but also two types of *Salmonella*, the bug that the Rajneeshees had used in their restaurant attacks:

Bacillus anthracis (anthrax), 100 grams
Pasteurella tularensis (tularemia), 20 grams
Venezuelan equine encephalomyelitis virus (encephalitis), 20 grams
Coccidioides immitis (valley fever), 20 grams
Brucella suis (brucellosis), 2 to 3 grams
Brucella melitensis (brucellosis), 2 to 3 grams
Mycobacterium tuberculosis (tuberculosis), 3 grams
Salmonella typhimurium (food poisoning), 10 grams
Salmonella typhimurium (chlorine-resistant food poisoning), 3 grams
variola virus (smallpox), 50 grams
staphylococcal enterotoxin (food poisoning), 10 grams
Clostridium botulinum Type A (lethal food poisoning), 5 grams
paralytic shellfish poison, 5 grams
Bungarus candidas venom (lethal snake venom), 2 grams
Microcystis aeruginosa toxin (intestinal flu), 25 milligrams
toxiferine (paralyzer), 100 milligrams

The senators blasted the spy agency and army biologists for ignoring the president's order. "We must insist that these agencies operate strictly within the law," Senator Church said.

Bill Patrick, though never called to testify, was unapologetic about the arsenal. In interviews, he said he had assisted the CIA but seldom knew specifics. "If you were asked to prepare a product that had certain characteristics, you'd know damn well what it was for. But I would never be told, 'We want this because we want to assassinate somebody.' "

A CIA document released at a hearing said the characteristics of its stockpiled germs included forms "suitable for dusting of clothes, pillows, etc.," which conjured up visions of dictators being killed in bed.

Patrick was not alone, it turned out, in developing agents to maim and kill individuals. Such work had long predated his arrival at Fort Detrick.

The story of America's only known biological attack came out piecemeal in the hearings and made no headlines, unlike the Church committee's other disclosures. The first hint emerged on September 16, 1975, when William E. Colby, then director of central intelligence, told the senators that the poisonous cache had been created partly because a predecessor agency, the Office of Strategic Services, had used such toxins effectively. In fact, he said, the OSS had conducted "a successful operation using biological warfare materials to incapacitate a Nazi leader temporarily." He gave no details. And publicly at least, the senators demanded none.

Senator Edward M. Kennedy of Massachusetts reported at the end of a lengthy follow-up hearing in May 1977 that some CIA documents had arrived only "a few minutes ago" which disclosed that spies had attacked the still unidentified top Nazi "to prevent his appearance at a major economic conference during the war. They used food poisoning, which was rather successful."

But Senator Kennedy mentioned no toxin and identified no victim. It took a CIA document, buried in the back of the hearing record and all but lost publicly, to finally do so. Even this report, however, avoided the embarrassing truth.

The target of the clandestine attack was Hjalmar Schacht, the financial brains of the Third Reich. After Hitler's rise to power in 1933, Schacht was appointed president of the Reichsbank. Sometime in the early 1940s, OSS agents had obtained the requisite poison and administered it to Schacht, apparently in his food.

Technically, the attack was said to be successful. The chosen toxin was from the staphylococcus bacterium, one of the same poisons selected for the strike against Cuba. The symptoms that Schacht experienced most likely included chills, headache, muscle pain, coughing, and high fever. Perhaps most important, the assault remained secret.

The problem was that Schacht turned out to be an anti-Hitler conspirator, or at least he appeared to be one. When plotters unsuccessfully tried to assassinate Hitler in July 1944, Schacht was thrown into a concentration camp. American troops found him at Dachau in April 1945 and released him. The Nuremberg tribunal acquitted him of all war crimes.

As congressional investigators probed the American germ program, the U.S. intelligence community disagreed on whether the Soviet Union was secretly forging ahead on biological weapons. The disputes began in 1975, the year the germ treaty took effect. (The time allotted for the destruction of global germ arsenals had been three years, the time it took the United States to scrap its own biological weapons.) Several government officials wrote a secret paper arguing that the Soviet Union was illegally continuing an offensive program. Their conclusions were dismissed as speculative. At about the same time, a State Department official asserted that the Soviets' new intercontinental missile, the SS-11, was designed to carry a biological warhead. His bosses, one colleague said, found the idea ludicrous. A senior Soviet diplomat stationed at the United Nations defected to the United States in 1978 and warned that his country's military had embarked on a secret biological program, a claim that was viewed as intriguing but unsubstantiated.

Gary Crocker was a believer. A junior analyst at the State Department's intelligence arm, Crocker had been questioning the Soviets' biological intentions since shortly after joining the department in 1974. He had begun his career, as had many of his colleagues, studying the Soviet nuclear program. But Crocker had quickly gravitated to the arcane specialty of biological and chemical weapons, known colloquially inside the government as "bugs and gas." The United States employed thousands of people to follow the twists and turns in Moscow's strategic programs. Phalanxes of experts tracked each new missile and warhead. These glamorous jobs tended to be held by men.

The bugs-and-gas crowd was much smaller and included an unusually large number of women. For much of the 1970s, the group assigned to assess the world's biological and chemical weapons could fit comfortably in a small room. That meant that a single analyst, man or woman, could change a policy—if that person's voice could influence the wider bureaucracy. Crocker would spend most of his career representing the State Department at the monthly meetings. "We all knew each other and worked together," he recalled. "We were kind of a little clique fighting the big guys, the big powers in nuclear weapons. We were this little crowd coming along and saying, 'Hey, look out, there are chemical and biological weapons, too.' "

IN October 1979, a Russian-language newspaper in Frankfurt, West Germany, that regularly carried news from Soviet émigrés, ran a sketchy report about what it called a major germ accident. The death toll was put in the thousands. By January 1980, the newspaper had fresh details: the catastrophe had occurred in April of the previous year. An explosion at a secret military base near the Soviet city of Sverdlovsk, an industrial complex in the Ural Mountains that built tanks, rockets, and other weapons, had sent a cloud of deadly microbes wafting over a nearby village. An estimated one thousand people had eventually died from anthrax, a disease that without treatment is usually fatal in two weeks. The paper said the Soviet military had seized control of the area, covering the contaminated ground with fresh topsoil.

The story gained international prominence in February as the *Daily Telegraph* in London and the *Bild Zeitung* in Hamburg ran pieces on what they described as a calamity of agonizing deaths, cremated bodies, and widespread decontamination work.

U.S. intelligence analysts had noticed nothing unusual that previous April. Now they began scouring archives of spy satellite photos and intercepted communications from the weeks at issue. Pictures of the area around Compound 19, a military part of Sverdlovsk, seemed suspicious, showing roadblocks, newly paved streets, and what appeared to be decontamination trucks. A report was uncovered that showed that the Soviet defense minister had visited Sverdlovsk just after the incident was said to have taken place. The signs pointed toward a military accident.

That assessment was plausible. *Bacillus anthracis* is an ideal biological weapon, prized by Soviet and American germ warriors of the 1950s and 1960s for its hardiness and lethality. Lederberg calls it a "professional pathogen." The microbe's life cycle begins when livestock graze in a spore-laden pasture. The anthrax bacillus can enter the human body through multiple pathways. It attacks the lungs when people inhale spores flecked off an animal hide; it burrows into the digestive tract of those who eat contaminated meat; or it can infect the skin if it comes into contact with scrapes or open sores.

Anthrax is a very rare disease in modern societies, afflicting a handful of wool workers and others exposed to large numbers of animal hides. The disease follows an aggressive course. Bacteria from spores germinating in the lungs produce several toxins that attack human cells. The symptoms are initially benign, a cough and fatigue. Patients often experience a brief recovery, the so-called anthrax honeymoon, as their immune system valiantly fights the disease. The respite is almost always temporary, and it is almost invariably followed by collapse of the respiratory system and death. The disease follows a similar pattern in people who eat tainted meat.

The reports about Sverdlovsk came at a moment of deepening tensions between the superpowers. The Soviet invasion of Afghanistan the previous December had brought an abrupt end to a decade of détente, and President Jimmy Carter was taking a harder line toward Moscow. In March 1980, his administration privately queried Moscow about the Sverdlovsk incident. The next day, before the Soviets could reply, the State Department announced there were "disturbing indications" that a biological agent had been released. The issue had become a cudgel in the propaganda wars. Moscow denounced the allegations as an "epidemic of anti-Soviet hysteria." The outbreak at Sverdlovsk, it said, was caused by nothing more nefarious than tainted meat.

A group of intelligence analysts and outside experts convened by the administration dismissed the Soviet explanation. They reviewed the satellite photos and other evidence and closed ranks around this scenario: a mishap at the Sverdlovsk facility had released a plume of deadly anthrax spores into the wind; some people had died after inhaling the aerosolized microbes, and others had fallen ill weeks later after eating the meat of infected cattle. The group felt that it needed two etiologies because the epidemic had gone on for seven weeks. The disease, it concluded, came first by inhalation, then by consumption of stores of infected meat.

The CIA called in Matthew Meselson. The Harvard biologist and proponent of the Nixon ban was eager to see the secret intelligence. He stayed at the home of a CIA official for ten days. "I got hooked on this," he recalled. "And I became somewhat skeptical of their explanations." If the CIA's hypothesis was correct, Meselson reasoned, one should find anthrax deaths caused by both contaminated air and tainted meat. But the agency said its best medical evidence on the outbreak—secondhand reports from Soviet doctors—described only inhalation cases. Meselson felt the lack of corresponding evidence on intestinal anthrax "cast doubt" on the veracity of the sources, and thus the CIA's developing theory. Perhaps the agency's informants were unsophisticated. Perhaps they were unable to "reliably distinguish" between the two forms of the disease, and, hence, between a natural outbreak and one caused by germ weapons.

Around this time, intelligence analysts began piecing together what they viewed as an even more disturbing story of human suffering in a distant part of the world. Hmong refugees from Laos were reporting that helicopters flown by Soviet-backed government forces had sprayed them with a mysterious substance that left horrific burns and lesions on their skin. The refugees called it "yellow rain." To investigate, the army sent a medical team, which interviewed thirty-eight Hmong tribesmen in October 1979. The symptoms and accounts, the officers reported, seemed to suggest attacks by as many as three different toxic agents, including one that caused internal bleeding.

Ronald Reagan took office in January 1981, determined to confront what he would eventually call the "evil empire." Weeks later, the administration was handed a yellow-rain contaminated plant sample collected by the Thai military near the Cambodian border. A private American laboratory analyzed the material for Washington and reported that the leaf and stem were contaminated with unnaturally high levels of mycotoxins—harshly toxic substances that occur in low concentrations in some crops or can be extracted from fungi. Administration analysts believed the sample advanced the argument that the Soviets were using biological weapons, which was already being built around blood samples and eyewitness testimony. Still, Crocker and some of the bugs-and-gas experts urged caution. The results were unconfirmed, and other samples and blood awaited analysis. The Thai sample was important, but short of a smoking gun, Crocker warned.

Reagan's new secretary of state, Alexander M. Haig Jr., pressed ahead. On September 13, 1981, he announced at a conference in West Berlin that the United States had "physical evidence" that Moscow's surrogates in Southeast Asia were using deadly toxins to wage biological warfare. If true, the charges were stunning, suggesting a blatant violation of the germ treaty that went well beyond what might have happened accidentally at Sverdlovsk. Haig's allegations touched off a contentious debate that would echo through the rest of the decade, dividing scientists, government officials, and even America's allies.

Chief among the critics was Meselson, the champion of the germ treaty. His high-level security clearances as an army consultant and stature as a scientist gave him additional clout. Over several years, Meselson and his allies developed an alternate theory that many experts, and some government officials, found plausible. The yellowish samples from Southeast Asia were actually dried bee feces, not the residue of biological attacks. And the presence of mycotoxins was an illusion, a lab error.

Meselson began to poke holes in the government's arguments on Sverdlovsk as well. He discovered a witness overlooked by the intelligence agencies, an American professor living in the city on an exchange program, who said nothing unusual had happened in April 1979. The debate grew harsh. Some Democrats charged that the Reagan administration was manipulating the intelligence to win support for its stance against Moscow and its trillion-dollar military buildup.

From the start, Lederberg was called into the federal clashes over Sverdlovsk and yellow rain. In 1978, he had been named president of Rockefeller University in New York City, one of the nation's premier facilities for biomedical research. But eager to reach beyond academe and play a role in Washington, he had joined in 1979 the Defense Science Board, a group of top scientists and industrialists who advise the military. Intelligence officials were eager to hear the Nobel laureate's assessment of Sverdlovsk, but he was agnostic. "I couldn't really make up my mind," he acknowledged. "It looked very, very suspicious, and the Russians were not cooperating to provide access to the kind of information that could clear it up."

Lederberg thought that there might be a benign explanation, even if Moscow's infected-meat story turned out to be a lie. An accidental release of anthrax spores, he said, could have stemmed from an attempt to test defensive equipment. The treaty provides broad latitude for such research. "There's nothing that says you can't do aerosol experiments with anthrax on a modest scale," he said. "If you're trying to test a vaccine against aerosol-induced anthrax, you might have to do experiments" in which test animals were exposed.

Lederberg said that his foray into the intelligence world left him unimpressed. "They would pick up any rumor and try to make something of it," he said of analysts in the Reagan years.

The arguments over the Soviet germ program were rooted in a clash of cultures between the biologists and the intelligence analysts.

Scientists are trained to work solely from empirical evidence, preferably drawn from experiments they can observe and document firsthand. Almost everything else is suspect. "We're in the business of being skeptical," Meselson said of scientists. "If you've got to have an answer and you're willing to have a forty percent chance of being wrong, ask somebody else."

Intelligence analysis is as much an art as a science. The evidence—the reports from spies, the snippets of overheard conversations, the photos from satellites hundreds of miles overhead—is inevitably incomplete, sometimes contradictory, and always open to varying interpretations. The best analysts weave a coherent narrative from the conflicting reports, drawing careful distinctions between what is proven and what is suspected. Some of the arguments are circumstantial. Crocker, for example, gave great weight to the similarity of eyewitness testimonies from Laos, Cambodia, Afghanistan, and Yemen. Primitive tribesmen, separated by thousands of miles, all seemed to be describing the same sorts of attacks by the Soviets and their allies. The case that the Soviets had continued their biological weapons program was a mosaic, he argued, of which yellow rain was but a piece.

The conclusions of intelligence analysts are comparable to grand jury indictments, offering charges, not proof beyond a reasonable doubt. They spark debates more often than they settle them. The disputes can rage for years, settled only when intelligence officers recruit a spy or defector who delivers the inside story. The need for human intelligence, HUMINT as it is known, is particularly acute in the study of biological weapons programs, which can be easily hidden within legitimate research establishments. From the outside, research on defenses against germs does not look much different from the development of germ weapons.

CROCKER and his colleagues in the secretive world of bugs and gas soon began to gather evidence that the Soviet Union was embarking on a two-track strategy: the production of old-fashioned germ weapons like anthrax as well as futuristic research into bioengineered pathogens—the very dangers Lederberg had warned of in his speech to the United Nations years earlier. Sverdlovsk, they were sure, was just a glimmer of something much bigger. Their charges set off yet another bitter clash between intelligence analysts and scientists that persisted throughout the decade.

The fears of recombinant weapons were deepened by the practical applications of the scientific breakthroughs pioneered by Lederberg and others. In theory, the gene revolution gave any researcher the tools to make bad pathogens worse. But by this time several companies had produced products in which DNA was manipulated for profit and public health. As the companies did for peace, government analysts feared, the Soviets might do for war.

With the right equipment, military scientists could make a pathogen much hardier or even more lethal. Researchers might use the new techniques to turn harmless germs into killers. Overall, the advances threatened to tip the balance between offense and defense decisively in favor of the attacker. Genetic manipulation made it possible to redesign bugs like anthrax so they could evade vaccines, one of the best protections against a biological weapon.

A 1981 paper by the army's Dugway center in Utah, "Recombinant DNA and the Biological Warfare Threat," warned that designer germs could be used to make highly concentrated poisons. A 1983 army study suggested that bioengineering could turn anthrax into a better weapon for the battlefield by making its spores much more vulnerable to sunlight and natural decay. (One of the drawbacks to attacking an enemy with anthrax is that the spores can persist for decades.) Growing concern that an enemy might make such a superbug prompted the Reagan administration in 1984 to raise the Pentagon's germ budget so the military could better assess the threat. The gene breakthroughs, officials judged, "could be applied to the creation of novel biological agents, or to the production of specific agents (such as protein toxins) in quantities that far exceed their natural levels."

In April 1984, the Reagan administration went public with its worries. Its annual assessment of the threat posed by Moscow, *Soviet Military Power,* said the Soviet Union was conducting gene engineering aimed at developing new weapons. Days later, the *Wall Street Journal*'s editorial page began a nine-part series that made bolder accusations. Based largely on interviews with Soviet émigrés and liberally quoting Lederberg on the dangers of recombinant technology, the series charged that Moscow was pioneering new kinds of evil. Its scientists allegedly had put genes from cobra snakes into otherwise innocent bacteria and viruses, which would then produce deadly venom in a victim's body. "The revolutionary techniques of recombinant DNA," the *Journal* said, were being distorted "to create a new generation of germ-warfare agents."

CIA spies pursued the story and heard many of the same allegations. But the agency's senior Soviet specialists found it hard to believe that Moscow was investing huge sums of money in biological weapons. Douglas J. MacEachin, head of the agency's Soviet analysis office from March 1984 to March 1989, was one of the most determined skeptics. A former Marine Corps officer, MacEachin found the notion of germ weapons implausible. His idea of war was rifles, artillery, bayonets. He accepted the reality of nuclear weapons and strategic missiles, but the germ threat was science fiction to him. MacEachin found the bugs-and-gas crowd overzealous, drawing firm conclusions from scant evidence. The harder they argued, the less he believed them. Through the Reagan years, MacEachin fought a rearguard action against what he termed the "evil empirists," analysts who believed the reforms begun by Mikhail S. Gorbachev would never bring real change. MacEachin had no quarrel with the view that Sverdlovsk stemmed from the accidental release of germ weapons. The question was whether the Soviet program had continued into the 1980s. Meselson and Lederberg played a significant role in undermining the credibility of the government's bugs-and-gas analysts. "We had these esteemed scientists briefing us," MacEachin recalled. On one side stood "these Armageddon preachers, and then you've got Matt and Josh, and they're saying, 'This is all concocted.' "

Whatever the doubts behind closed doors, the Reagan administration battered the Soviets publicly and privately over germ weapons for much of the decade. It was perfect stuff for a propaganda war. A presidential commission charged in June 1985 that Moscow was investigating exotic arms that could cause "sudden panic or sleepiness in defending forces." In 1986 the Defense Intelligence Agency issued an unclassified, twenty-eight-page report that accused the Soviets of "rapidly incorporating biotechnological developments into their offensive BW program."

Douglas J. Feith, a senior Pentagon official, told the House Intelligence Committee in August 1986 that Soviet scientists had begun rearranging germs to develop "new means of biological warfare." The "stunning advances" in the field of biotechnology, he added, were inaugurating a new era in which designer pathogens could be made in hours, while antidotes like vaccines might take years. The United States filed formal diplomatic protests with Moscow over its germ activities in October 1984, February 1985, December 1985, August

1986, July 1988, and December 1988. Among the suspected sites were high-security compounds near Sverdlovsk and Zagorsk, an old town outside Moscow famous for its monastery, religious icons, and fairy-tale churches topped by onion domes of blue and gold.

The 1980s marked one of the most rapid military buildups in American history. The Pentagon was flush with money for new weapons systems, including billions for development of Star Wars anti-missile weapons. In an area of the budget where a small amount of money went a long way, funds for scientific research on biodefense doubled, reaching a high of $91 million annually. Mid-level officials believed that the United States was moving into an undeclared biological arms race with the Soviet Union, and they pushed America's best researchers to explore the frontiers of biotechnology, hoping to find new technologies that could help them understand the threat of new types of pathogens for war.

Several major American scientists contributed to the research, including Stanley Falkow, a Stanford University biologist and member of the National Academy of Sciences who was later elected president of the American Society of Microbiology. Working on a contract, Falkow and a colleague moved a dangerous gene into *Escherichia coli,* the common microbe of the human gut, creating a superbug that attacked human cells. The donor pathogen was *Yersinia pseudotuberculosis,* a less virulent cousin of *Yersinia pestis,* the microbe that causes plague and in the fourteenth century decimated the population of Europe in the Black Death. Writing guardedly in the unclassified scientific literature, Falkow and his collaborator, Ralph R. Isberg, said in September 1985 that they had named the gene section of *Y. pseudotuberculosis* that produces its virulence INV, for invasion. The germ's aggression, they asserted, "can be successfully mimicked." They said nothing of weapons or war. Instead, they argued that their study could help science understand how infections develop and how to block them.

Less openly, military scientists improved ways of transforming benign germs into dangerous ones. At an army institute in Washington, D.C., researchers reported that they had developed methods to modify the genetic makeup of normally harmless bacteria such as *E. coli* to produce any desired "level of pathogenicity."

The Reagan administration interpreted the 1972 biological weapons treaty as permitting all research with deadly pathogens, as long as the intent was peaceful. "We are doing research in genetic engineering and related disciplines to understand what's possible," acknowledged Richard L. Wagoner, an assistant secretary of defense and one of the program's managers. "But it's for the purpose of understanding how to design a defense, not to design an offense." An analysis of Pentagon research between 1980 and 1986 illustrated the shadowy line between the two. In the name of protecting soldiers and citizens, the American government had paid for fifty-one projects aimed at making novel pathogens, thirty-two at boosting toxin production, twenty-three at defeating vaccines, fourteen at inhibiting diagnosis, and three at outwitting protective drugs.

This research ended doubts about whether the production of recombinant germ weapons was possible. The uncertainty had arisen because scientists in academe and industry, instead of producing waves of superbugs, as some experts at Asilomar had feared, found themselves struggling to keep their designer germs alive. Most man-made bugs turned out to be weaker than their natural rivals. As a result, by the mid-1980s, regulators lifted most curbs on DNA research and approved the release of some bioengineered organisms into the environment. But the military's research demonstrated that what was rare by accident was feasible by design. If superbugs were hard to make *accidentally,* they could be made *deliberately.* It was simply a matter of sustained effort, one largely kept secret.

The military found it easy to recruit some of the nation's best scientists to work on such challenges, especially when the research extended the frontiers of genetics. But as a rule, top scientists were uninterested in working on mundane preparations for germ defense—the procurement of vaccines, medical treatments, and battlefield detectors. The Reagan administration had funded the Pentagon lavishly. But spending on biological protections was limited. And the military, even when it had the money, found it difficult to buy talent in fields that lacked scientific allure.

ON Christmas Eve 1984, David Huxsoll, the head of the United States Army Medical Research Institute of Infectious Diseases, and Richard Spertzel, a top germ specialist at Fort Detrick, worked late into the night because the Pentagon urgently wanted to know what supplies were needed for germ defense. The two men produced a crash paper that recommended the stockpiling of enough anthrax and botulinum vaccine to inoculate two million soldiers against attack.

Their assessment was part of a wider study of biological defenses led by General Maxwell R. Thurman, the army's vice chief of staff. A wiry, intense man with a buzz cut and a short temper, Thurman was known affectionately as "Mad Max." He was popular, a soldier's soldier. In 1985, he did what the military called a "functional area assessment" of germ defense, gathering several dozen top generals and civilians in a room to critique training, doctrine, and equipment. Thurman's report found deficiencies in all three. The military was seriously underprepared. American forces would be, in essence, naked if attacked by a conventional agent like anthrax. Bill Richardson, a Pentagon civilian with decades of experience in chemical and biological defense, admired Thurman's drive but felt that his confrontational tactics were doomed to fail. "Thurman pushed extremely, too extremely," Richardson recalled. "The system grinds on regardless, and vice chiefs come and go." Thurman's report was largely ignored.

As the document circulated, the army's medical officers asked pharmaceutical manufacturers to submit bids on what it would cost to build the stockpile recommended by Spertzel and Huxsoll. One company offered to open a new factory that would produce millions of doses of vaccine against anthrax, botulinum, and other agents. The price was high, several hundred million dollars, and army medical officials could not persuade the Pentagon to make the investment.

Tired of what he saw as the timidity and inaction at Fort Detrick, Bill Patrick quit his army job in 1986 soon after helping the FBI investigate the Rajneeshee assault in Oregon. He was fifty-nine but called himself an "old fossil" wedded to germ weapons. Sure that the threat was rising, Patrick set up a germ-warfare consulting firm that targeted federal agencies and private clients. His business card bore a skull and crossbones. Atop his stationery was a representation of the grim reaper, the black scythe labeled BIOLOGICAL WARFARE and the figure's outstretched arm sowing a rain of germs.

One of the key deficiencies of military preparedness identified by Thurman was a lack of vaccines against what Patrick called the "oldie moldies"—the traditional agents. It was a problem that Huxsoll, Spertzel, and Patrick had all agitated about at Fort Detrick, to no avail.

In the late 1980s, a young army major named Robert Eng decided to do something about the problem. Eng was serving in the office of the army surgeon general as the medical defense officer for NBC—nuclear, biological, and chemical weapons. Soon after he took the job, he learned that the military had little vaccine stockpiled to protect soldiers against anthrax or botulinum, the two best-known biological agents. Eng had read the intelligence reports about the biological warfare programs of the Soviet Union and other nations and recognized that if it came to war, military commanders would want adequate supplies of both vaccines. And so he set out to arrange the purchase of 500,000 doses of vaccine, enough to give 160,000 soldiers the first three shots of the six-shot series.

Eng turned for help to Anna Johnson-Winegar, a civilian army official with a doctorate in microbiology. Johnson-Winegar understood the destructive power of the anthrax bacillus firsthand. She had specialized in the pathogen while working at Fort Detrick as a researcher before moving to the army command that handled medical programs. Johnson-Winegar was one of the few people in the United States government to have received the anthrax inoculation, which was developed in the 1950s and licensed by the Food and Drug Administration in 1970. The vaccine was administered in six shots spaced out over eighteen months. Doctors had little experience with it. In nearly two decades, the vaccine had been given only to a small number of technicians and researchers at Fort Detrick and several hundred workers at wool factories. Some of those receiving the shots complained of swelling and other localized reactions after receiving the injections. There were no reports of long-lasting harm.

The army understood the shortcomings of the anthrax vaccine. A few years earlier, in 1985, it had invited pharmaceutical manufacturers to develop an improved product. The existing formulation, it said, was "highly reactogenic," or likely to cause reactions, and "may not be protective against all strains of the anthrax bacillus." No major drug makers bid on the contract. The disease was rare, the army's commitment to buying large quantities uncertain.

America's major pharmaceutical manufacturers were fleeing the vaccine business, driven out by low profits and a rising number of lawsuits. Universal vaccinations for children had vanquished diseases such as whooping cough. But a handful of patients had suffered severe reactions after taking the shots, including brain damage and death. The scientific evidence tying these cases to immunizations was in dispute, but juries were sympathetic to the young victims, and had delivered some staggering judgments against drug manufacturers. By the mid-1980s, shortages of vaccine were emerging.

There was a single, licensed manufacturer of anthrax vaccine in the

United States: an antiquated laboratory operated by the Michigan Department of Public Health in Lansing. Established in 1925, the lab had developed the first diphtheria-tetanus-pertussis vaccine. Some of its buildings dated to the Depression; others had been built in World War II by the army.

Officials at the Michigan lab told Johnson-Winegar that they could not possibly meet Eng's goal. The facility had been making small batches of the anthrax vaccine about once every four years, yielding between fifteen thousand and seventeen thousand doses. On September 29, 1988, the army signed its first-ever contract to buy large quantities of anthrax vaccine, asking Michigan to produce and store three hundred thousand doses.

Under the contract, the army agreed to pay for bigger fermenters, in which germs are grown for vaccine making. Much of the lab's equipment was more than twenty years old and, as one army document put it, "certainly not state-of-the-art." Recognizing the difficulties of speeding up production, the army gave Michigan five years—until September 1993—to deliver the three hundred thousand doses.

THE number of analysts working on bugs and gas had grown rapidly during the Reagan years, and the intelligence agencies began to collect new information about germ weapons and to issue reports on their spread to Third World countries. In June 1988, they said Iraq was well on its way to building "a bacteriological arsenal" under the cover of legitimate scientific research. The classified study was produced at Fort Detrick, by the Armed Forces Medical Intelligence Center.

The center had grown in the early 1980s from a single analyst who monitored the world's biological programs to a staff with many different specialists, including a microbiologist with advanced training in genetic engineering. The center's 1988 report on Iraq focused on the oldie moldies. The analysts said Baghdad's scientists had already produced weapons from *Clostridium botulinum,* which produces a deadly toxin ten thousand times more lethal than nerve gas. It said Iraq was also trying to produce "research and development" quantities of weapons from anthrax and other deadly microbes.

Intelligence reports are often larded with caveats, but this one included several alarming claims. It said Iraqi scientists were developing germs for assassinations. And it reported that Hussein Kamel, President Saddam Hussein's son-in-law and the head of Baghdad's powerful intelligence agency, was personally supervising the biological research. These details, which suggested that the program was supported by the most important elements of the Iraqi dictatorship, were accompanied by a revelation that should have caught the eye of senior American officials. Iraqi scientists were buying their starter germs—the foundation of any biological-weapons program—from an American company. "The Iraqis are very good organizers and excel at purchasing needed offensive supplies using procedures which effectively disguise their intent," the report said. "They are currently purchasing bacterial strains from the American Type Culture Collection."

A scientific supply company then located in the Maryland suburbs of Washington, the American Type Culture Collection, the same company from which the Rajneeshee cult in Oregon had bought some of its germs, was a natural focal point for Iraq's clandestine effort. It housed the world's largest collection of germ strains, including particularly virulent variants of anthrax and botulinum discovered in the mid-1950s as part of the American germ-warfare program. The collection served as a global lending library for scientists beginning their own research in the field of microbiology and was considered an important tool in the fight to improve global health. Overseas customers were required to obtain a license from the Commerce Department to export the most virulent strains, but in 1988 this was largely a formality. Applications, from Iraq or anywhere else, were seldom denied. Two years earlier, in May 1986, the company had sold an assortment of germs to the University of Baghdad, including three different types of anthrax, five variants of botulinum, and three kinds of brucella, which causes an animal disease, brucellosis, that is incapacitating but rarely fatal. Further orders were planned.

There seemed little reason to stop them. In the late 1980s, American officials viewed Iraq as more of an ally than a possible foe. For most of the decade, American policy had been to "tilt" toward Baghdad in its long war with Iran, and the United States even provided the Iraqi army with spy satellite photographs of Iranian troop formations. The American military's contingency plan for the Persian Gulf presumed that American forces would be called upon to defend the region's oil fields against a Soviet thrust into Iran and that they would join forces with Iraq to fight the Soviet army. American officials had been disturbed by Iraq's use of nerve gas against its Kurdish minority. Nevertheless, the relationship continued.

The intelligence report on Iraq's biological efforts went to the State Department, the CIA, and various parts of the military, including the Central Command, which planned and directed American operations in the Middle East. It prompted no action. No one called the American Type Culture Collection to warn against further sales of germs to Baghdad. No one suggested that the Commerce Department withhold licenses on germ exports bound for Iraq.

On September 29, 1988, three months after the report was completed, eleven strains of germs, including four types of anthrax, were sent from Maryland to Iraq. The order was placed by the Iraqi Ministry of Trade's Technical and Scientific Materials Import Division, or TSMID, which American intelligence officials had already identified as a front for Baghdad's purchases of germ weapons materials. One of the microbes, strain 11966, was a type of anthrax developed by Fort Detrick in 1951 for germ warfare.

Belatedly, the intelligence on Iraq prompted American officials to prohibit the transfer of more germs to Baghdad. On February 23, 1989, the Commerce Department banned sales of anthrax and dozens of other pathogens to not only Iraq but Iran, Libya, and Syria, which were also suspected of trying to develop germ weapons. The ban, said the announcement, grew out of "heightened concern" about the biological threat.

As the bugs-and-gas experts worried about new foreign threats, Lederberg turned his attention to new microbial dangers at home. Weapons were only the most dramatic part of the threat, he declared. In fact, human beings were beginning to lose the battle against infectious diseases. The bugs, through the natural processes of genetic mutation and evolutionary selection, were proving increasingly resistant to medicine's arsenal of antibiotics and drugs—as witnessed by rises in AIDS, drug-resistant tuberculosis, and new varieties of *E. coli* that were dangerous or could even kill. Old diseases once thought to be defeated were showing new strength, and the authorities found themselves poorly equipped to anticipate or manage these new kinds of infections. In 1988, Lederberg and a small group of virologists and tropical-medicine specialists plotted strategy. In May 1989, they acted.

"Nature isn't benign," Lederberg told a scientific audience at a three-day meeting at the Hotel Washington, just down the street from the White House. "The survival of the human species is *not* a preordained evolutionary program."

Lederberg's warning had broad support among knowledgeable doctors and scientists. The number of Americans who died of infectious disease was rising rapidly. AIDS alone was a raging epidemic. The meeting was sponsored in part by the federal government's National Institutes of Health, and a follow-up report by the Institute of Medicine concluded that the danger was real.

But new money to fight the rise in infectious disease proved to be extraordinarily hard to attract. In 1951, near the height of germ-warfare fears, it had been easy to found the Epidemic Intelligence Service, which trained thousands of disease detectives. Now, however, antibiotics were seen as triumphant. Public health and microbiology had become backwaters, and the billions of federal dollars spent on biology tended to go into research aimed at cancer and the illnesses of old age, such as heart disease. Public health was messy and old-fashioned. It had a weak constituency and little presence on the public agenda. Trillions of dollars were spent on weapons. For lack of a tiny fraction of that amount, millions of children throughout the world died of infectious diseases each year.

LEDERBERG made little headway. In the last years of the Reagan administration and the early years of George Bush's presidency, Washington had limited enthusiasm for countering any kind of germ threat, whether from nature or human adversaries. The cold war was ending, the propaganda war against Moscow winding down. Independent experts increasingly questioned whether anyone was really building deadly biological arms. The treaty was working, they said.

The doubts reached into the ranks of the nation's military officers, most of whom had never liked germ weapons in the first place. Many approved of the ban. By the late 1980s, germ warfare was on the fringes of American military strategy. And the gene revolution just made things worse, with its vision of more terrifying agents that could outwit vaccines. When American national-security officials weighed the military balance with the Soviets, they worried about conventional weapons, nuclear missiles, and Moscow's huge arsenal of chemical munitions. The Soviet Union's extensive civil-defense preparations against biological attack, including shelters with specially designed air filters, were seen as just another example of Moscow's paranoia. After all, Soviet troops did not conduct exercises in the use of germ weapons. The problems resulting from a biological attack were too staggering to contemplate. What could a NATO force defending Europe really do if it were doused by anthrax or some other germ? The airplanes, airfields, and gear would all be contaminated. The soldiers would all be dead or gasping for breath. Nothing in the warrior culture of the American military accustomed planners to thinking about battle as a medical competition.

General Brent Scowcroft, a retired air force officer who had served as national security adviser to Presidents Ford and Bush, summed up the prevailing attitude in an interview years later. "I have a bias," he said. "To me, biological weapons are not practical or militarily useful. As terror weapons, they're fine. But in the context of warfare, they're too imprecise, the reactions too delayed. They're not likely to be employed."

The United States military had no real plans for how it might defend itself against a germ threat. "There's an in box, an out box, and a too-hard-to-do box," recalled one officer. "We saw it as a threat, but we didn't want to deal with it, to put together a war plan. It was too difficult." The same "too-hard-to-do" logic marked the military's response to nuclear threats. The weapons were so quick and devastating that, after decades of hard study and research, officials had found that no defenses were possible, despite President Reagan's dream of a Star Wars shield. If anything, the problem of germ protection was more complicated. An adversary could hide germ weapons amid factories that were ostensibly civilian, making it hard for a defender to draw up realistic threat assessments.

The United States had no plans to protect the civilian population from biological attack. A few specialists, including Bill Patrick and Josh Lederberg, had begun warning that terrorists might try germ weapons. But the Oregon salad-bar attack was forgotten, and most senior government officials who thought about the problem at all saw germ weapons as an issue for the military. Why would any group want to unleash an epidemic when explosives could make a similar point without endangering the attackers?

In the late 1980s, skepticism over the germ threat rose as civilian specialists assailed the military's financing of recombinant experiments in the name of biological defense. It was pointless, they charged. Charles Piller, an investigative journalist, and Keith R. Yamamoto, a molecular biologist at the University of California at San Francisco, obtained a stack of documents describing the research through the Freedom of Information Act and in 1988 published *Gene Wars: Military Control over the New Genetic Technologies.* The authors condemned the burgeoning effort as needless and illusory. Germs in their natural state were already horrible enough. Moreover, superbugs could never be turned into viable weapons, they argued, since any country developing them would have to proceed in secrecy and be unable to test them properly. The threat was overstated, Piller and Yamamoto wrote. "Imaginations have run wild."

Their arguments gained political currency. On May 17, 1989, the Senate Governmental Affairs Committee held a hearing on the Pentagon-financed research. Yamamoto testified, citing new evidence of the military's recombinant DNA research, including development of a type of botulism germ that could evade vaccines. "I see no evidence for a BW biotech research race," Yamamoto told the senators, referring to possible biological warfare strides by the Soviets. "But if there is one, the United States is far ahead."

Senator John Glenn, an Ohio Democrat and the committee's chairman, sent Congress's investigative arm, the General Accounting Office, to review the Pentagon's germ research. The investigators' report, released in December 1990, questioned the military's choice of projects. American intelligence agencies had a formal process for producing a "validated threat list"—the list of pathogens believed to be in development around the world as weapons. GAO investigators found that forty-nine of the government-financed projects, totaling $47 million, involved exotic germs not on the list, suggesting that the research lacked a defensive rationale. "The Army," the report concluded, "unnecessarily expended funds on research and development efforts that did not address validated threats."

Eileen R. Choffnes, a Glenn aide with a master's degree in genetics and doctorate in toxicology, helped organize the 1989 hearing and subsequent curbs on the new wave of military research. "Yamamoto and Piller were right," she said. "It was even worse than they thought."

Philip K. Russell, the head of the army's medical research and development command at Fort Detrick, was staggered by the GAO report and the Glenn hearing. Russell had ordered studies of the exotic viruses after reading intelligence reports that said the Soviets were aggressively pursuing them. Putting a new agent on the threat list was a time-consuming, bureaucratic exercise. Russell did not feel he should wait. "We knew the Russians had sent people to Africa trying to collect Ebola and Marburg viruses," he recalled. "That was good enough for us."

Senator Glenn believed that the army had veered dangerously close to offensive research. A retired Marine Corps officer, Glenn was no foe of defense spending. He just thought the United States government had no business toying with superbugs in its laboratories. It was repugnant, even if permitted under a lawyerly interpretation of the treaty. Glenn pressed to restrict funds. Russell was frustrated. He felt certain that the Soviets were making progress on viral weapons, yet the Senate had blocked him from research that could help defenses. "It was a sad set of events, a tragedy," he said years later. "It set our virus programs way back."

To the public, the case that Moscow was making biological weapons seemed cloudier than ever. Meselson and such allies as Thomas D. Seeley of Yale University had undermined the government's claims on yellow rain, making their arguments in a series of carefully documented articles over several years. One offered a side-by-side comparison of bee pollen and yellow-rain samples as viewed through electron microscopes. They looked identical. A 1987 article in *Foreign Policy,* "Yellow Rain in Southeast Asia: The Story Collapses," noted that government laboratories in Britain and the United States had looked for mycotoxins in more than one hundred samples and failed to find them. Quoting documents obtained through the Freedom of Information Act, the article disclosed that an American team had scoured Southeast Asia for two years in a failed effort to find evidence supporting the charges. According to the documents, the team re-interviewed the original eyewitnesses, several of whom recanted and said their accounts were, in fact, secondhand.

The controversy over yellow rain faded even though some American officials continued to believe that the Soviets had used unconventional agents in Southeast Asia. Lederberg remained suspicious. The evidence on mycotoxins was undeniably weak; the government had left itself open to ridicule with its flawed science. But he was bothered by the persistent reports that the Soviets and their allies had used germ or chemical weapons. The case, Lederberg felt, remained open.

Questions also remained about Sverdlovsk. The Soviets had repeatedly rebuffed the requests from Meselson and others to allow an independent team to visit the city and investigate the outbreak. But Moscow eventually said it had new studies showing that the deaths were caused by tainted meat. In April 1988, Meselson arranged for two Soviet scientists to come to the United States to make their case at public meetings in Cambridge, Baltimore, and Washington. The meetings were polite, with no pointed challenges to the bad-meat claim. Meselson said he found the Soviet account "completely plausible and consistent with what we know from the literature and other recorded experiences with animal and human anthrax." Still, he thought more investigation was needed inside the Soviet Union.

Russell was unpersuaded and felt that the Soviets had held back crucial evidence. Anthrax contracted from airborne spores can be distinguished from the intestinal variety. A patient infected through the lungs typically suffers severe, distinctive damage to the mediastinum, the area between the lungs just below the sternum. The Soviets presented autopsy slides of several people who died in Sverdlovsk. But none showed the key tissues.

In an article for an arms-control journal a few months later, Meselson said the American charges about a germ warfare accident at Sverdlovsk were clearly "in need of careful and objective review." Further details from the Soviets, he wrote, could eliminate "any remaining reasonable doubt that the outbreak was a natural occurrence." At a Senate hearing in May 1989, Meselson articulated his views before a more influential audience. "The burden of the evidence available," he said, "is that the anthrax outbreak was the result of a failure to keep anthrax-infected animals off the civilian meat market, as the Soviets have maintained, and not the result of an explosion at a biological weapons factory, as previously asserted by the United States."

Meselson told the senators that he believed that the germ treaty had been a success around the globe: "Today, to the best of my knowledge, no nation possesses a stockpile of biological or toxin weapons."

A FEW months later, in October 1989, a top Soviet biologist defected to Britain and told a very different story. Vladimir Pasechnik described to interrogators a top-secret world in which tens of thousands of specialists were laboring to perfect and deploy germ weapons. Yes, Pasechnik said, the Soviet Union had produced long-range missiles to deliver germs. He had been the director of the Institute for Ultra-Pure Biological Preparations in Leningrad, one of scores of false fronts scattered throughout the Soviet Union. His institute of four hundred scientists, he said, had been doing research on modifying cruise missiles to spread germs. The low-flying robotic craft, which can outwit early-warning systems, were adapted to spray clouds of aerosolized pathogens over unsuspecting foes.

Never before had Western experts penetrated so deeply into the Soviet germ program, and what they found seemed to go beyond the most hawkish views of the bugs-and-gas experts. American intelligence analysts had guessed that the Soviets might be working on plague. But Patrick and other experts had dismissed the idea, noting that the United States had never produced an aerosolized plague microbe that could survive outside the laboratory. Pasechnik told his British handlers that his institute had succeeded where the Americans failed. It had even created a new, genetically improved version of the Black Death.

The superbug that he and his colleagues had created was more resistant to heat, cold, and antibiotics, Pasechnik said. The Soviets had put the discoveries of Cohen and Boyer to lethal use. One of the Americans' early experiments had created a penicillin-resistant strain of *E. coli,* the common bacteria. The Soviets had used the same techniques to perfect their plague weapon. Plasmids, the tiny rings of DNA crucial for gene engineering, also controlled a germ's natural defenses against microbial foes. The Soviets had tinkered with the plasmids of plague, making the new strain impervious to existing vaccines and antibiotics. It was a disconcerting milestone.

Pasechnik's disclosures gave Moscow's work on classical agents and engineered pathogens new importance. Soviet scientists had already married the old and new approaches to create an entirely new class of deadly weapon, Pasechnik alleged. Superplague, he explained, was no mere laboratory curiosity. The Soviets had packed a dry, powdered form of the germ into bombs, rocket warheads, and artillery shells, and it was only one of many improved agents that the Soviet Union had developed.

Intelligence officers view defectors with suspicion. Such men tend to say what their new masters want to hear. But Pasechnik made an immediate, positive impression on the two British experts who debriefed him. One of them, David Kelly, a microbiologist and top British specialist on germ weapons, concluded that Pasechnik was a meticulous witness, careful to distinguish between what he knew from personal experience and what he had heard from others. "He didn't waffle," Kelly said. "When he didn't know, he said so."

By this point Douglas MacEachin, the skeptical CIA analyst, had been assigned to support the Bush administration's diplomatic maneuvering over the Soviet germ program. In early 1990 he learned from a friend in the directorate of operations, the agency's clandestine service, that British intelligence was debriefing a highly placed defector from the Soviet germ program. The name was omitted. "I said, 'Okay.' I knew this was going to be a big problem," he recalled. He asked a technical expert on his staff to review earlier studies of the Soviet program as well as satellite photos and communications intercepts. Nearly everything the mysterious defector said could be independently corroborated. The agency even sent Josh Lederberg to London to debrief Pasechnik, government officials said. The biologist who had issued some of the earliest warnings about the new biology reported that the defector appeared to be genuine.

MacEachin was now ready to believe that it was possible that the Soviet Union had been maintaining a vast, secret germ-weapons program, just as the bugs-and-gas analysts had been charging for a decade and a half. In February 1990, MacEachin briefed the UNGROUP, the Bush administration's secret committee for dealing with arms-control initiatives. He said, "We have extremely credible information from an extremely credible source," adding, "We've gotten tons of corroboration." The group immediately decided to confront the Soviets with the allegations. The charges seemed to be wild. But the group agreed that they had to be resolved.

Within a month, the British had raised the issue with Dmitri Yazov, the Soviet defense minister. He replied, "I know where you're getting that. It's from Pasechnik." A month later, Secretary of State James A. Baker III was in Russia for a meeting with his Soviet counterpart, Eduard Shevardnadze. MacEachin, who was part of the delegation, drafted a summary of the charges that Baker wanted to give to Shevardnadze as they went sightseeing. He asked one of Baker's aides where the group was spending the afternoon. They're visiting the monastery in Zagorsk, he was told. The irony was huge. Right there, outside Zagorsk, behind high walls and barbed wire, was where the Soviet Union had set up one of its secret germ facilities.

MacEachin acknowledged that Pasechnik's disclosures had prompted him to see years of evidence in a new light. The zealots, he candidly acknowledged in a recent interview, turned out to have had a point. "They were right. There was this program." And it turned out, he added, that the Soviets "were investing in it."

American intelligence officials began to look closely at foreign germ efforts, especially those involving the new biology. In June 1990, analysts zeroed in on Iraq's research center at Al Tuwaitha, near Baghdad. The facility employed some of Iraq's best military scientists and in 1981 had been preemptively bombed by Israel, which suspected it was making fuel for nuclear weapons. Now American experts at the Defense Intelligence Agency studied its buying patterns for advanced supplies and grew to suspect that Al Tuwaitha was doing "molecular biology and genetic engineering" for new classes of biological weapons, the same thing Pasechnik had warned that the Soviets were doing. This was plausible, the analysts wrote in a report, because the center had produced "no publications in these areas." There were, of course, innocent explanations for the purchases. But Al Tuwaitha's history and its high security precautions worried the analysts. Was Iraq, like the Soviet Union, trying to develop a lethal edge?

4

Saddam

In September 1990, a researcher breathing though a respirator and clad in protective gear—head-to-toe suit, two pairs of gloves, and two pairs of boots—walked into a laboratory at Fort Detrick and injected a rhesus monkey with enough anesthetic to render it unconscious. As the monkey's breathing slowed, she moved the monkey to a Plexiglas box, placed its head in a rubber-lined enclosure, and turned on an aerosol device, which released a carefully measured dose of *Bacillus anthracis,* the microbe that causes anthrax. The germs were Vollum 1B, the same strain sold to Iraq. The particles of the aerosol mist measured about one micron—ideal for lodging in the lungs. One by one, sixty monkeys were infected and returned to their cages. The experiment, one of the most ambitious ever attempted at Fort Detrick, had an unusual practical urgency. Saddam Hussein's army had invaded Kuwait on August 2, 1990, and American intelligence agencies believed that his arsenal included germ weapons. The experiment was designed to find out which treatments offered the best protection after an anthrax attack: antibiotics alone, vaccines, or a combination?

Years of neglect had left American forces vulnerable. None of the soldiers or reservists serving in the military had been inoculated against anthrax or other well-known biological threats, despite Sverdlovsk and the accumulating evidence of wide Soviet preparations for germ warfare. The Pentagon had no plans for how it might "surge" production of vaccine in a crisis and had little of it on hand. Even more significant, the armed forces lacked equipment to detect attacks with germ weapons such as anthrax, which takes hold in the body over several days. By the time a patient infected with *Bacillus anthracis* becomes feverish, the first sign of the illness, it is often too late. The scramble at Fort Detrick to study possible treatments of the disease was only one sign of how much was lacking.

The experiment took place at Building 1412, not far from where American scientists had spent the 1950s and 1960s perfecting germ weapons. A specially designed ventilation system kept the air pressure slightly lower than the outside air, an arrangement known as negative pressure. If any microbes were accidentally released, they would be sucked back into the lab. Technicians followed stringent rules for the handling of deadly organisms, obeying practices required in bio-safety level 3 labs. Only "hot" agents like Ebola, for which there is no treatment, are handled more carefully.

The researchers knew surprisingly little about how to combat anthrax. The studies leading to the Food and Drug Administration's approval of the vaccine in 1970 involved only a handful of people afflicted with the inhaled form of the disease. Studies suggested that short courses of antibiotics were ineffective, but Arthur M. Friedlander, the chief of the bacteriology division at Fort Detrick, had a hypothesis that large doses of antibiotics administered for at least a month would vanquish the disease.

Researchers divided the infected monkeys into six groups. One group got the vaccine alone; another received the vaccine and antibiotics; and three groups were treated for thirty days with antibiotics—penicillin, doxycycline, or ciprofloxacin. A control group of ten monkeys received saline injections every twelve hours. Days after they inhaled their first spores of *Bacillus anthracis,* the monkeys in the control group began to lose their struggle with the disease. Their final hours offered a grisly proof of the microbe's lethal power. By day eight, all but one were dead. Two days later, eight of the ten monkeys given the vaccine alone had died as well. By contrast, the monkeys treated with daily antibiotics survived into the fourth week.

The experiment was exhausting, a staggering exercise in logistics. One of the antibiotics in the study could be administered by injection, but the other two could be given only in oral form, which meant the monkeys had to be anesthetized twice a day and fed the drug through tubes threaded down their mouths and into their stomachs. Every dead monkey was immediately autopsied. Friedlander marveled at the intensity of the teamwork. Researchers are solitary sorts, preferring to work alone or with a cadre of trusted colleagues. This experiment involved more than sixty scientists and technicians. The monkeys were anesthesized more than 3,700 times, and the technicians took more than a thousand blood cultures to chart the progress of the disease. "It was an extraordinary effort, a very frantic time," Friedlander said. The military kept a close eye on the experiment. "There were people calling on a daily basis from the Pentagon asking, generals asking, 'What happened today?' "

After a month, the results were in and they deepened the dilemma facing General H. Norman Schwarzkopf, the field commander of the allied army assembling in Saudi Arabia. Thirty days of antibiotics had prevented anthrax in most of the monkeys. But the anthrax spores, which can hide for weeks in the lymph system, proved resilient, and there were some differences among the antibiotics. Three of the ten monkeys who received penicillin died after the treatment was halted. Those given doxycycline or ciprofloxacin fared better, with only one monkey in each group dying of anthrax after the treatments stopped. The combination of vaccine and doxycycline performed best, with all monkeys surviving through the months after the experiment ended. The study included a significant proviso: every monkey had begun treatment within twenty-four hours of exposure.

The message seemed clear: a month of antibiotics improved a patient's odds but were not a complete solution. Even the newest and most effective of them, doxycycline or ciprofloxacin, eventually failed in about 10 percent of cases. That would potentially mean tens of thousands of casualties if a large group of soldiers was attacked. The best defense against germ attack, the experiment suggested, was inoculating the troops before they set foot in the Persian Gulf region. If that wasn't possible—and it almost certainly was not, given the shortage of vaccine—allied forces could feel confident in the face of the anthrax threat only if they could administer antibiotics and vaccine within a day of attack. They needed three items: detectors that could warn immediately of exposure, large reserves of vaccine, and enough antibiotics to treat every soldier for at least a month.

Antibiotics could be found on the global market. But the technology to detect airborne anthrax spores was in its infancy and would almost certainly not be available before the January 15, 1991, deadline set by the United Nations for Iraq's withdrawal from Kuwait. Prospects for boosting supplies of vaccine by that date were equally clouded. The facilities at the Michigan Department of Public Health could not come close to bridging the gap. American industry had met such challenges before, but speeding up production of a drug was not like ordering a defense contractor to punch out more tanks or mortars. A new vaccine factory would need FDA approval and would have to comply with strict rules for making drugs, a process that takes years.

SCHWARZKOPF had begun to worry about the Iraqi threat months earlier, soon after he took over the Central Command, or CENTCOM, which was responsible for defending the Persian Gulf. In a war game in late July, his commanders rehearsed how they might respond to an Iraqi invasion of Kuwait and Saudi Arabia. The exercise, codenamed Internal Look90, made clear the logistical problems an American force would face fighting a war halfway around the world. Saddam Hussein's chemical weapons rated a brief mention. None of the participants explored the question of how the war might be affected by an Iraqi biological arsenal.

Days after the Iraqi invasion, Schwarzkopf and a delegation of Americans asked King Fahd of Saudi Arabia to accept the deployment of hundreds of thousands of American troops on his soil. The king agreed, and as the American delegation flew over Egypt on its way back to the United States, Schwarzkopf's mind was racing over the challenges of assembling a credible force in the desert. He picked up a secure telephone and placed a call to his chief of staff, General Robert B. Johnston, who was in the headquarters of the Central Command in Tampa, Florida. Schwarzkopf ran briskly through his list of pressing issues. He was moving the command's headquarters to Saudi Arabia to manage the operation. "In all directives and orders, include that we must be prepared to fight in a chemical environment," he said. Johnston took Schwarzkopf's order as shorthand. "Chemical" in military parlance meant chem-bio, chemical *and* biological weapons.

Intelligence analyses of Iraq's biological capabilities began to circulate more widely at the Pentagon. On August 6, 1990, the navy sent its commanders a message warning that Iraq might have germ weapons that could be effective against ships at distances of up to twenty-five miles. The initial assessment, the message said, was that the Iraqis "would deploy these agents if needed." Analysts at the Armed Forces Medical Intelligence sent around a report summarizing the threat. "Iraq has a mature offensive BW program," they wrote. "Substantial amounts of botulinum toxin likely have been produced and are probably weaponized. In addition, bacillus anthracis (anthrax), vibrio cholerae (cholera), staphylococcus enterotoxin (SEB) and clostridium perfringens bacteria or its toxin may be in advanced stages of development or available for weaponization."

The document noted that Baghdad had several options for delivering germ weapons, from aerosol generators that could be carried on trucks, boats, or helicopters to cluster bombs, spray tanks for high-performance aircraft, artillery shells, and missiles. The gloomy predictions included one bit of reassurance. "It is assessed that Iraqi forces will use BW only as a last resort."

At about the same time, the CIA weighed in with a report, entitled "Iraq's Biological Warfare Program: Saddam's Ace in the Hole," which waffled on the issue of whether Baghdad had yet turned its noxious germs into usable weapons. "Iraq's advanced and aggressive biological warfare program (BW) is the most extensive in the Arab world," the study said. "By the end of 1990, the Iraqis will probably have deployed significant numbers of biologically filled aerial bombs and artillery rockets."

CIA analysts quoted Saddam Hussein as telling a delegation of U.S. senators earlier in the year that he would use chemical weapons in retaliation for a chemical or nuclear attack on Iraq. "Iraq will treat BW similarly, in our judgment," the report said, "and probably save its biological weapons as a retaliatory option." The analysts raised the possibility the Iraqis might try a sneak attack. "Iraq may have contingency plans to use biological agents covertly," they wrote. "Botulinum toxin and, to a lesser degree, anthrax bacteria lend themselves to covert dissemination because even small amounts placed in the food supplies are sufficiently toxic to kill large numbers of people. Iraq also could covertly use spray tanks or aerosol generators purchased for its chemical warfare program to create large toxic clouds of bacterial agents upwind of a target area."

Stripped of technical language, the analysts delivered a simple message: American troops could be badly hurt by germ weapons. Top officials at the White House and the Pentagon struggling to assemble an international coalition against Baghdad were scarcely aware of the issue. But in the less traveled corridors of the Pentagon, the men and women assigned to supply the troops and protect their health understood immediately. On August 8, a day after the first American forces began moving to Saudi Arabia, experts from the army surgeon general's office briefed General Gordon Sullivan, the army's vice chief of staff, on the germ threat. They recommended immediate inoculations against anthrax and botulinum with the stocks on hand. As more vaccine became available, more soldiers could be immunized. Three days later, a Pentagon logistics officer called Fort Detrick to ask how much anthrax and botulinum vaccine could be sent to Saudi Arabia. The query was seemingly routine. America's armed services stockpile every imaginable item, from fuel to food to ammunition to body bags. Surely the government had ample amounts of vaccine against the two germ agents Iraq was known to be producing. The answer was not reassuring. The United States had vast amounts of vaccine against several diseases, including some obscure viruses, but almost none for anthrax or botulinum.

A few days later, army health officials spelled out the dimensions of the problem to General Schwarzkopf's aides. The army had only 10,000 doses of the anthrax vaccine on the shelf. Even if the supplies were stretched by giving each soldier only two or three shots, the army had enough to protect at best a few thousand soldiers. Army officials rushed to Michigan to assess whether the Department of Health could step up production. The laboratory had been pushing to fill the Pentagon's 1988 order for anthrax vaccine even before Iraq invaded Kuwait. It had 140,000 doses that could be ready to ship by late September, enough to protect about 46,000 soldiers. The pace was ahead of schedule, but it was nowhere near enough to immunize the hundreds of thousands of soldiers who would be assembling in Saudi Arabia.

The military's choices for protecting its troops against an attack with botulinum were even more limited. Botulism is extremely rare, and American drug companies, lacking a market, had never paid the considerable costs of putting the vaccine through full clinical trials. Tests suggested that it was safe for humans, but the larger-scale studies, which researchers rely upon to uncover less common side effects, had never been done, even though the vaccine had been around for more than two decades. The drug was experimental. Under federal rules, it could not be given to patients unless they gave "informed consent" after learning of all possible risks. Stocks of the vaccine totaled just 34,000 doses, enough for no more than 10,000 soldiers.

On August 20, Pentagon health officials pulled together a committee of experts. Its members amounted to the government's brain trust on biological weapons, an assortment of doctors, intelligence specialists, and mid-level military officers who understood the issue. Their title was appropriately wonkish: the Ad Hoc Working Group for the Medical Defense Against Biological Warfare. The site of their meeting was revealing. Proximity to power is a status symbol in military culture, with top defense officials competing for the coveted offices in the Pentagon on the E Ring, near the secretary of defense. The ad hoc committee gathered at the army surgeon general's office in a building on traffic-clogged Leesburg Pike, miles from the Pentagon.

The meeting quickly got down to business. The intelligence analyst who had drafted the June 1988 report on Iraq's purchases of germs from an American company reported that Baghdad's most advanced germ efforts involved anthrax and botulinum. Army medical specialists reviewed the difficulties in diagnosing inhalation anthrax, which begins with "nonspecific" symptoms like fever and fatigue. The disease is "virtually 100 percent fatal" if left untreated, they said. Botulinum is easier to recognize—victims suffer severe neurological problems within hours—but harder to protect against. There are seven strains of the toxin; Iraq had obtained the "seed cultures" for the F strain, a variant not covered by the small amounts of vaccine the United States had on hand. The committee moved briskly through the material, speaking in the clipped shorthand of scientific professionals. Everybody knew what needed to be done. By 4:00 P.M., when they adjourned, the officials had agreed on a far-reaching recommendation: the United States should immediately begin vaccinating troops sent to Saudi Arabia, using available supplies to protect those "most at risk."

Out in the desert, General Schwarzkopf and his troops were spooked by a more practical problem. An army major scouting out landing sites for the air force's giant C-130 transport planes came across the carcasses of as many as a thousand dead sheep and goats in varying states of decomposition. Was it a germ attack or just some very sick livestock? The army did not take any chances. On August 30, General Schwarzkopf's staff sent a three-man team for a firsthand look at the rotting remains. A Saudi chemical officer accompanied the team to conduct his own investigation and to translate for the Americans. He proved superfluous; the local Bedouins had hired shepherds from Bangladesh to mind their flocks, and the interviews were conducted in English. The team quickly concluded that the sheep and goats had died from natural causes. The shepherds said they themselves were in fine health. The problem was simple: animals in the region were not properly vaccinated against indigenous diseases.

Reports of the false alarm reassured officials in Washington, where the proposal to inoculate soldiers was gaining support. The top medical officers of the army and the air force endorsed the plan, as did Enrique Mendez, the Pentagon's senior civilian official on health issues. The Armed Forces Epidemiology Board urged anthrax vaccination "as soon as possible," reflecting the medical consensus. Mendez drafted a memo for Dick Cheney, the defense secretary, ordering the inoculations to begin.

Cheney and General Colin L. Powell, chairman of the Joint Chiefs of Staff, and Schwarzkopf all believed that the recommendation needed more study. The plans envisioned keeping any vaccination program secret. But Saudi Arabia was crawling with reporters, and the inoculation of large numbers of troops was certain to leak out. The vaccinations would also complicate the administration's most important diplomatic initiative: assembling an international coalition against Saddam Hussein. The United States did not have nearly enough vaccine to protect its own soldiers, let alone those of its allies, the American civilians living in the country, or Saudi civilians.

On September 21, 1990, Powell met with the commanders of the nation's armed services in the "tank," the sealed, spyproof conference room in the Pentagon reserved for the regular meetings of the heads of the army, air force, navy, and marines. The Joint Chiefs decided it would do more harm than good to begin vaccinations immediately. Contradicting the intelligence analysts, they concluded that the threat was "tenuous"—there was no confirmation that Iraq had figured out how to turn the germs it was producing into weapons. A memo written after the meeting succinctly summarized the outcome. "The decisions were no longer medical in origin. Rather, they were political, social and military/operational." This did not mean that Powell was taking the threat of biological warfare lightly. A career infantryman, Powell was far more worried about Iraq's germ arsenal than its chemical weapons. American soldiers, he felt, were prepared to defeat a chemical attack. They had trained for it, had practiced combat operations in suits and masks. "You can deal with CW," Powell said in an interview years later. "Chemicals don't linger. You can get out of their way. I knew how to handle it." The allied germ defenses, by contrast, were improvised, drawn up on the fly. "BW was a wild card," he recalled. "Something I had less ability to deal with." The situation did not surprise Powell. "I'd been kicking around thirty-odd years," he said. "I was quite aware that we did not have protective equipment or vaccines."

Powell picked Brigadier General John Jumper, a one-star general working for the Joint Chiefs of Staff, to serve as his personal emissary on the issue, to push the production of vaccine and detectors.

"Jumper, I'm worried about BW defense, and I want you to deal with it," Powell said.

"Sir, I think you've got the wrong guy," he replied. "I'm just a dumb fighter pilot from Paris, Texas. I've got a degree from VMI in electrical engineering. I don't know much about this." Don't worry, said Powell, you just have to know how to fix things.

Jumper set out to educate himself. He quickly learned that the Michigan Department of Public Health was the nation's only licensed producer of vaccine against both botulinum and anthrax. An antidote to botulinum, known as an antitoxin, could be extracted from the blood of horses or other animals exposed to the disease. But the United States only had one horse, named First Flight, that could produce the medicine, an experimental drug that had been given to only a handful of people. Jumper found a picture of the horse and mounted it on an easel for one of his briefings of the military chiefs. This, he declared, is the entire industrial base for production of botulinum antitoxin in the United States. The generals reacted with disbelief. Jumper received briefings from expert after expert. What surprised him most was the military's lack of a doctrine for biological defense. There had been no planning for how to handle casualties, where to put detectors, or who would be inoculated.

It wasn't even clear that there would be any detectors. For decades, the army had been trying to design equipment that could sniff an aerosolized cloud of germs and provide immediate warning of germ attack to soldiers. Scientists had not come close to conquering the technical challenges. The air is filled with biological material, from pollen to germs. No one could figure out how to design a machine that could distinguish between the normal "background" and germ weapons. A 1970s army system, a gigantic air-sniffing device coupled with a crude detector, had been shelved because of its limited accuracy. Shortly after the Iraqi invasion, Powell ordered the Defense Nuclear Agency to improvise a detector that could be mounted on an airplane. The initial tests were encouraging, but the plans were shelved when the prototype airplane was destroyed in a crash.

The vaccine shortage worried the Pentagon's top civilian health officials. Peter Collis, an emergency-room physician who was deputy assistant secretary of defense for health affairs, pressed the ad hoc working group to find other sources of vaccine production. He decided the group needed a more inspiring name. Collis, who had studied infectious diseases as an army doctor, remembered from medical school that anthrax spores were unusually persistent and that the illness could be contracted from household items, even a shaving brush. The best shaving brushes were made from the hair of badgers. The Ad Hoc Working Group for the Medical Defense Against Biological Warfare became Project Badger. "You can't do that," an aide said, reminding him of the Pentagon's procedures for code names. "I just did," said Collis.

The team included Anna Johnson-Winegar, the civilian official who had tried to build the stockpile of anthrax vaccine through the late 1980s. She was a natural choice for Project Badger. Pentagon officials were baffled by why the pace of production could not be accelerated. She tried to explain. The conversations were frustrating to both sides.

"Just call them up and tell them to deliver five hundred thousand doses next week."

"Sorry, can't do it."

"Well, money's no object."

"It's not a matter of money."

"They can make three times as much as they used to by putting on three shifts of workers a day."

"What do you want them to do, stand there and watch the bacteria grow?"

Johnson-Winegar explained the basics to her uniformed colleagues. The only way to speed up production was with multiple assembly lines, which would have to be certified by the FDA. The vaccines had to be processed, tested for potency and sterility, and packaged. The schedule could not be truncated.

Every major pharmaceutical company in the United States was approached for help. The responses were discouraging. Producing anthrax vaccine posed special problems. Its spores were so dangerous and so long lasting that buildings and fermenters in which the vaccine was made could never be used for anything else. Making botulinum vaccine was an even greater production challenge. The microbes had to be grown in an oxygen-free environment (anthrax, in contrast, is an aerobic germ). There were seven different strains, each requiring its own fermenter and production line.

Botulinum antitoxin, refined from the blood of horses or other animals, posed its own problems. It took months of repeated exposure to the biological poison to stimulate the immune systems of the animals. Project Badger set out to supplement the production of the army's lone horse, First Flight, by placing orders for one hundred horses that were to be kept in paddocks at Fort Detrick. The experts knew it would be as much as a year before the first lots of antitoxin could be refined and purified from the army's herd.

Of the 150 companies Johnson-Winegar approached to make anthrax vaccine, only 15 owned the appropriate equipment, and only Lederle agreed to try producing it. Collis tried calling the CEOs of pharmaceutical companies himself, telling them that the secretary of defense had personally authorized him to make the request. The results were no better. Part of the problem was the deal Collis and the Project Badger team was offering. The army had decided not to indemnify the companies against lawsuits, even though vaccine makers faced huge potential liabilities from people who suffer severe reactions to the drug.

At a meeting in early October, the Project Badger team promised the army that it would have significantly more vaccine available by April 1. Colonel Harry Dangerfield, a senior member of the committee, viewed the commitment as optimistic. Any vaccine for American troops would be inspected by the FDA, which often rejected entire lots. Producing mass quantities of anthrax or botulinum vaccine would take close to a year, not a few months. "We were faced with a problem that couldn't be solved in the time frame it needed to be solved," he said later. "The people setting the goals just didn't understand what the issue was."

WITHIN weeks of the invasion, the intelligence reports about Iraq's biological program came to the attention of more senior officials in the Pentagon. One who felt particularly frustrated by the report's carefully hedged language was I. Lewis Libby, a trim, boyish lawyer and one of the Pentagon's top policy officials. His job, as deputy to Paul Wolfowitz, the undersecretary for policy, was to think two or three steps ahead on the international chessboard. Where was the next hot spot? What could be done to head off trouble? What were policy makers overlooking as they coped with the rush of day-to-day crises? To Libby, the words "probably" and "possibly" jumped out of the reports about Iraq's biological program. He thought the analysis was cogent, as far as it went. But it left open the most important questions about Iraq's intentions. Libby knew the United States had few spies inside Saddam Hussein's secretive regime who could provide so-called real-time intelligence. The reports on biological weapons drew largely on satellite photographs of suspect sites, snatches of intercepted telephone conversations, and interviews with Western businessmen who had sold equipment to Iraq. Libby told colleagues that intelligence analysts had an unfortunate habit: if they did not see a report on something, they assumed it did not exist. Or, as another veteran intelligence officer put it, absence of evidence is not evidence of absence.

A few weeks after American troops began gathering in the deserts of Saudi Arabia, Libby assembled his top aides to coordinate defense against biological warfare. His instructions were simple: "If we go to war and Saddam uses biological weapons, I want to be able to look in the eyes of the soldiers and their families and say: 'We did everything we could.' " He asked an aide, a career navy officer named Captain Larry Seaquist, to look more closely at whether Saddam really had a usable germ arsenal. Seaquist, who had commanded the battleship *Iowa,* proved an inspired choice. Navy captains are trained to be generalists who can master the interlocking complexities of complicated engineering systems. Seaquist's skills as a quick study served him well. He was not afraid to dive into the unknown. Learning about biological weapons was no different from what he had done all his life in the navy. In his view, you just had to start asking the right questions and keep at it.

In late September 1990, as Powell was slowing the pell-mell rush to begin vaccinations, Seaquist and another officer who reported to Libby, Colonel George Raach, flew to Wright-Patterson Air Force Base in Ohio for their education in biological warfare. Their teachers were a couple of aging scientists who had worked with Bill Patrick to make germ weapons for the United States. They started with the basics: how the agents worked, how they could be easily disseminated, how many people could be killed in a single attack. They moved to more advanced topics, such as LD-50, the amount of agent needed to administer a lethal dose to 50 percent of a target population. By the end of the day, Seaquist and Raach were reeling. Biological warfare could be just as deadly as a nuclear attack, if not more so. The implications for the conflict in the gulf were immediately clear: Saddam Hussein had the means to inflict catastrophic, Hiroshima-sized harm on the allied forces with a single well-planned attack. The two military officers drove their rental car back to the airport, struggling to digest what they had learned. Their conversation continued in the departure lounge, where they became so absorbed that they missed their return flight.

Seaquist asked intelligence officials in Washington more detailed questions. The scientists at Wright-Patterson had said biological weapons could be easily disseminated from aerosol sprayers used to apply pesticides. Did Iraq have an ample supply? Yes, they replied. In the spring of 1990, the Iraqis bought forty top-of-the-line "aerosol generators" from an Italian company. Each one was capable of dispersing 800 gallons per hour. They were compact enough to fit on the back of a pickup truck, small boat, or single-engine aircraft and could disperse either liquid or dry biological agents. Seaquist looked over the weather charts for the Saudi desert. On three to five nights a week in the fall, he noticed, there was a temperature inversion, meaning that a layer of hot air sat atop the cool air near the ground. Seaquist understood the significance. Inversions make biological weapons much more deadly because they trap a cloud of agent over the target. He did some more calculations: Assume the United States had 200,000 troops camped out in Saudi Arabia. If the Iraqis could slip a single small boat equipped with an aerosol generator into the Persian Gulf, it could unleash a cloud of anthrax that would, in appropriate weather conditions, kill 90 percent of the soldiers.

Seaquist had made a crucial leap, one that would elude some of his more senior colleagues. The distinctions in the intelligence reports between producing large quantities of an agent and turning it into a weapon were not as important as they appeared. In germ warfare, the germ *was* the weapon. Every country that had tinkered with biological warfare had worked hard to figure out how to make a bomb or missile that would deliver the agent. (The blast from high explosives destroys most of the germs.) But Saddam's engineers did not need to invent a workable biological missile or shell. The aerosol generators they bought in the spring of 1990 would do just fine.

And then there was the terrorist threat. Seaquist and Libby hauled in the intelligence officials for additional briefings. What if Saddam Hussein announced just prior to war with the United States that he had five containers that would release biological weapons in Washington? Could he do that? You bet, came the answer. Seaquist spelled out his conclusions in a highly classified memo for Libby and Paul Wolfowitz. The United States was vulnerable both at home and on the battlefield.

Libby believed it was possible, but not likely, that Iraq would use its germ weapons. Saddam was under less pressure than it appeared. In Libby's view, he could survive a defeat in Kuwait. The day after the fighting ended, Libby told colleagues, Saddam would still command the only army in Iraq. To deter the Iraqi leader from a germ or chemical attack, the United States needed to threaten something that Saddam Hussein treasured—his hold on power. Cheney was briefed on Seaquist's finding and senior officials began discussing how to send an unmistakable message to the Iraqi leader.

Josh Lederberg was also worrying about civilians. Saddam Hussein's statement earlier in the year that he would burn Israel with his "fire" was widely taken as a threat to use chemical and biological weapons. As a member of the Defense Science Board, Lederberg had access to some of the classified intelligence on the Iraqi biological program. "I knocked on every door, got to see Brent Scowcroft, got to see the intelligence people, the FBI," he said. He also asked a simple question: "What are we going to do about the civilian population?"

No one had thought through which government agencies would respond to a domestic germ attack. Lederberg urged Scowcroft, Bush's national security adviser, to begin planning civil defenses. Saddam Hussein's most likely use of germ weapons, he argued, would be through a surrogate, a terrorist group. White House officials quietly began assembling an emergency-response team centered at the Department of Health and Human Services. The staffer who handled his meeting with Scowcroft was a rising young star named Condoleezza Rice, a Soviet specialist whom Lederberg had known when she was a student at Stanford. The preparations were modest: a training exercise or two and some antibiotics stockpiled in the Washington area. But Lederberg felt he had made a contribution. The nation's capital had been made somewhat less vulnerable to terrorist attack.

AMERICA's top military commanders were beginning to understand the magnitude of the threat. Briefed in late November about the lack of biological detectors, General Carl Vuono, the army chief of staff, was stunned that American troops would be used as canaries in a potentially lethal coal mine. He pushed the bugs-and-gas experts to tell him why the soldier was the "first indication of BW agent use." Wasn't there some other intelligence system or equipment that could warn of an impending attack? Vuono said that many countries might follow Iraq's example and acquire biological agents. "The U.S. should never be put into this position of not being able to vaccinate the force," he told a senior aide. He demanded a "long-term solution," a stockpile of vaccines or some other idea.

In Riyadh, General Schwarzkopf felt trapped. He was persuaded that the biological threat was real. But the supplies of vaccine were well short of what was needed. Someone was going to have to play God by deciding who would be protected. Schwarzkopf believed such a decision should be made in Washington and kept pressing the joint chiefs to deliver enough vaccine to immunize everyone. "It was a nightmare," he said years later.

Schwarzkopf was also frustrated by the carefully worded assessments coming from the intelligence agencies. He needed facts, not surmises. What was Iraq's doctrine for using germ weapons? Did their missiles have germ warheads? The bugs-and-gas experts in Washington were chagrined by their own inability to provide clearer answers. But there was just not enough firsthand information available on a program that was one of Iraq's most closely guarded secrets. An October 22 report by the Armed Forces Medical Intelligence Center was typical. "Large quantities of anthrax bacteria and botulinum toxin likely have been produced," it said. "We believe that these agents have been weaponized and that biological and toxin munitions already exist."

The analysts mentioned another potentially disturbing bit of news. The Baghdad offices of a Chilean company that was supplying cluster bombs to Iraq had been overheard discussing "the parameters for aerosol testing of botulinum toxin with the home office." Did this mean that Iraq had such a weapon? The analysts were not sure. The evidence, they wrote, would allow only the "inferential assessment" that the Iraqis were trying to buy such weapons, not that they had them. On the crucial question of what Iraq might do with its weapons in a war, its "use doctrine," the experts admitted what Schwarzkopf already knew: there were "significant intelligence gaps." A DIA report from the same period said that Iraq "is assessed to have weaponized anthrax and botulinum. The type and number of weapons or dissemination systems is not known." The DIA officials acknowledged, "There is no reliable information on how Iraq might use their BW weapons. The most suitable way is in a clandestine manner prior to the outset of hostilities."

The impasse over when to begin inoculations continued, and Powell returned to the "tank" to confer with the chiefs of the four branches of the American military. Only the navy supported immediate vaccination. A few days later, on October 21, Charles Freeman, the American ambassador in Riyadh, raised his own objections. Vaccinating soldiers but not civilians was a bad idea, Freeman warned in a secret cable. It could sow panic, fracture the coalition. Besides, there were American citizens in the region, working on oil projects. What about them? There would be a public-relations disaster, Freeman warned, if the United States protected only its soldiers.

Intelligence officials began trying to figure out what might prompt the Iraqi leader to use his germ weapons. "If Saddam concluded his personal position was becoming hopeless," CIA analysts warned in November 1990, "this could convince him to use biological weapons to shock the coalition into a cease-fire. In such a situation, the use of anthrax against a coalition military installation or major Saudi oil facility might seem an attractive option."

AIR force officers drawing up allied plans for a massive bombing of Iraq thought they could resolve some of the uncertainties about Baghdad's germ program with a few well-placed explosives. American intelligence had identified what it believed were Iraq's main research and production sites for biological weapons, as well as more than a dozen refrigerated bunkers. The analysts were not sure what Iraq was doing with the bunkers. Air force officers designing the air campaign planned to demolish them and any other structures that might be linked to the Iraqi biological program.

That stirred a new set of uncertainties among intelligence analysts back in Washington. No one was sure what would happen if you bombed a building or bunker packed with anthrax or botulinum. Would a plume of agent scatter illness and death across the region? Some preliminary calculations by yet another hastily assembled committee of experts, dubbed the Interagency Fusion Group, suggested that casualties inside Iraq could be staggering. Attacks on ten separate sites in Iraq, they estimated, could kill more than 150,000 Iraqi civilians—more than the number of people killed by the Hiroshima bomb. The British weighed in with an equally gloomy assessment.

General Charles Horner, the Riyadh-based American commander who was planning the air assault on Iraq, believed that the estimates were exaggerated. Horner pressed for more facts and learned that sunlight killed anthrax spores. His aides proposed several ways of minimizing the risk. The bunkers could be hit with bombs that would cause them to implode rather than explode. Pilots could drop incendiary devices, creating a blast of heat that would kill the spores. The missions could be flown just before dawn so that if a plume of anthrax were inadvertently released, it would be weakened by the heat of the day.

In December Horner briefed Cheney and Powell on his plans for bombing the bunkers. He described the revised calculations of how much anthrax might waft across Iraq after a raid. "I can't promise you we are not going to have casualties in this," he told them. "There has got to be a penalty for a country that manufactures and stores these horrible weapons. So, if there is a fallout and if there are civilian casualties from this, maybe that point needs to be made to all the other countries who conceive to do this."

A few weeks later, the fusion group acknowledged that it had miscalculated, and it revised its estimate of casualties sharply downward. Because they did not know where Iraq was hiding its germ weapons, the analysts had put together a "worst-case" forecast in which it was assumed that each site housed Iraq's entire arsenal. There was a second, crucial mistake. The fusion group had initially estimated that 1 to 10 percent of the spores would survive an explosion and float into the air. Consultation with several veterans of the American germ program, including Bill Patrick, suggested that the actual number would be much, much lower—between .001 and .1 percent.

The new analysis still offered sobering predictions. It said the allied bombing was likely to cause hundreds, perhaps thousands, of Iraqi civilians to suffer painful, lingering deaths from inhaled anthrax. Senior officials weighed the risks and—assuming that was to be the cost of taking the biological option out of Saddam's hands forever—accepted them.

THE general whom Powell had assigned to work the vaccine issue, John Jumper, sent a message to Schwarzkopf in early December suggesting a way to stretch the limited vaccine supplies. The commanders could administer antibiotics to all 500,000 troops and keep the vaccine in reserve to treat anyone attacked with anthrax. A combination of doxycycline and vaccine had kept alive all the monkeys in the Fort Detrick study.

Jumper's cable puzzled Schwarzkopf's medical officers. Anthrax multiplies rapidly, and the researchers had begun treating the monkeys within a day of exposure. Because the detector problem remained unsolved, the first signs of attack could appear days later, when soldiers began complaining of flu. Sending troops on antibiotics into battle was risky, Schwarzkopf's medical advisers felt. The drug recommended in the cable, ciprofloxacin, was a powerful antibiotic that sometimes caused serious diarrhea and other side effects. It also killed normal bacteria in the gut, leaving only resistant germs. A soldier shot in the stomach would stand a greater chance of contracting an untreatable infection.

The combination of cipro and vaccine had not been tested on the monkeys. But Pentagon officials were recommending it because cipro was a newer drug and thus more likely to be effective against genetically engineered anthrax. Intelligence officials had recently warned that the Iraqis were capable of creating such a strain. Still, no one knew if they had done so.

The doctors met with Schwarzkopf on December 6 to review Jumper's cable. Colonel James David Bales Jr., an infectious-disease specialist assigned to CENTCOM just a few days after the invasion of Kuwait, leaned over the commander and watched his eyes flicker across the page. As Schwarzkopf read the passage about the need to administer the vaccine and cipro within twenty-four hours, Bales said: "And how are you going to know that?" The more experienced officers in the room braced for an explosion. Schwarzkopf hated interruptions. But he agreed with Bales. The antibiotic approach was too risky without reliable detectors.

The next day, Schwarzkopf sent a private "eyes-only message" to Powell rejecting the widespread use of antibiotics. He preferred to vaccinate, even if there was not enough for everyone. A decision to immunize, he reiterated, had serious implications for the coalition and could be made only in Washington.

The impasse frustrated Colonel George E. Lewis, a microbiologist who was working for Jumper and helped prepare the cable on antibiotics. Military officers made life-and-death decisions every day. They calculated how many men would be sent to attack a particular target or how much air cover was required for dangerous operations. Why was it so hard to decide whom to protect against biological attack?

The January 15 deadline set by the United Nations for Iraq to withdraw from Kuwait was just five weeks away, and Lewis knew the allies had already waited too long. To obtain even partial immunity, inoculations should have begun in mid-November. Each day's delay raised the danger. The Pentagon calculated that soldiers who received their first shots on December 10 would not be minimally protected against anthrax until January 21 and against botulinum until April 1.

Project Badger's efforts to make more vaccine had fallen well short. The Michigan lab had produced enough anthrax vaccine to protect about 150,000 of the half million American troops in the region. There was none to share with allies or civilians in Saudi Arabia. Defenses against botulinum were even more limited. Despite considerable effort, not a single additional dose had been produced. There was still only enough for 10,000 soldiers.

On December 9, the British added to the pressure. General Peter de la Billière met with Schwarzkopf in Riyadh and told him that he intended to begin immunizing his troops unless the United States had "strenuous objections." British officials, unlike the Americans, believed Iraq had turned anthrax and plague into weapons, and they intended to protect their troops against both. The British bought an American vaccine against plague that was still experimental and was known to cause temporary but debilitating flulike side effects.

The Saudis learned of the British plans and asked Schwarzkopf whether he could provide some anthrax vaccine to protect their troops. The American commander said he did not have enough for his own soldiers. A senior Saudi general asked whether Washington could at least set aside some shots for the royal family. It was an awkward request. Saudi Arabia's willingness to allow American forces on its soil had made Operation Desert Storm possible. But Schwarzkopf had none to share.

On December 9, Schwarzkopf spoke with Powell by secure phone and suggested a bolder tactic to avert biological attack. The United States should announce that it would severely punish Saddam Hussein if he used his chemical or biological weapons. Powell said he was pressing the White House to warn Iraq that the United States would use its own "unconventional weapons" if Iraq attacked with chemicals, a reference to nuclear retaliation. The next afternoon, Schwarzkopf was on the phone with Powell again, rehashing the decision to bomb the biological bunkers and production plants. He told the chairman that it "would be an unforgivable sin" not to attack the germ targets. It was the only way to guarantee that the weapons would not be used.

Schwarzkopf began to plan who would be immunized with the limited supplies. Months earlier, he had asked his chief medical officer, Colonel Robert Belihar, and other top aides to assess how and where Iraq might attack with germs. Think as Saddam Hussein would, he told them. Belihar said he lacked information. Would the Iraqis use anthrax as an aerosol, or would they pour botulinum into the water supplies? Would they spread germs on the front lines, or use it as a terror weapon in the United States? A key question was whether Iraq had vaccinated its soldiers against anthrax or other agents. The Americans arranged for blood samples to be taken from Iraqi deserters being held by Saudi Arabia. The blood was flown to Fort Detrick for analysis in early December.

Schwarzkopf and his aides eventually concluded that anthrax would most likely be aimed at rear areas. A germ weapon that takes effect over several days, they reasoned, would not be very useful in a frontal assault. Botulinum dissipates quickly and was thus a threat to the forward positions. Schwarzkopf was trying to make the best of a bad situation. It was entirely possible that his guesses were wrong, that the Iraqis would use anthrax against the front lines. If that happened, there would hardly be any vaccine left to treat those who were ill.

His proposal for rationing the vaccine stirred up interservice rivalries. At a meeting of air force generals in December, one officer complained that the recommendations by Schwarzkopf, an army general, reflected an "infantryman's concerns for his fellow infantrymen." The air force had done its own analysis, which concluded its crews would be a primary target for germ weapons. Air force officers drew up a list of 30,000 airmen who should be vaccinated.

A December 11 assessment prepared by army health officials outlined three possible approaches. "Course of Action #1" was to begin vaccinating U.S. troops as quickly as possible. This would have a "favorable domestic reception" and could deter Iraq. "Course of Action #2" was to share the limited supplies with allies, which would "promote perception of shared risk," support the unity of the alliance, and might dissuade Baghdad from using germ weapons. The third option was to defer vaccinations until there was enough on hand for a "comprehensive program," while sharing the antibiotics with allies. This would heighten the dangers for everyone, but the "shared risk" would promote allied unity and keep on hand a supply of vaccine in case of attack. Pentagon officials wrote that the last approach offered another benefit: It "will temper media reaction."

On December 17, 1990, Powell finally delivered his recommendations to Secretary Cheney, calling for inoculations to begin immediately. It had been four months since the Pentagon's medical experts had made the same proposal, months in which the service chiefs and other top military and civilian officials had undergone a crash course in the science, tactics, and ambiguities of biological warfare.

Cheney and Powell took their plan to the White House. At the table were President George Bush and several of his top aides. Some were surprised by the seriousness with which the Pentagon presented the Iraqi biological threat. Was it really likely that the Iraqis would use such weapons? What were the risks of the vaccines? White House officials had made a pact among themselves. They would not micromanage; they would not overrule the commanders on purely military issues. The president approved the plan without revisions. "I'll be honest: if I had been making the decision on the anthrax, I probably would not have used it," Brent Scowcroft, Bush's national security adviser, said in a recent interview. "I didn't object to it, if the military thought it was the thing to do, but I probably would not have recommended it."

As Powell's aides were drafting the memo outlining the vaccination program, the Pentagon received an encouraging bit of news. The blood samples taken from Iraqi deserters detained showed no sign of immunizations against anthrax or botulinum. Schwarzkopf's medical team did not feel particularly relieved. The Iraqi soldiers in Saudi hands were conscripts, cannon fodder whom Iraqi generals would view as expendable. Perhaps Saddam Hussein had vaccinated his elite troops, the Republican Guard. Maybe the Iraqis would attack the front lines with germs, killing both the unvaccinated Iraqi conscripts and the Americans.

Back in Washington, the army was pressing the FDA to permit it to give the botulinum vaccine to soldiers without obtaining the "informed consent" normally required for patients given experimental, unapproved drugs. Botulinum's effects were felt within hours of exposure, which meant the weapon might be useful in an attack on frontline troops. There would be no time for medical niceties, military officials argued. The FDA reluctantly agreed.

By December 29, Powell notified Schwarzkopf that the decision had finally been made to immunize the troops against anthrax and botulinum. Days later, Schwarzkopf reviewed the plan for rationing the vaccine with his top aides and demanded some changes, overruling a proposal to give botulinum shots to officers at the Riyadh headquarters. What little vaccine there was, he thought, should be reserved for frontline troops. Schwarzkopf also said every soldier assigned to receive the experimental drug should be given a release form to read and sign. Anyone who wanted to refuse the shot could do so.

In early January, teams of medics began immunizing the troops against anthrax and botulinum; every soldier was also issued a five-day supply of cipro. A Marine Corps medical official told General James M. Myatt, the leader of the First Marine Division, that there were only 7,000 doses of anthrax vaccine for the 22,000 soldiers under his command. Like Schwarzkopf, Myatt was reluctant to play God. If there was not enough vaccine to go around, nobody would get it. Adherence to Schwarzkopf's orders on the botulinum vaccine was spotty; some soldiers were given the shot without being told what it was. To quell fears about the vaccines, Schwarzkopf himself took an anthrax shot. "Just make sure that I'm one of those nine monkeys that survived," he told the medics.

DETECTORS were still a major headache. By December, senior army officials had devised a plan to use an untried, laser-based technology to detect germ attacks. The Los Alamos National Laboratory had been experimenting with airborne lasers that might be able to pick up the release of an aerosol cloud. The lasers were to be placed aboard C-130 aircraft that would fly continuously over the gulf. Schwarzkopf's second in command, General Cal Waller, hated the idea. "Thanks for your proposed deployment of experimental developmental BW defense assets," Waller wrote to the Pentagon. "The fielding of a fully supported system of operational worth on such short notice is ambitious." Waller said he was reluctant to divert his limited intelligence aircraft from "priority missions" to an untried, uncertain system of biological defense. The air force also hated the plan, fearing that the lasers might blind pilots. They were not deployed.

Instead, the army relied on a dozen detectors its technicians had improvised. The system was built around a sniffer that forced huge quantities of air into a liquid reservoir. Servicemen would test for biological attack by dipping specially treated paper into the liquid at regular intervals. In theory, the paper would change color in the presence of suspect microbes, such as the anthrax bacillus. Then the liquid would be tested with more sensitive lab equipment. No one had much confidence in the system.

In the late fall, Bush and his top national-security aides, known as the "gang of eight," considered whether the United States should explicitly threaten Iraq with nuclear retaliation for a germ or chemical attack. The consensus was against such a move. Washington was trying to hold together a fragile coalition that included Iraq's Arab neighbors. Bush made his preferences clear. He would not be the second president in American history to order a nuclear strike. If the United States needed to punish Iraq, it could expand its conventional bombing campaign. The decision was kept secret. On January 9, 1991, Secretary of State Baker flew to Geneva to give Tariq Aziz, the Iraqi foreign minister, a letter from President Bush to Saddam Hussein. The White House had decided against threatening Iraq with nuclear attack. "Your country," the letter said, "will pay a terrible price if you order unconscionable acts." Aziz refused to accept the letter, and Baker, taking an expansive view of his instructions, set out to persuade the Iraqis that the United States was poised for nuclear retaliation. "God forbid," he said, "chemical or biological weapons are used against our forces—the American people would demand revenge." Baker added, "This is not a threat."

The president's inclination against a nuclear strike was hypothetical, and Pentagon officials who had done some preliminary planning of nuclear options believed that they could reopen the issue if Saddam Hussein did the unthinkable. The day after a biological strike, political pressures would force any president to reassess. "If we had a loss of 200,000 soldiers, how could we not nuke them?" asked Larry Seaquist, whose report for the Pentagon's top policy officials had raised the specter of such losses. Scowcroft acknowledged that the issue would have been revisited if Americans had died in a germ attack. But he added that he would have adamantly opposed nuclear retaliation. Scowcroft said he would have strongly argued for changing the aims of the war to, say, toppling Saddam Hussein. Using atomic weapons against Iraq would have been "another case where it's OK against the Japanese, OK against the Arabs, not against white folk."

In the days before the air war began, President Bush grew increasingly anxious about Iraq's biological arsenal. On January 13, he asked Powell to review the plans for bombing the refrigerated bunkers. What were the risks? Was it worth it? Powell reassured the president that the bombing would probably destroy the germ agents. Some might be released, but it was "a gamble we had to take." In his memoirs, Powell wrote that Bush "was already agitated, and this added worry did not soothe him." Two days later, the chief of the British Defence Staff, Air Vice Marshal Sir David Craig, called with the same questions. "Bit risky that, eh?" Craig remarked. Powell said he believed that the benefits outweighed the costs. "If it heads south," he told Craig, "just blame me." The United Nations deadline expired on January 15. Officials later said that in the meeting in which President Bush sent American forces to war against Baghdad, he had only two concerns: a last-minute diplomatic move by France and the threat from biological weapons.

AT 1:30 A.M. on January 16, the first Tomahawk cruise missiles were launched from vessels in the Red Sea and the Persian Gulf, marking the start of the largest sustained bombing campaign in the history of warfare. In Washington, the officials who had tried to build a credible defense against biological attack held their breath. Seaquist believed that the United States' best defense was Iraq's fear of nuclear retaliation. But the Bush administration had undermined that strategy in the weeks leading up to the war; while administration spokesmen studiously maintained the ambiguity, Vice President Dan Quayle said that he could not imagine the president ordering a nuclear strike. (The next day, Quayle said, "We don't rule options in or out.")

Iraq's air defenses were destroyed in the first days of the war, and allied bombers pounded what intelligence officials believed were the key biological targets. Salman Pak, a research facility where intelligence analysts suspected that Iraq had begun its nuclear program and done some germ experiments, was leveled. Nearly all of the refrigerated bunkers were slammed with high explosives. On January 21, the pursuit of biological targets became a major public-relations problem for the allied coalition as bombers slammed a factory in Abu Gharib that intelligence officials believed was a "backup" plant for germ production. Iraq said the factory produced infant formula, and CNN's Peter Arnett was taken to the site for a tour and chance to broadcast pictures of a container bearing, in English, the notation BABY MILK.

Replying to some skeptical questioning from Schwarzkopf's aides after the raid, intelligence officials said they had spotted little day-to-day activity at the plant since it had opened in 1983. It produced little "baby milk" and was camouflaged just prior to the war, suggesting to the American analysts who were picking the targets that it was linked to biological warfare. (The connection of the factory to biological weapons remains controversial.)

In January, as the allies prepared for the ground war that would culminate with the invasion of Kuwait, Schwarzkopf's aides confronted a grisly question: If soldiers were killed by a germ attack, what should the military do with their remains? Botulinum toxin decays rapidly, so victims of that microbe could be buried in temporary graves and then exhumed for interment in the United States. Anthrax, with its highly persistent spores, was another matter. "The presence of living anthrax spores in the remains can only be confidently excluded if the bodies are incinerated," a memo from Fort Detrick said. If that was not possible, the remains would have to be soaked in powerful bleach to kill all the spores. The bones could then be flown back to the United States for burial with minimal risk to morticians.

American intelligence officials scoured the region for signs of a biological release, particularly on days after germ-related targets were hit. The DIA reported after the bombing of Salman Pak that there had been no outbreak of disease in Baghdad. Allied bombers hit another suspected germ facility, at Taji, on January 21 in a raid that American officials had feared could release as much as five hundred grams of anthrax, killing as many as five thousand people in Baghdad.

"It has been three days since the attack," the report noted, "and casualties should already have started." The findings were immediately reported to Cheney. It looked like the allied bombing was not going to unleash a plague on the Middle East. There were some anxious moments. A half dozen British soldiers stationed near the front lines reported flulike symptoms. They were sent to a Saudi hospital and closely monitored. Their symptoms, it turned out, were caused by the flu. The balky detectors continued to pick up signs of anthrax and botulinum, all of which were found to be false alarms. The ground attack was ready to go, and Saddam Hussein was running out of time to use his ultimate weapons.

5

Secrets and Lies

LARRY Seaquist felt relieved as the sun rose over Washington on February 28, 1991. After weeks of bombing and one hundred hours of one-sided ground combat, Iraq was surrendering, its elite Republican Guard fleeing Kuwait at top speed. The doomsday scenario he had outlined for Pentagon officials—nearly 200,000 allied soldiers dying from a single, devastating Iraqi anthrax attack—had turned out to be just another of those wartime contingencies. No aerosol cloud had drifted over the American forces in the desert. No suitcase-sized bombs had been detonated in the subways of Washington or New York. None of the antibiotics set aside to treat victims of a germ attack in the nation's capital had been needed.

The mood among American military and national-security officials in the months after the war ended was triumphant. Allied troops had demonstrated their superior technology and training, smashing a well-armed foe while sustaining hardly any casualties. "Smart" bombs, guided by a wisp of laser light, had homed onto targets with precision accuracy from thousands of feet away. American ingenuity appeared to have neutralized even Saddam Hussein's best terror weapon, the Scud missile, as the Patriot defense system seemingly blasted incoming rockets out of the sky.

With President George Bush's popularity soaring above 90 percent, the only controversy was over whether the allied forces had killed too many Iraqi soldiers. The biological arsenal that had worried General

Powell and the president before the bombing began in January seemed a footnote, a historical curiosity. American intelligence officials poring over satellite photos of Iraq believed that the allied bombing had demolished Saddam Hussein's germ operations. A secret March 1991 study by the Defense Intelligence Agency said it would cost up to $200 million and take as long as eight years for Iraq to rebuild its germ factories and laboratories. The bombing had destroyed "all known" Iraqi facilities for researching or producing bioagents, the DIA said.

Over time a more accurate picture emerged. The footage of "smart" bombs turned out to be misleading; many targets had been missed. The antimissile defenses were not nearly as effective as they had appeared on CNN; most, if not all, of the Iraqi Scud attacks had evaded the Patriot system. And, perhaps most important, Iraq had built and concealed a germ arsenal that was far larger and far more lethal than anyone had understood, a sprawling complex of buildings and laboratories that had survived the war largely intact.

Iraq's failure to use its biological weapons gave powerful new arguments to those who believed the germ threat was overstated. Deterrence appeared to have worked. Secretary of State James Baker's veiled threat of nuclear retaliation had proven an effective defense, staying Saddam Hussein's hand. Hardly anyone inside or outside government knew the full story about the extent of the Iraqi threat or the lack of effective defensive preparations. The prewar intelligence reports about Iraq's program, the Pentagon's frantic, unsuccessful attempts to produce enough vaccine for its troops, General Schwarzkopf's fury over the choices he was forced to make—all remained closely guarded secrets. Key documents spelling out the military's vulnerabilities—such as Seaquist's report—were never provided to Congress or widely distributed within the government. Some details, such as precisely how little vaccine the United States produced during the war, are still classified as secret a decade later.

IN the months after the war, the Bush White House had larger reasons to avoid fanning public fears about germ weapons. Intelligence about Moscow's illegal germ-weapons program continued to seep out of the Soviet Union, confirming the revelations of Vladimir Pasechnik, the Soviet defector who had provided the first inside account of Moscow's germ program. A team of American and British experts, visiting the Soviet laboratories in January 1991, filed a secret report which found the Soviets were operating a "massive, offensive biological warfare programme" that included genetic engineering and, ominously, research on smallpox. In comparison, the Iraqi efforts seemed rudimentary.

Britain and the United States stepped up diplomatic pressure on Moscow, but the Soviet leadership, which was publicly embracing openness, continued to deny that it was producing germ weapons. The impasse posed a serious problem for Bush and Prime Minister John Major of Britain, who were trying to bolster the nascent reform movement in the Soviet Union. Either Soviet chairman Mikhail S. Gorbachev and his foreign minister, Eduard A. Shevardnadze, were lying, or they were being misled by their own military on a crucial issue. The British were eager to expose the Soviet lies, perhaps by arranging for Pasechnik to make his charges public on a television documentary. Robert Gates, Bush's deputy national security adviser, was horrified by this idea, fearing it would humiliate Gorbachev and hamper his reform efforts. "We were really pushing the Soviet leadership very hard to deal with this problem," Gates recalled. "By that time, Gorbachev was pretty embattled. The way I characterized it to Brent Scowcroft and the president was: 'He has to pick his fights. This one, as important as it might be to us, had to be considered against the overall safety of the world.'

"Gorbachev had much higher priorities in terms of where he was prepared to spend political capital," Gates said. "There was a feeling that he already was on a tightwire. We didn't want to do anything that would result in him being forced out. It's a classic situation where an issue comes up to the president and you don't have any good options. You end up choosing the one that is least bad."

Bush's aides concluded that secrecy was the key to finessing the issue. If word leaked out that the Soviet Union had flouted the germ weapons treaty for nearly two decades, the delicate dance between Gorbachev and his military, as well as that between East and West, would be seriously disrupted. "The information was tightly held," Gates recalled. "And the Bush administration had a pretty good reputation for keeping secrets."

ON April 4, 1991, the United Nations Security Council pledged to lift the international ban on Iraqi oil sales as soon as it confirmed that

Baghdad had destroyed its biological, nuclear, and chemical weapons. The arrangement seemed more than fair to American officials. Iraq had already agreed to give up its unconventional weapons in the resolution ending the war. Everyone would come out a winner. The profits from Iraqi crude would help Baghdad repair the damage caused by the allied bombing. American and European companies would win lucrative contracts to rebuild Iraq. The region would be more secure. And the Iraqi leaders, who benefited personally from each oil sale, would have more money to divide among themselves.

What seemed clear in Washington and London was less obvious in Baghdad. Iraq's influence in the region stemmed from its ability to bully its neighbors, from Jordan to Kuwait to the tiny states scattered across the gulf. To the Iraqis, the benefits of keeping their germ, chemical, and nuclear programs alive were substantial, even at the risk of losing billions of dollars in oil revenues. The bugs were their atom bomb. Even if the biological weapons were never used, the rumors of their existence could make little nations take note and might give Western powers reason to pause.

Soon after the war ended, Iraq's foreign minister, Tariq Aziz, ordered top officials to prepare plans for hiding the weapons infrastructure from the United Nations. Iraq would not disclose its nuclear or biological programs or its stocks of VX, an advanced nerve agent. It would conceal all evidence of the germ operations and of the crash nuclear program, which had come much closer to building a crude nuclear bomb than Western intelligence officials suspected. Iraqi officials destroyed documents, built fake walls in key buildings, and rehearsed the lies they planned to offer United Nations arms inspectors. On April 18, 1991, Iraq delivered its first official statement to the U.N. on its unconventional weapons. It acknowledged limited production of chemical weapons and did not mention work on either nuclear or biological weapons.

The job of verifying that statement fell to a new group of experts, the United Nations Special Commission. Known by its acronym, UNSCOM, the organization was unlike any the world had ever seen, a mix of diplomats and specialists who had been Cold War adversaries. Retired Soviet military officers worked with their counterparts from the American and British military. The U.N. Security Council picked Rolf Ekeus, a Swedish diplomat and former judge, as UNSCOM's leader.

Ekeus felt he did not have much to go on in pursuing Iraq's biological weapons. There were no signposts, no hints. Baghdad had made some significant admissions about its chemical and missile programs. Ekeus's initial view was that Iraq had filed a "relatively honest declaration" that needed to be verified. He immediately wrote a letter to fifty nations around the world from which Iraq might have bought chemical, nuclear, or biological equipment, asking for help. Only Britain and the United States replied with detailed information.

Few members of the first group of inspectors recruited by UNSCOM had much experience with biology, but the team did have one genuine specialist: David Kelly. A microbiologist from Porton Down, Britain's Fort Detrick, Kelly was one of two British officials who had initially debriefed Pasechnik in 1989. Weeks before he joined UNSCOM, he had inspected Soviet germ sites as a key member of an American-British team. Kelly was a skilled interrogator. In January, he had badgered a Soviet scientist into acknowledging that the Soviet Union had experimented with smallpox at Vector, a laboratory that was part of Moscow's germ warfare program. The admission confirmed Western officials' worst fears: that the Soviets were trying to make a weapon from one of the world's most dreaded diseases.

The American-British trips through the Soviet Union were models of how arms inspectors could work closely with intelligence agencies. Kelly and his colleagues were guided by accounts from Pasechnik and by other, less precise information painstakingly assembled by British and American intelligence. They had an idea of where to look and what to look for. UNSCOM, on the other hand, had much less information to go on. American officials provided a briefing to the U.N. team in early May. But the CIA was reluctant to share its best information with an international group that included former Soviet officials and that was wide open to compromise by Iraqi spies. Although British and American inspectors were privately given more information, the intelligence agencies were circumspect about what they said, even to allies.

On August 2, 1991, UNSCOM's first biological-inspections team arrived in Baghdad. The inspectors were greeted by Hossam Amin, a colonel who had worked before the war in the Military Industrialization Corporation, Iraq's purchasing agent for weapons. Amin handed the inspectors a one-page statement acknowledging for the first time that Iraq had performed biological research for "military purposes." The experiments, the statement said, were purely defensive and involved only ten scientists. Kelly found the statement implausible but was intrigued by a list Iraq had drawn up of its "dual use" facilities that had possible military applications. The document said fermenters were being stored at "Al Hakam warehouse." What, exactly, was Al Hakam? The team was already scheduled to visit the bombed-out remains of Salman Pak, the laboratory where Iraq had performed its initial biological-weapons research. Kelly asked the UNSCOM leadership for permission to broaden his itinerary and send some inspectors to Al Hakam for a closer look. Robert Gallucci, the American diplomat appointed as Ekeus's deputy, turned down the request. Another U.N. team, he told Kelly, was already in Iraq pursuing solid allegations about Baghdad's nuclear program. UNSCOM could not support more than two missions at the same time. Ekeus assumed Al Hakam was not a biological-weapons factory. After all, the coalition had never bombed it.

KELLY'S instincts were sharper than he knew. In May, just as UNSCOM was setting up shop, the CIA had received a startling report from a source with access to some of Iraq's darkest secrets. Baghdad's germ weapons had survived the war and were being hidden. The source said Iraq had manufactured two different kinds of biological agents at "the Al-Hakam facility in Jur Al-Shakar," according to a summary circulated among CIA officials in May 1991. Iraq had buried its biological bombs and missile warheads "one meter underground." The "production lines" still existed.

In August, as Kelly arrived in Iraq, CIA analysts sent the agency's director, William H. Webster, a fuller account of what Iraq had produced and where it was being hidden. The report said Baghdad had buried seventy-five bombs containing "Agents A, B, C" near Airfield 37, a major military base. The Iraqis were concealing fifteen germ warheads for their missiles in two locations, one of which was near the Iraqi army base in Al Mansuriah. More than one hundred bulk containers were hidden at yet another military base.

The CIA's report added new details about the Iraqi germ production at Al Hakam. It said the facility had five large laboratories where Iraq had produced toxins from bacteria in 1,400-liter fermentation tanks.

And it said that Iraqi scientists had made weapons from fungi at a factory in Al Fadhaliya. The report noted that Iraq had "no plans to move any of the chemical or biological agents which have been hidden."

The source's account and other new evidence gathered by American intelligence operations prompted the CIA to reassess its view of the damage done to Iraq's program by the coalition bombing. In fact, most of Iraq's biological operations appeared to have survived the war. In September 1991, the agency identified eight sites at which Iraq was suspected of producing germ weapons. Only two had been bombed. On the list was Al Hakam, which Iraq said was a warehouse. The report said the facility appeared to have made botulinum, anthrax, and *Clostridium perfringens,* a bacterium that causes gangrene and attacks the body's internal organs.

The report about Iraq's buried germ weapons, as well as the CIA's assessment that much of Baghdad's biological program was intact, does not appear to have made its way to Ekeus and other senior UNSCOM officials. Major Karen Jansen, an American military officer who was serving as a U.N. inspector, recalled being told by a CIA briefer in late 1991 or in early 1992 that germ weapons were hidden at Airfield 37 and at several other locations. Jansen said she had told the CIA that she was grateful for the tip but that UNSCOM needed precise locations or sites. The inspectors could not dig up an airfield. A senior CIA official said in an interview that the agency believes it also relayed the reports to two other UNSCOM officials in 1991 or 1992. Neither could recall such a briefing. In any event, none of the information was put into UNSCOM's files. Nor was it made available to David Kelly, the British official who led the initial inspections.

In September, a month after the CIA's new assessment of the Iraqi program was completed, Kelly and the UNSCOM team set off for their first visit to Al Hakam. The site, about thirty miles southwest of Baghdad in the middle of the desert, seemed an odd choice for a civilian facility. The buildings were surrounded by an impressive security fence and spaced more than two miles apart. The Iraqis had already abandoned their initial story about Al Hakam. It was not a warehouse but a factory for making animal feed, they said. The inspectors did not contest that explanation. The equipment installed in the cavernous buildings was being used to make animal feed. But the same fermenters could just as easily produce germ weapons. It all depended on what was cultivated. "Although at this point there is absolutely no evidence of participation in a biological weapons program, the team was concerned that it might feature in development of such a program," the inspectors reported. UNSCOM inspectors left Iraq divided over whether germ weapons had ever been produced at Al Hakam.

The U.N. team was stumbling through Iraq, unaware of what it was seeing: a vast war machine hidden in plain sight. In the fall, one group of inspectors toured the Al Walid air base in western Iraq to examine some R-400 bombs with black marks painted on their sides. Before the inspections began, the Iraqis gave UNSCOM a misleading explanation of the color codes on its unconventional weapons. Yellow paint meant a shell, warhead, or bomb was filled with mustard gas; black was for sarin nerve gas; red for tabun. Iraq would eventually acknowledge that it had lied and that a thick black stripe denoted a germ weapon. The inspectors walking through Al Walid had seen bombs containing biological agent, not sarin.

Shown copies of the CIA's 1991 reports years later, several UNSCOM inspectors could only wonder what might have been. The information, they said, must have come from a top official inside Iraq, someone with knowledge of the entire biological program. Even the nomenclature was correct. The Iraqis referred to bioweapons by different designations: A was for botulinum, B for anthrax, C for clostridium. (The Iraqi air force used the code names "tea," "coffee," and "sugar.")

It would take the United Nations teams nearly four years and countless trips to Iraq to piece together what the CIA had figured out by the fall of 1991. The failure to exploit the spy's report, however it occurred, slowed the inquiry. In the years that followed their first trip to Al Hakam, inspectors disagreed about whether Iraq even had a bioweapons program. Kelly, in particular, saw missed opportunities in the CIA reports. The first UNSCOM inspection of Airfield 37, in November 1991, was a cursory look for bunkers. Having the complete report from the spy in 1991 would have prompted the field teams to ask much tougher questions, to look harder. "It would have affected the investigation profoundly," he said. Ekeus, who refereed the fights among UNSCOM inspectors over Iraq's biological capabilities, could only shake his head when shown the CIA material ten years later. "This is scandalous," he said. "Yes, it definitely would have mattered."

As UNSCOM inspectors struggled to decipher the Iraqi germ program, the Pentagon began work on rebuilding America's biological defenses. The Project Badger team that had struggled to increase vaccine supplies in the fall of 1990 circulated a classified report on what should be done. The military could not depend on the nation's pharmaceutical industry to meet demand for vaccine in a crisis. Serious steps were needed to guarantee the supply, the report said. American soldiers should never go to war again without adequate protection against biological attack.

Days after the fighting ended, army health officials sent a memo to the secretary of the army, Michael P. W. Stone, saying that "the biological defense program has suffered from minimal support since the termination of the offensive BW program in 1969." The memo noted that "U.S. forces are immunologically vulnerable to the full spectrum of enhanced BW agents even though we have the capability to immunize against many infectious diseases." The United States, it argued, needed to begin a program of peacetime vaccinations for troops stationed in high-threat regions.

That enthusiasm proved difficult to sustain. The army, which was in charge of biodefense, noted a few weeks later that an expanded program to inoculate the troops would be expensive and might raise fears among civilians by putting the "sensitive BW topic in the public eye." It could also spur potential adversaries to create vaccine-resistant bugs. Army officers also raised the possibility of stockpiling enough vaccine to protect the civilian population of the United States. This, they maintained, would show potential adversaries that "we are prepared for BW." Army planners pointed out that this option carried "the highest total cost." The idea was rejected.

Army officials were persuaded by their Gulf War experiences that neither private industry nor Michigan could be relied upon to make sufficient quantities of vaccine in times of war. The Defense Department, the army said, should immediately begin work on its own vaccine plant. Coalition warfare was the way of the future, and the United States should have enough vaccine to supply not only its own needs but those of "select allies." A contractor could be hired to run the factory.

On April 9, 1991, Colin Powell and the Joint Chiefs of Staff agreed to a first step, recommending that the army stockpile enough anthrax vaccine to protect the entire active duty force of 1.6 million men and women. Powell outlined the proposal in a secret memo to Defense Secretary Cheney. He termed it a "minimal goal."

The army's plan to vaccinate a limited number of soldiers in peacetime proved far more contentious. By the fall of 1991, the demise of communism was stirring serious demands for a peace dividend. As the White House scoured the Pentagon for budget cuts, the army, air force, and navy all fought to save their favorite projects. Every dollar for biological defense was a dollar taken away from other programs that had long enjoyed greater stature. The army's vaccine factory would cost more than $1 billion over several years.

At an October 1991 meeting of the chiefs, the air force asked whether the threat was "real" and whether the Pentagon could afford such an ambitious plan. Cheney's aides had qualms about some aspects of the army proposal but insisted to military leaders that "two components of the plan (peacetime vaccinations and adequate vaccine production) should be accomplished immediately." The services did not agree. "There is a lack of consensus on the threat," one official wrote in February 1992. "Navy and Air Force do not see the same threat to their forces as ground forces." As a result, there was "no vaccination policy."

The air force and navy suggested inoculating troops on the eve of hostilities and only if they were going to war against a foe known to have germ weapons. Health officials said such a plan was medically unsound. They reminded Powell that it takes as long as six months to build up full immunity. The chairman began quietly pushing the plan to begin peacetime vaccinations of selected units.

Powell found a more receptive audience when he met with the military's special operations commanders in early 1993. He asked how the Delta Force and other units were preparing to deal with terrorist attacks involving WMD, weapons of mass destruction. The military's counterterrorist units had been training for years to storm airplanes and rescue hostages. But Powell believed that the threat of the future might come from a terrorist group that had obtained a nuclear, chemical, or biological weapon. The special forces commanders understood the issue and began to rethink their training and strategy, focusing first on the dangers posed by radiological bombs—conventional explosives packed with radioactive material.

British and American efforts to force Moscow to own up to its biological past appeared to be making progress. Boris Yeltsin, who became president of Russia after the Soviet Union ceased to exist on December 31, 1991, told President Bush at a February 1992 summit meeting at Camp David that he had personally reviewed the issue. "I got your paper," Yeltsin said, referring to the American document detailing the violations of the biological treaty. "I'm absolutely certain it's true." Yeltsin, the Communist Party chief in the Sverdlovsk region in 1979, at the time of the anthrax release, said he had long doubted the tainted-meat story. He said the KGB and military had lied when they assured him that the anthrax deaths had a benign explanation. "I will issue a decree and clean this up," he promised Bush. In passing, Yeltsin mentioned some alarming news to his American hosts: his own military was holding back from his aides information about the illicit program. As the summit concluded, a Russian spokesman publicly announced Yeltsin's pledges to dismantle the Soviet germ empire.

A confrontation, however, soon erupted over whether Moscow had been candid about its germ history. Britain, the United States, and Russia had agreed to make declarations to the United Nations about their respective programs. Russia's proposed draft, circulated in late April 1992, disappointed officials on both sides of the Atlantic. It did not acknowledge any weapons production and omitted many of the facilities Pasechnik had disclosed. The account, one official said, was the diplomatic equivalent of saying "I smoked but I didn't inhale."

On May 27, in an interview with a Russian newspaper, Yeltsin acknowledged publicly what he had already told Bush—that the 1979 outbreak of anthrax in Sverdlovsk stemmed from an accident at a military facility. Yeltsin said the Soviet intelligence agency, the KGB, had privately admitted "our military development was the cause." He offered no further details.

Days later, a scientific team led by Meselson, the Harvard professor, arrived in Russia to investigate the Sverdlovsk incident. After years of requests, Moscow was finally permitting an independent review of the evidence. The team interviewed health officials and families of the victims. Russian scientists showed the visitors autopsy slides of the area between the lungs—the key evidence missing from the Soviet presentations in the United States in 1988. The tissue bore the telltale signs of damage from inhaled anthrax.

In the summer of 1992, Meselson and his team presented their preliminary findings to an audience of experts at the Brookings Institution in Washington. The team's pathologist, David Walker, said the autopsy slides were conclusive evidence that the anthrax had been contracted from inhaled spores, not tainted meat. But Meselson said what had caused the outbreak was still unclear.

The impasse between Russia, Britain, and the United States over the United Nations declarations became public in July, when the *Washington Times,* a small newspaper with solid military and intelligence sources, broke the story. Russia was depicted as having lied about its past just as the administration was pushing for an election-year treaty on nuclear arms. Once again, the biological issue had taken on domestic political significance. The Bush administration demanded an immediate meeting with the Russians to resolve the issue and salvage its arms proposal.

British, American, and Russian officials gathered in Moscow for talks that quickly grew tense. Addressing a room crowded with more than fifty people, Valentin Yevstigneev, a burly Russian general, defended his country's declaration to the United Nations as complete. He said the Sverdlovsk incident was omitted because Russian experts were unable to agree on whether it was caused by tainted meat, defensive research, or sabotage by anti-Soviet forces. Pasechnik was a liar, another Russian official said. The Soviet Union had never weaponized plague. The talks dragged into the next day, and in the final hours, the Russians delivered an ultimatum: Moscow would agree to visits to suspect biological facilities, but only if the United States and Britain agreed to open their pharmaceutical companies to the same scrutiny. The British and American officials hated the offer because it made all three sides appear equally culpable. But diplomacy is the art of the possible; they accepted the deal, which was announced with great fanfare and sold to Congress as a diplomatic accomplishment. Within weeks, as the negotiations for the mutual inspections plodded ahead, the United States unexpectedly learned even more about the extent of Moscow's biological deceit.

THE CIA men walked Bill Patrick into the middle of his yard in rural Maryland, away from possible surveillance in the house. They told him that they wanted him to interview a Soviet defector. He got into a government car and was driven around for a long time to elude any tail. Finally, at a secluded spot near Washington, they arrived at the safe house. Patrick was surprised that the Russian defector looked so Asiatic. The chauvinistic Soviet system usually reserved top posts for ethnic Russians. Even so, the short man, born in Kazakhstan, was allegedly a scientist who had worked for seventeen years inside the Soviet biological-weapons program and had risen to become the second in command of something called Biopreparat, which American intelligence agencies had been tracking for years. If real, he was better than Pasechnik, a leader with a detailed overview of the whole clandestine effort. Now it was up to Patrick to test his credibility and see what he might have to offer.

Patrick handed the defector his business card, the one with the skull and crossbones. The defector couldn't read a word of English, but he laughed. The ice was broken. "We got right down to it," Patrick recalled. The two men discussed how the Soviets handled seed stocks, drying, concentrations—everything. "It was like a glove," Patrick said. "You take any of our so-called experts. They don't know the stability of agents, or what you have to do to protect agents during freezing and drying." This defector was the real thing, someone who knew the intricacies of the art. "I won't say we fell in love," Patrick recalled. "But we gained an immediate respect for one another."

Kanatjan Alibekov, the defector who later changed his name to Ken Alibek to make it sound more American, had been through two months of debriefings with American experts. He had been surprised to find these sessions like academic seminars in which he often had to explain the basics. "It wasn't until Bill Patrick walked through the door," Alibek said, "that I felt someone understood what I was trying to say."

What Alibek had to say was horrifying. Moscow, he reported in grim detail, had secretly produced hundreds of tons of anthrax, smallpox, and plague germs meant for use against the United States and its allies. The amounts dwarfed anything American experts had ever imagined. Alibek also described a germ empire that stretched from the Soviet Council of Ministers to the Soviet Academy of Sciences, through the Ministries of Defense, Health, and Agriculture, and into Biopreparat, his own ostensibly civilian pharmaceutical agency. In fact, Alibek said, Biopreparat was a biological war machine that employed tens of thousands of people at more than forty sites spread across Russia and Kazakhstan. Most important for Patrick, Alibek related details of the Soviet scientific work. As Pasechnik had first revealed, the germ recipes developed by Moscow drew on many different breakthroughs of the biological revolution that had begun just after Washington's own germ program was shut down in 1969. Patrick buried his head in his hands. The Soviets had made significant advances.

The secret debriefings, begun in late 1992, went on for one year. Alibek's assertions were greeted with some skepticism by the broader intelligence community. It wasn't that anybody doubted that the Soviet Union had built a large germ arsenal, but there were questions about the details, especially Alibek's assessment of the Soviets' successes in recombinant research. His statements to the Americans in time included both firsthand knowledge and his interpretation of what Russian scientists were publishing in the open literature. Alibek was a medical doctor, some analysts grumbled, and out of his depth on hard science. But with Patrick as his new ally, he continued to raise the alarm about Russia's germ achievements, first in the secret councils of government, then eventually in public.

BILL Clinton took office in January 1993 determined to keep foreign affairs in the background, which quickly proved impossible. The Soviet menace was gone but the end of the Cold War had brought new dangers. Clinton was a quick study and he understood when aides identified a new kind of threat: the thousands of Soviet scientists left unemployed and bitter by the collapse of their country. Many had spent their lives producing biological, chemical, or nuclear weapons. Their recruitment by a rogue nation or terrorist group could threaten the United States and its allies.

The new president had been in office just thirty-five days when he was confronted with an attack on American soil. Islamic radicals exploded a bomb under the World Trade Center in New York City, shaking it with the force of a small earthquake, collapsing walls and floors, igniting fires, and plunging the city's largest building complex into smoke and darkness. The blast killed six people and injured roughly a thousand.

President Clinton and his aides came to believe that the United States was vulnerable to even more devastating assaults. The attack on the Twin Towers by a loosely organized group of people who had fought against the Soviets in Afghanistan marked a fundamental shift in the nature of terrorism. For much of the twentieth century, ethnic groups had used bombings and assassinations in their fight for independence and their own nation-states. The World Trade Center attack reflected a new paradigm—a holy war, with the United States cast as the enemy. Some of the more violent groups were underwritten by a wealthy Saudi exile, Osama bin Laden, who hoped to spread his radical vision of Islamic rule to Muslims everywhere and drive the United States out of the Middle East and Africa.

The White House initially pushed to improve defenses against the sort of truck bomb that tore into the World Trade Center. But at the Department of Health and Human Services, a handful of officials began to worry about the damage a terrorist group might be able to inflict with a germ weapon. In November 1993 two officials at HHS, Frank E. Young and William Clark, organized an exercise to see how New York City might handle an anthrax attack. A microbiologist who directed the office of emergency preparedness, Young believed that the nation's defenses against a domestic biological attack were woefully inadequate. The Gulf War had vividly demonstrated the superiority of America's battlefield technology. Future adversaries would be looking for easier targets.

The exercise, code-named Civex '93, had been planned in the final months of the Bush administration. But Young and Clark hoped that it would get the attention of a White House increasingly preoccupied by the terrorist threat. In the scenario hundreds of thousands of panicked New Yorkers fled their homes as the disease spread. Federal and New York City officials were impressed by the havoc a germ attack could unleash.

Clinton's choice as defense secretary, Les Aspin, also understood the biological issue. A respected military thinker, Aspin had served for years as chairman of the House Armed Services Committee and had studied the Persian Gulf conflict closely. The central lesson, he believed, was that American military power would not deter dictators like Saddam Hussein from seemingly irrational acts. Aspin argued that American forces would have to be ready to counter whatever an adversary could invent, including chemical and biological weapons.

Early in Aspin's tenure, army officials pressed their case for building a modern, government-owned pharmaceutical facility where the military could produce large quantities of several different vaccines. Surveys of the nation's pharmaceutical companies, they wrote, had found "no commercial interest" in making vaccine for biodefense. The officials added that the burgeoning research into recombinant bacteria and viruses raised the threat of new kinds of weapons. Construction of the plant was slated to begin in early 1994. If all went well, the facility—equipped with six separate production "suites" for making different vaccines—would produce its first lots in 1998. Army officials hoped it would eventually make seventeen different vaccines.

A fight soon erupted over who would be in charge of making the military's vaccines. For years, biological defenses had been divided between the army's chemical corps, which handled suits and detectors, and the army's medical research and development command at Fort Detrick, which managed vaccine programs. Aspin's aides believed biodefenses would be stronger if all the projects were guided by a single hand. It made no sense, they felt, to buy biological detectors in one corner of the army and vaccine in another.

The medical command, the army "docs" as they were called, lobbied against the idea, arguing that while the merger looked logical on paper, it would be disastrous in practice. While some in the military referred to "chem-bio" as a single threat, the two disciplines were very different. Because the bulk of the spending on biodefense went for detectors, the leader of a new joint office would inevitably come from the chemical corps. Vaccine production, the docs argued, should be overseen by someone with a biology background. The former head of the army's medical research command, Philip Russell, asked Lederberg to intercede with John Deutch, the former MIT chemistry professor who was the Pentagon's top procurement official. Deutch was unmoved. The chemical corps officers were more experienced at buying military equipment. They were perfectly capable of managing vaccine projects, Deutch believed. In June 1993, the Pentagon created the joint program office for biological defense.

As the military made plans for stronger germ defenses, Josh Lederberg grew increasingly anxious about the threat posed by impoverished germ scientists, especially in Russia and the former Soviet republics. They might be tempted to sell their expertise to rogue states or terrorist groups, he worried.

A breakthrough came in 1993. During the previous decade, the National Academy of Sciences, the nation's preeminent scientific group, had established a committee of top American scientists that met regularly with Soviet counterparts to discuss security issues. Lederberg was a member. A meeting of the joint committee in Moscow in 1993 had a surreal air as the Russian biologists denied that they had ever done anything wrong—contradicting Pasechnik, Alibek, and even Yeltsin. Finally, Lev S. Sandakhchiev, the director of the State Research Center of Virology and Biotechnology, known as Vector, stood up. "Let's cut the crap," he said. The past had to be acknowledged, he declared, alluding to what he and his colleagues had long denied. Sandakhchiev, of course, knew what he was talking about. His own giant institute in remote Siberia had specialized in turning the world's deadliest viruses into weapons.

Lederberg was stunned. The candor, he recalled, continued in the months ahead as both sides discussed the deteriorating state of the former Soviet germ warfare centers and what could be done to keep their secrets and scientists from falling into unfriendly hands.

Lederberg proposed closer ties. His idea was to use Defense Department funds to pay former Soviet biowarriors to do peaceful research at home in cooperation with American scientists. Under a law sponsored by Senator Sam Nunn, Democrat of Georgia, and Senator Richard Lugar, Republican of Indiana, Congress for two years had been providing defense funds to help former Soviet states dismantle their nuclear warheads. Lederberg and his allies wanted Nunn-Lugar funds extended to biologists and the academy to organize the cooperative program.

The Pentagon resisted. Military officials argued that biological projects would be inherently for dual use—aiding the development not only of medicine but also of weapons. The uneasiness was most intense among American military scientists, especially some at Fort Detrick, who feared that their collaborations with the Russians would be attacked as aiding Moscow's development of germ weapons. But Lederberg strongly disagreed. He argued that the projects could be screened carefully to minimize the military risk. "He had great credibility at the Pentagon," an academy official recalled. "He could address those kinds of issues with more authority than anybody else."

Lederberg won, and the Defense Department began to finance cooperative projects. For instance, it paid for a Detrick scientist to work with Russian scientists at Vector. The team analyzed the genome of the monkeypox virus, a cousin to smallpox. Lederberg became chairman of an academy committee that advised the Pentagon on the safety of the Russian research proposals. Among the committee's fourteen members was Matt Meselson, the Harvard biologist. While the two men were developing somewhat different views on Moscow's biological past, they agreed on the value of the cooperative projects.

In the summer of 1993, Lederberg began urging the Pentagon to broaden its biodefense agenda to include civilians. The Defense Science Board, the influential advisory panel on which he served, was studying what it called "technical military capabilities for future contingencies." Lederberg drafted a sharply worded section on the vulnerability of civilians to biological attack. "BW is a weapon of mass destruction. But no agency has done any serious planning about how to defend against a BW attack on our own cities or those of our allies," it said. "We urge DOD to take the initiative, together with the Centers for Disease Control and Prevention, in formulating a comprehensive plan for civil defense against BW attack. If such an attack should occur, the military establishment will be blamed for the failure in national defense, regardless of the purported mandate—and above all, we will blame ourselves."

The military was having enough trouble addressing its own needs. On November 26, 1993, the Pentagon committed to its first peacetime vaccination program against biowarfare agents. The order was signed by William J. Perry, then the deputy secretary of defense. It directed the nation's armed services to inoculate personnel stationed in "high threat" areas or who were assigned to units that would be mobilized first in a crisis. The Pentagon, it said, must improve existing vaccines, develop new ones, and stockpile supplies to protect soldiers against "all" possible biological threats identified by the intelligence agencies. Powell had finally persuaded military leaders to begin inoculations against biological attack before a war broke out.

Producing the vaccine would be challenging. A design for the army's proposed factory was complete and sites were being scouted at Pine Bluff, Arkansas, and Fort Detrick. But congressional opposition was growing. In the fall of 1993, lawmakers ordered the army to stop work until it completed a study justifying the need for a new government-owned facility. The issue came before John Deutch, who was then the undersecretary of defense for acquisition. Several civilian officials argued that the military had no reason to get into the drug business. But the army docs pleaded with Deutch to keep the factory alive, even if that meant a fight with Congress.

Deutch was strongly inclined to leave vaccines to the private sector. The army docs, he felt, had a poor track record in moving from research to manufacturing. The factory was a substantial investment, $200 million, but Deutch saw the question as a straightforward procurement issue: How could the military most efficiently buy its vaccines? He had spoken with pharmaceutical executives who were willing to take on the work if the government insulated them from legal liability.

The army's plans for a vaccine factory were shelved. Pentagon officials began searching for a "prime contractor" to manage the development of new vaccines. For the foreseeable future, the military would rely on the Michigan lab for its anthrax shots.

The army projected that it could stockpile enough to protect every serviceman by 1997, but only if Michigan doubled its production without a hitch, and only if the FDA accepted a proposal to lower the number of shots required from six to five.

JOSH Lederberg continued to raise the alarm about civilian vulnerabilities. In May 1994, the New York City health commissioner, Margaret A. Hamburg, arranged for him to brief mayor Rudolph W. Giuliani on the biological threat. A former prosecutor, Giuliani had begun to grasp the political impact of public-health issues when the New York City tabloids broke the story of an eight-year-old killed by a rare form of virulent streptococcus. Some parents panicked; there were calls to close every school where a child had strep throat—a move, Hamburg noted, that would have shut down every school. The incident, she recalled, marked the first time that Giuliani "understood how unsettling an invisible, infectious disease can be to a population."

The mayor was behind schedule, so the briefing began in the early evening. Lederberg offered a tutorial on the life cycle of the anthrax bug. Hamburg was dismayed. At the end of a long day, the mayor could not possibly be interested in such particulars. "The mayor's going to kill me," she remembered thinking. "But Giuliani loved it. They went into all kinds of detail—how many spores could you get into a lightbulb, that kind of thing. It went on for two hours."

A few weeks later, Lederberg was seated next to Anthony Lake, President Clinton's national security adviser, at a dinner meeting in New York at the Council on Foreign Relations. He told Lake that the administration was not doing enough to defend Americans against germ weapons.

"Do I really have to worry about this?" Lake asked. "I've got enough on my plate right now."

"Yeah, I really think we do," Lederberg replied. He followed up the exchange with a letter on June 11, 1994, recounting his meetings with the Bush White House during the Gulf War. "The result was at least a skeletal framework for coordination," he wrote, "that has largely evaporated." Lederberg said that biological weapons were already "widely proliferated" and that "New York City and Washington are the world's juiciest targets for such attacks."

As Lederberg worried about the future, Meselson and his team addressed one of the central mysteries of the Soviet past. On November 18, 1994, they published their final conclusions on the Sverdlovsk anthrax outbreak. Its cause, they wrote, was airborne spores leaking from a military facility, not bad meat. "What we had thought and said to be plausible," Meselson recalled, "was actually entirely wrong."

His position had evolved as the inquiry gathered new evidence. A turning point came in late 1992 when the team discovered some long-forgotten research that seemed to explain a central mystery of the case: how an epidemic of inhaled anthrax had continued for seven weeks. British scientists discovered in the 1950s that monkeys exposed to tiny doses of anthrax did not become ill for many weeks, a delay consistent with the duration of the outbreak in Sverdlovsk.

The team obtained data on the direction of the winds and mapped the whereabouts of seventy-seven victims, sixty-six of whom had died. Most were clustered on a line between the military base and the edge of the city. If the victims had eaten infected meat, they would have been more randomly distributed. The pattern of illnesses was also consistent with the wind direction on the day of the suspected release.

The team's definitive paper, which appeared in *Science,* the respected journal, marshaled an overwhelming case that inhaled anthrax had caused the tragedy. The team estimated that a tiny amount of anthrax had escaped from the military facility, between a few milligrams and a gram, or one thirtieth of an ounce. The size of the leak was potentially significant. The Meselson team's estimate left open the possibility that the lab had been doing defensive research that went awry.

The larger estimates favored by American intelligence officials suggested that the accident was a catastrophic mistake by a weapons factory. Bill Patrick, who had studied the complex mathematics of dispersing biological agents, continued to argue that the accident involved pounds of anthrax. Recalling a lecture in which Meselson presented his low estimate, Patrick said, "We hooted." Ken Alibek, too, had few doubts about what had happened at Sverdlovsk. Compound 19, he later said, had been the Soviet program's busiest production plant for anthrax, making industrial quantities. The work went on around the clock, in three shifts.

In Iraq, the efforts to understand Baghdad's biological history were not going nearly as well. Taking a page from Russia's book, the Iraqi's denied that they had ever produced a single germ weapon, though they conceded that having done some research. American officials encouraged UNSCOM to send its experts to Bill Patrick, who walked U.N. trainees through Building 470, the towering factory at Fort Detrick where the United States had made anthrax weapons in the 1960s. While he was impressed by their enthusiasm, he felt that they were unprepared and stood little chance of success.

The United States sent Colonel David R. Franz, a senior scientist at Fort Detrick and a friend of Patrick's, to strengthen the UNSCOM team in March 1993. A veterinarian who had specialized in toxins, Franz had never worked in a germ-weapons factory, but he had a good idea what one should look like—or so he thought. He joined an UNSCOM team that inspected Al Hakam, which Iraq continued to insist had only produced animal feed and "biopesticides," natural substances that kill insects.

The buildings at Al Hakam were vast, hangarlike spaces with metal doors and ordinary ceiling fans for ventilation. A laboratory in the United States handling anthrax or botulinum would have much more stringent safety precautions, including negative air pressure to keep stray germs inside the building. Safety was an obsession at Fort Detrick. As he craned his head to look up at the fans, Franz thought, "They couldn't be producing human pathogens in here; it's just too dangerous." Al Hakam looked nothing like a comparable facility from America's germ program. The United Nations team was led around the building by Rihab Taha, an Iraqi scientist who answered all their questions politely. Nothing remarkable, she said, had ever happened at Al Hakam.

Back in the States, American officials tried to help by recruiting a veteran of America's germ program to serve as an inspector. Lisa Bronson, a Pentagon official who had helped Alibek defect, asked Franz whether he knew anyone with the requisite experience who might want to visit Iraq in the summer, all expenses paid. Franz suggested his friend Richard Spertzel, the Fort Detrick scientist and associate of Patrick's who had decades of experience with germ weapons.

Spertzel, then sixty-one, was a kindly gentleman whose bearing, shock of white hair, and gracious manner can only be described as grandfatherly. But looks can be deceiving. In fact, Dick Spertzel was a preternaturally stubborn human being who would infuriate the Iraqis and, at times, try the patience of his colleagues. Spertzel had an instant, almost photographic recall of names, dates, and places. Some people are put on earth to be dancers, some baseball players. Spertzel may have been one of humanity's few natural-born weapons inspectors. He had spent nearly three decades in the army, rising to deputy director for research at Fort Detrick before he retired in 1987. He was ready for a change.

Spertzel began work at UNSCOM's headquarters in New York on March 31, 1994. He quickly understood that the investigation had reached an impasse. Senior U.N. officials suspected, but could not prove, that Iraq was lying about its biological weapons. Spertzel closeted himself with the files. They revealed a discouraging pattern: UNSCOM officials would travel to Iraq and pose a long list of questions about bioweapons. Baghdad would delay and provide guarded answers. Months later, the same questions would elicit slightly different replies.

Spertzel turned to Bill Patrick for help, adding him to the UNSCOM mission that was leaving for Iraq on May 28. Patrick was taken aback by Iraq's petty and effective harassment of the inspectors. The U.N. team was billeted at a military hotel where phones rang at all hours with anonymous calls and the showers produced only steam.

Patrick focused on Al Hakam. On his visit there, he shone a flashlight inside the fermenters, which had been produced in Iraq. The interior walls weren't as reflective as a mirror or glass, but they were sufficiently gleaming to meet Western standards and certainly good enough to make the cleanup of deadly spores relatively easy. Al Hakam had top-of-the-line centrifuges, high-quality water, and an intricate array of well-sealed pipes and valves. Patrick felt certain the facility had been built to make more than animal feed.

UNSCOM was divided over whether Patrick was right. Several months later, in October, Rolf Ekeus questioned whether the biological team would ever prove that Iraq had made germ weapons, touching off a tense showdown. Ekeus said that several years of inspections had not yet yielded any convincing evidence that Iraq had weaponized its germs. Tempers flared, and the chief germ-weapons inspector, Annick Paul-Henriot, a French lawyer, said she would quit. Spertzel, dejected, checked out of his hotel in New York, expecting to return to his home in Frederick, Maryland. But cooler heads prevailed. "We had more discussions with Ekeus," Spertzel recalled. "He wasn't convinced. But he promised to devote as much time each day as we needed to lay out our case." The evidence was circumstantial. The inspectors could show that Iraq had researched germ weapons and had imported high-tech equipment that could be used for making them. They had nothing more concrete.

Ekeus acknowledged the Iraqis were hiding something. He suggested a compromise: if the U.N. could not unravel the Iraqi deceptions, it could at least keep an eye on the laboratories and factories in which Iraq might make biological weapons.

Spertzel and the biological inspectors traveled to Iraq in November with orders to put in place a regimen for monitoring any potential weapons laboratories or production sites. The U.N. had been working for months to set up video cameras at any site that had both civilian and military uses. Ekeus said the team was free to use its spare time to conduct interviews with Iraqi scientists. But he warned them that they would have to find evidence of wrongdoing soon or give Iraq a clean bill of health.

Spertzel and the inspectors decided to do two jobs at once, setting up the cameras and monitoring equipment by day and questioning Iraqi officials by night. Iraq claimed that the nine scientists who conducted germ research had all been transferred from scientific labs to Al Hakam during the war. The inspectors asked: Why would a highly trained team of microbiologists and technicians be sent to work at an animal-feed factory?

"They told absolutely incredible tales," Spertzel recalled. "One of these guys was a fermenter technician. We asked him, 'What did you do?' He said, 'Nothing.' We couldn't get anything out of him beyond 'nothing.' It was obvious they were lying."

One of the first mysteries Spertzel wanted to resolve was a report that Iraq had tried to order an air-handling system from a British company for a mysterious facility its procurement agents called Project 324. The particular system Iraq wanted would create negative pressure, a crucial requirement for working with dangerous pathogens such as anthrax. The British company had refused to make the sale, but Spertzel suspected that Iraq had gone ahead with the building. There were rumors that it might be somewhere in the western desert, but there was no confirmation. If the inspectors could figure out which Iraqi facility was Project 324, Spertzel reasoned, they would have a much better idea of where Baghdad had produced its germs.

At last the scientific detectives got a break. Combing through Iraqi records, the team noticed that most of the orders for germ supplies had been placed by the Technical and Scientific Materials Import Division, which was part of the Ministry of Trade. An Iraqi clerk told the UNSCOM team that TSMID, as it was called, was actually an arm of Iraq's intelligence services. The U.N. quietly asked member nations for any records of exports to TSMID. A few weeks later, an Israeli intelligence officer turned over documents to UNSCOM showing that British and German companies had sold TSMID nearly forty tons of microbial food—nutrients needed to produce mass quantities of germs. Scientists around the world use the food, called growth medium, every day for lab experiments that involve culturing germs, but Iraq had imported far more than it needed for research and medical treatment.

Once again, the key clue had been in the hands of American intelligence for years. By the late 1980s, the United States had identified TSMID as the covert purchasing agent for Iraq's biological program. A 1990 classified document written by the bugs-and-gas experts said the United States had been investigating "TSMID's search for anthrax cultures" since 1988. Another American intelligence report in early 1991 speculated about certain cooling equipment recently bought by Iraq. "It may be possible these are the refrigeration units involved in Project 324," the analysts wrote, referring to the suspected biological weapons facility they were monitoring. The purchase could "tie these buildings to TSMID and BW." The United States even knew about the purchases of growth media. A November 1990 analysis by naval intelligence officers noted that Baghdad bought "microbial media sufficient for the production of at least 74 billion human lethal doses of botulinum."

Spertzel had some strong suspicions about where the tons of growth media bought by TSMID might have gone. Just a few weeks earlier, he had ordered an inspection team at Al Hakam to sample a production line that was producing biopesticides. The results added another piece of circumstantial evidence to Spertzel's argument that Al Hakam had once been a biological weapons factory. Particles taken from the spray drier measured one to five microns, quite small for biopesticides but perfect for bioweapons. Spertzel asked the laboratory to analyze the DNA of the pesticide it produced, a natural pathogen called *Bacillus thurengensis,* or BT. The bacterium belongs to the same family as *Bacillus anthracis.* It produces a toxin that is harmless to people but kills Japanese beetles or grubs, among others.

The lab's report baffled American intelligence officials and UNSCOM. The strain of BT taken from Al Hakam lacked the genes for producing such toxins, making it useless for killing insects. So, why were the Iraqis making it? Spertzel racked his brains for an explanation. The production techniques for making BT and aerosolized anthrax are similar. Perhaps the Iraqis had produced the toxin-less pathogen as a practice run.

Spertzel found more evidence of Al Hakam's true purpose during a trip to Iraq in January 1995. He asked one of the architects who had worked on the project how Iraqi officials had referred to it during construction. The answer was intriguing: "Project 900." Spertzel tracked down another Iraqi engineer—she had designed the sanitation and water systems for the facility—and asked the same question. "Project 324," she replied. Trying to keep a casual expression, Spertzel put the question to Taha, the Iraqi scientist assigned to answer the inspectors' questions. Was this the correct name for the facility at Al Hakam? Oh, yes, she explained helpfully, Project 324 was the name for the entire construction site; Project 900 was a subunit. The two Iraqi production sites seemed to be one and the same.

That night, Spertzel shared his suspicions with his colleagues. UNSCOM knew that before the Gulf War Iraq had attempted to order high-powered ventilation devices from a British company for Project 324. No animal-feed factory in the world needed such equipment. If the inspectors could prove that Al Hakam was, in fact, Project 324, it would be another argument that the animal-feed factory had been set up as a germ plant.

Spertzel remembered that the intelligence report about the attempt to buy the air-handling equipment mentioned that it was to be installed in buildings identified by the letters E and H. Was that part of Al Hakam? Iraq had turned over the blueprints of Al Hakam to UNSCOM; now another U.N. inspector found a copy and held it up. Buildings E and H were clearly identified. Spertzel had cracked the code, making a connection that had long eluded American intelligence analysts. The mysterious Project 324 was Al Hakam. The Iraqis had concealed their most important germ factory in plain sight.

On the same trip, U.N. inspectors decided to try to bluff the Iraqis into admissions. In Baghdad, an Australian inspector, Rod Barton, sat across the table from four Iraqi generals, the documents provided by Israel in hand, and demanded to know why they had ordered so many tons of microbial food. The Iraqis mumbled denials. Barton began flicking through documents, reading off the numbers of a letter of credit used to pay for the purchase. "Do these refresh your memories?" he demanded.

A scientist with considerable experience in biological defense, Barton was skeptical when he joined UNSCOM in September 1994. "My perception was, maybe they had a research program, but as to producing, well, they probably didn't get that far." The proof on Al Hakam, he felt, was not conclusive. "I was sitting on the fence. There were quite a lot of people who had been there, well-qualified people who did not believe it."

After the confrontation with Barton, Iraq changed its story yet again. Iraq admitted to buying the media but said they were for civilian purposes. It offered UNSCOM documents purporting to show that the bacterial food had been distributed in 1988 and 1989 to hospitals and laboratories. That assertion could be tested. After four years of evasions, the documentation of the growth-media purchases had backed the Iraqis into a corner. It was just a matter, Barton thought, of demolishing the Iraqis' story, piece by piece.

As UNSCOM made progress in understanding Baghdad's program, the inspectors and Western intelligence analysts began to worry that some Iraqi germ biologists might seek work abroad. The inspectors were told, for instance, that Amir Medidi, a top scientist of Baghdad's germ effort, was no longer in Iraq. Had he fled to Libya? No one seemed to know. Apprehensions about an Iraqi exodus paralleled those about Russian germ biologists, whom Lederberg and the National Academy of Sciences had tried to engage in peaceful research.

As Washington sought to limit the spread of dangerous expertise from Iraq and Russia, the apartheid regime in South Africa abruptly collapsed, deepening the threat.

Wouter Basson, a top cardiologist and military officer, had founded and led South Africa's secret program to make germ weapons, Project Coast. Through a maze of front companies and covert science labs, he had built an apartheid arsenal of anthrax, botulinum toxin, Ebola, Marburg, and human immunodeficiency virus, the cause of AIDS. Many of these agents were intended to cripple or kill apartheid foes. Chocolates were laced with anthrax, beer with botulinum, sugar with salmonella. According to testimony before South Africa's Truth and Reconciliation Commission, Basson's germ program claimed many lives and was even considered for use in crippling Nelson Mandela, the antiapartheid leader, in jail.

After white-minority rule ended in 1994 and Mandela became president, Western intelligence learned that the Libyan leader, Muammar Qadaffi, was trying to hire South African scientists, including Basson. He had made repeated trips to Libya, and American officials feared he might leave South Africa permanently.

White House officials said a politically delicate covert operation was mounted to block Basson's emigration. Now unemployed, he could disappear at any time. And the problem went beyond Libya. As the creator of South Africa's germ arsenal, Basson had developed a global network of covert and overt contacts. British and American representatives met with President Mandela, described the dangers, and persuaded him to return Basson to government service, where his actions could be more easily tracked. Even then, a White House official added, Basson and his allies were seen as posing a danger of spreading weapons secrets. Basson subsequently went on trial in Pretoria, charged with murder and other crimes. Relaxed and smiling, he denied all accusations.

6

The Cult

ON Sunday evening, March 19, 1995, wire services began sending urgent bulletins from Japan, saying thousands of people had been sickened in the subways during their Monday-morning commute by a mysterious gas or toxin. Richard A. Clarke, the White House official who handled terrorism issues, immediately called Frank Young of the Department of Health and Human Services for a scientist's assessment of what might be unfolding in Tokyo. Clarke described the early, sketchy accounts to Young, a microbiologist who headed his department's office of emergency preparedness. No group had claimed responsibility for an attack. But Young surmised that someone had released sarin, a potent nerve gas, on the subway platforms. Young decided to check his impressions with Josh Lederberg, whom he had met nearly four decades earlier as a graduate student. He knew just how to find him. Lederberg was part of a secret federal program known as Reach Back, a group of medical experts who wore beepers so they could be instantly consulted in the event of a terrorist attack involving chemicals or germs.

Awakened from sleep in a San Antonio hotel, Lederberg was not surprised by reports of the poisonings in Tokyo. "I hope it's not too bad," he told Young. Lederberg wondered whether the terrorists in Tokyo had also tried out biological weapons. "If they were playing around with this kind of fire," he told himself, "how could they not be lighting that box of matches, too?"

The next day, Clarke called a meeting of the Counterterrorism Security Group, the administration's terrorism specialists. They gathered in the White House situation room. The reports from Tokyo were becoming clearer: as many as a dozen people were dead and thousands injured by an attack with sarin, the nerve gas Young had identified. Suspicions immediately focused on an obscure religious cult, Aum Shinrikyo. For much of the past two years, Clarke had been trying to bring some order to the government's counterterrorism programs, which sprawled across more than forty departments. He was an intense, obsessive man, the sort of person who, one colleague said, liked to relax in the bath while reading a few back issues of *Aviation Week*. Clarke had a talent for getting things done and making enemies. Some in the FBI distrusted him, worried that he was too involved in the specifics of their operations. Others admired his energy. Clarke already knew a lot about the dangers of biological weaponry. He had worked in the State Department's intelligence arm in the late 1980s and was among a small group privy to the secret reports about the Soviet germ-weapons program.

At the White House meeting, Clarke was flanked by Young and career officials from the Pentagon, the FBI, and the Central Intelligence Agency. The more junior aides were perched in the "strap hanger" chairs around the edge of the wood-paneled room, taking notes for their bosses. The people at the table, the government's brain trust on terrorism, knew one another well. Just thirty-five days into Bill Clinton's presidency, they had gathered to coordinate a response to the World Trade Center bombing. Earlier in the year, the group had polished plans for the covert operation to arrest the mastermind of that attack, Ramzi Youssef.

It quickly became clear that neither the CIA nor the FBI knew much about Aum Shinrikyo or had any idea why it might try to kill thousands of Japanese commuters. The CIA's Tokyo station had sent word that the group had an office in New York at a midtown apartment building a few blocks from Times Square. The report surprised the FBI, which had found no references to the cult in its databases on dangerous groups. Pentagon officials immediately sent soldiers from the Technical Escort Unit, the army outfit that, among other things, is responsible for defusing chemical or biological bombs, to New York City. Perhaps the Aum members in the United States had stockpiled nerve agent in the building.

James Kallstrom, the head of the FBI office in New York, briefed

Mayor Giuliani and the New York City police commissioner on the situation. FBI agents began a twenty-four-hour surveillance of the apartment at 8 East 48th Street, just off Fifth Avenue. It was an anxious moment. The FBI had no idea how many people in New York belonged to the cult or where they might live. Following the group's members as they made their rounds only deepened the mystery. Each meeting added a new person to investigate.

FBI agents were eager to conduct a covert search of the group's apartment. There were, however, legal obstacles to such a break-in. Aum purported to be a religion, and its members were not known to have violated any laws in the United States. The group had not hidden its presence in New York City and had filed incorporation papers with the state for Aum USA Company Limited. The cult's operatives had even registered with New York State as a charity with the stated purpose of fostering "spiritual development through the study of and practice of Eastern philosophy and religion." The FBI was seeking the search warrant under the rules of the Foreign Intelligence Surveillance Act, a statute that gives investigators wide latitude in national security cases. Even so, Justice Department officials in Washington felt that the bureau did not have enough evidence to gain approval from the secret three-judge panel that rules on such cases.

Over the next two days, Louis J. Freeh, the FBI director, repeatedly called his counterparts in Japan, asking for help. None was forthcoming. Finally, FBI agents in New York obtained enough information about the group to justify the warrant. More than forty-eight hours after the first news reports from Tokyo, FBI agents quietly entered the building to search for dangerous substances and plant listening devices. The army's unit was not much help. Its best detection equipment, designed for the battlefield and mounted on a large truck, was useless in a covert search of an apartment building on a busy midtown Manhattan street. It was not clear what the FBI agents would have done if they had found chemicals or germs. They had not informed local health officials in advance and none of the agents wore masks or carried gear that could identify hazardous substances. Fortunately, no chemicals or biological agents were discovered in the apartment. The Aum branch in New York seemed to pose no threat. It had been used to procure computer software and other equipment, not to mount terrorist operations. But the incident had exposed some worrisome gaps in the government's preparations.

It took American officials several years to pry details about Aum

Shinrikyo's operations from the reluctant Japanese police. The group, officials would learn, was a religious cult with a $1 billion war chest and an apocalyptic theology that sanctioned mass killing. Lederberg's guess turned out be accurate. Evidence emerged over several years that scientists working for Aum had tried several times to make germ weapons from anthrax and botulinum. They had staged as many as a dozen unsuccessful germ attacks in Japan from 1990 to 1995.

The failure of the intelligence agencies to detect Aum's activities around the world was particularly disturbing to Clarke and his White House colleagues. The cult had tens of thousands of members in Russia and Japan, had published a statement threatening President Clinton, and hoped to hasten Armageddon by sparking a war between Russia and the United States. On a radio program it sponsored in Russia, it had broadcast statements touting the virtues of biological weapons.

The 1995 subway gassings changed the thinking of many American officials. Clarke's counterterrorism group had recently begun planning defenses against terrorists who might use chemicals or germs to achieve their ends. But until CNN flashed around the world the footage of Japanese commuters dying in Tokyo, it was seen as a remote possibility. Before Aum, the 1993 bombing of the World Trade Center by Islamic extremists was the model for modern terrorism. The technology to produce explosives from fertilizer could be mastered by a high school chemistry student. Few experts knew the full story of the salmonella attack carried out a decade earlier by disciples of the Bhagwan Shree Rajneesh. (Scientists, worried about copycats, were still arguing over whether it was appropriate to publish a detailed account of the Oregon poisonings when the Aum attack occurred.) The United States had set up a team to deal with nuclear terrorism in 1975 after an extortionist had threatened to explode a nuclear device in Boston. It was called the Nuclear Emergency Search Team. No similar unit existed to deal with chemical or biological terrorism aimed at civilians.

ON the morning of March 20, 1995, the *Honolulu Advertiser* carried a huge color photo of Japanese authorities ministering to the victims of the Aum attack. It caught the eye of General Charles Krulak, a general who was preparing to begin his tour as commandant of the Marine

Corps. "I asked the question, Is there anything we can do for the Japanese right now?" Krulak recalled. "The answer I got was that we really had nothing." The military's special-operations teams, such as Delta Force, were set up to foil a terrorist attack, not deal with its aftermath. Krulak began to sketch out plans for a Marine Corps unit that would react to chemical or biological incidents. "We wanted a force that could go in there, seal off the area, conduct decontamination, and treat people who were sick." Krulak hoped the new team could be ready for the 1996 Summer Olympics in Atlanta.

General Krulak's views were unusual. Most senior military officers had shied away from involvement in domestic terrorism defense, fearing that it would siphon money away from more traditional military missions and needs. Krulak disagreed. Dealing with bioterrorism was a way to get money, not lose it. The world was changing, and Krulak planned to use his tenure as commandant to reexamine the Marine Corps's basic assumptions. As he tried to build support for his antiterrorism unit, which bore the unwieldy name of Chemical Biological Incident Response Force (CBIRF, pronounced see-brif), Krulak discovered a powerful ally in the Pentagon. His name was Richard Danzig, the undersecretary of the navy.

Undersecretaries are typically assigned to deal with the day-to-day business of running their services: budgets, base closings, and maintenance. But Danzig, who had been a partner in a leading Washington law firm, had broader interests and some glittering credentials. A Rhodes scholar and Yale Law School graduate, he had clerked for Supreme Court Justice Byron White and had been named to a senior civilian post in the Pentagon during the Carter administration at the age of thirty-three.

In 1993, shortly after he was appointed undersecretary of the navy, Danzig began to ponder what threats the military was overlooking. He believed that civilians at the Pentagon should be focusing on the larger strategic issues. What kind of navy did the United States need in the post–Cold War world? What threats were being ignored? Danzig's attention turned almost immediately to the dangers of germ weapons. Despite the reforms pushed by Congress, biodefense was still scattered through several parts of the Pentagon. Nixon's 1969 renunciation of the offensive program had undermined the powerful motivation for building defenses. Without a visceral, direct sense of what the weapons could do to an enemy, military officers lacked compelling reasons to invest in vaccines, detectors, or better protective suits. The military, Danzig discovered, was largely detached from biology, the most rapidly evolving science on the planet. Josh Lederberg seemed to be the only biologist with any standing among Pentagon officials.

Danzig's staff scoured the intelligence agencies and government medical institutes for analysts and outside experts who understood biological weapons. He told them that he wanted to "get smart" on the intricacies of germ warfare, how the agents really worked, what it took to disseminate them, what kind of defenses the United States could deploy. He spent hours with these experts, learning the life cycle of *Bacillus anthracis* and the basics of cellular biology. He sought out I. Lewis Libby, the Pentagon civilian who had looked closely at germ-weapons issues during the Gulf War, and General John Jumper, the officer Powell had assigned to biodefense. They described how the Bush administration had scrambled to prepare for war and had fallen short.

Libby recounted one of the more frightening scenarios studied during the war. What if Saddam had responded to the air campaign by announcing, If you don't stop bombing my country and its civilian citizens, I'll use biological weapons against your civilians? Danzig immediately thought of another, more daunting strategy that would have allowed the Iraqi leader to distance himself from the threat. Saddam could say he had learned that terrorists, acting independently and against his wishes, were plotting to attack the United States with germ weapons unless the bombing was halted.

Danzig was not impressed with the military's preparations for germ combat or the effort to bring biodefense under a single office. The issue was still scattered throughout the bureaucracy. One part of the Pentagon handled training, another masks and detectors. Military leaders continued to view nuclear, biological, and chemical weapons as a single threat, as they had for decades.

Danzig turned to Lederberg, whom he knew casually from his days as a Stanford University law professor. The biologist recounted his repeated failed attempts to warn Washington of a growing threat. "Josh had been sowing seeds for decades in sterile soil," Danzig said. Lederberg was a brilliant analyst and conceptualizer—his ideas were often decades ahead of the conventional wisdom—but he was not much of a showman, Danzig thought. His explanations of the biological threat could be pedantic, dense with details that overwhelmed his audience.

Lederberg's main problem, Danzig concluded, was that nobody was "working the bureaucracy from the inside."

It was a job Danzig relished. He had begun his second stint at the Pentagon with a clear sense of how to change entrenched bureaucracies. You had to sell the leadership on new ideas, take them on an intellectual journey. The temptation in any large organization was to give orders, to impose your will through rank. Danzig was looking for converts. He wanted to walk military officers through the logic that had persuaded him of the danger of biological weapons. He understood that he would succeed only with the wholehearted support of the nation's senior military officers, the leaders of the four services. They would be a tough sell. Germ defenses were about planning for a contingency that might never happen. Meanwhile, the services faced expensive, real-world problems with immediate implications. The army's outmoded trucks needed to be replaced. Enlisted men were underpaid. The navy's fleet was shrinking.

From conversations with experts and intelligence officers, Danzig assembled a detailed presentation on the threat posed by biological weapons. He asked Lederberg to vet the scientific arguments. One of the first audiences was a group of about twenty Marine Corps generals. They were respectful but skeptical. Danzig left the room certain that he had not persuaded any of them.

Danzig sharpened the presentation and began meeting with top civilian and military leaders from all the services, ten or twelve people at a time. His talk was designed like a legal brief. It began with basic science—the ease with which a nation with a pharmaceutical industry and a few crop dusters could create a terrifying weapon. It moved to specifics: Danzig would project a slide showing the troop and ship deployments on a particular day in the Gulf War and walk the officers through an anthrax attack by a small plane or a boat. Using the Socratic method he favored as a professor, Danzig led the officers through a scenario in which germ terrorists attacked Washington, D.C. "There were a lot of what-ifs," recalled one official. Danzig made the case that investments in the millions, not billions, could help. The military needed anthrax inoculations, more intelligence, better analysis.

The seminars, which stretched through 1994 and into 1995, were jointly run by Danzig and Admiral William Owens, the vice chairman of the Joint Chiefs of Staff and another of the Pentagon's most innovative thinkers. The two Rhodes scholars had worked closely together on a number of issues, but Owens was not sold on the germ threat. He was particularly skeptical of the push to vaccinate select troops against anthrax.

Owens had an array of arguments against the program. The vaccinations would be militarily useless. If the United States inoculated its forces against anthrax, he told colleagues, adversaries would just shift to botulinum or some other agent. Anthrax vaccinations would complicate the delicate diplomacy of coalition building. How could the United States persuade allies with unprotected troops to join an international alliance as it had for the Gulf War?

More fundamentally, Owens believed that inoculations would touch off a biological arms race. He proposed a radical alternative. The United States should issue a public statement threatening nuclear retaliation against anyone who used germ weapons against Americans. The prospect of nuclear annihilation, he told colleagues, had certainly gotten Saddam Hussein's attention. Owens saw biological defense as folly. You can't protect everybody, he told colleagues. The military would never order the immunization of servicemen's families. Nor would American officials demand that every American take the anthrax shots.

Owens clashed with the medical experts who were pushing the program, particularly Stephen Joseph, the New York physician who took over as assistant secretary of defense for health affairs in early 1994. Owens believed that researchers were moving closer to producing a "multivalent" vaccine, a single shot that protects against multiple diseases, and wanted to defer vaccinations until the new vaccine was available. He repeatedly asked whether the existing anthrax vaccine was safe, whether it interacted with other drugs or caused dangerous side effects. The answers did not satisfy him. "I never felt we got the complete assurance that we were completely safe to vaccinate," Owens said later.

Joseph, a former New York City health commissioner, knew that it would be hard to persuade Owens and his colleagues in uniform. Biodefense was an alien concept to them. The threat was theoretical. In a time of shrinking budgets, germ defense took money away from tanks and trucks. But to Joseph and the Pentagon's health experts, the decision to vaccinate against anthrax was straightforward, a "no-brainer." He believed a multivalent vaccine was decades away, at best, and might never be produced. American servicemen needed protection against germ attack now. Joseph understood the reluctance of military officers to grapple with the issue. If they accepted the biological threat as real, their entire concept of warfare would have to change.

Danzig was challenging Pentagon officials to do just that, to begin incorporating the possibility of biological attack into their operational planning.

The military patrols the airspace above large public gatherings, such as the Olympics. Danzig suggested that the cordon be expanded to protect against the downwind release of a biological weapon by a crop duster or small plane. The Joint Chiefs invited him to watch an exercise in which commandos practiced an assault on a terrorist safe house in which biological weapons were stored. Danzig was impressed by the efficiency of the nighttime attack, which began with stun grenades and ended with a well-choreographed rush to the target. But he noted with dismay that some of the soldiers mounting the attack had trampled on test tubes that supposedly contained the agent, crushing them beneath their combat boots, releasing their contents.

ON April 19, 1995, a month after Aum's attack in Tokyo, the Clinton administration was staggered by the bombing of the Alfred P. Murrah Federal Building in Oklahoma City. The attack killed nearly two hundred people, including many children. Days later, the FBI began investigating a videotape that warned of a chemical attack against visitors to Disneyland over Easter weekend. The tape showed hands mixing chemicals. Federal agents swarmed over the amusement park. The person or persons who sent the tape were never found.

John Sopko, a senior Senate investigator, was not thinking much about terrorism when he went to the Pentagon's Defense Intelligence Agency for briefings later that month. A former federal prosecutor who had specialized in mob cases in Cleveland, Sopko worked for the Permanent Subcommittee on Investigations. He was at DIA to gather information on a subject that his boss, Senator Sam Nunn, had made a personal crusade: Russia's diminishing control over its nuclear arsenal. Scientists working for Russia's interlocking network of civilian and military research institutes were poorly paid, vulnerable to recruitment by aspiring nuclear powers. The DIA intelligence official he spoke with mentioned the I. V. Kurchatov Institute of Atomic Energy, named for the father of the Soviet Union's nuclear weapons program.

"You know that the Aum was there," the DIA official said, referring to the presence of Aum followers among the Russian scientists.

"The Aum?" Sopko said, puzzled.

"You know, the guys who did the subway gassing."

Until that moment, Sopko had not seen the Aum story as anything more than an international oddity. He pressed for more information. Why had a Japanese religious cult penetrated one of Russia's premier weapons laboratories? The answers were sketchy. Sopko returned to the Capitol with a new idea: the subcommittee's next round of hearings should examine the Tokyo attack. Nunn agreed, as did the Republican majority, and the panel's staff immediately began to investigate the issue, asking for briefings from the intelligence agencies and others in the government. The size and scope of Aum's operations stunned Sopko, and he was even more startled by the government's lack of information. Few government analysts knew anything about the cult. Sopko did his own research. He obtained police records from Australia and Japan, interviewed witnesses, and studied press articles. What he discovered alarmed him.

Aum, he learned, had been founded in the late 1980s by Shoko Asahara, a half-blind Japanese guru who had earned a fortune running a chain of meditation centers. A bearded man with flowing black hair and claims of supernatural powers, Asahara saw himself as a messiah who would lead his followers to safety as the end of the world drew near. The group's name was drawn from the Indian meditation chant *om* and a Japanese word that means "supreme truth." Its membership ran the gamut from lost souls to Japan's elite: biochemists, doctors, even a few Japanese police officers. Aum charged its members steep fees, and by 1995 its holdings in land, cash, and gold bars totaled more than $1 billion. Worldwide, Aum had fifty thousand members in six countries.

Sopko learned that Aum's sarin gas had been produced by its "health and welfare minister," Seichi Endo, who had done graduate work in chemistry and molecular biology. The group's ambitions went well beyond nerve gas. It had built sophisticated laboratories and tried to produce weapons from anthrax, botulinum, and Q fever. An Aum member had already confessed to the Japanese authorities that in 1993 the cult had tried to spray anthrax germs from an improvised aerosol device on the roof of a building Aum owned in downtown Tokyo.

The Aum case looked like a serious American intelligence failure. FBI and CIA officials privately told Sopko that they had not seriously begun collecting information on the cult until September 1995, months after Nunn announced that the Senate would hold hearings on the subway attack. The agencies, Sopko discovered, had missed clear warnings that something was afoot in Japan.

As early as 1993, a Connecticut company had notified the U.S. Customs Service that a Japanese publishing company called Aum Shinrikyo was trying to buy an interferometer, a device used to make very accurate measurements of small objects. The export of such equipment is restricted because it can be used for making nuclear bombs. And Aum ultimately abandoned the purchase. But no one in the American government had recognized the danger or pressed for an explanation of why a Japanese publishing company would want such equipment.

A year later, in 1994, Japanese newspapers carried stories about a poisoning with sarin gas in the provincial city of Matsumoto that left seven people dead and dozens injured. The incident fascinated Sopko because it seemed to be a harbinger of Aum's 1995 attack in the Tokyo subways. American intelligence agencies did not look closely at the Matsumoto gassing, which was blamed on a local resident who had supposedly produced sarin while trying to make home-brewed pesticides. The incident, however, obsessed the Japanese press and, beginning in late 1994, a Washington-based consultant who specialized in chemical weapons. The consultant, Kyle Olson, was hired by a Japanese television network to review the purported accident. He traveled to Japan, interviewed the suspect himself, and found the authorities' explanation scientifically improbable. Producing sarin is an exacting process, not something that can be done accidentally. On the long flight back to the United States in late December 1994, Olson hunched over his laptop and pounded out a warning for officials in Washington. The Matsumoto case was no accident, he wrote; it was terrorism. "It is not far-fetched to postulate a small group of persons (their motivations being beyond the scope of this paper) carefully tested and evaluated the effectiveness of a weapon based on technology with which they were unfamiliar," he wrote. "Such testing would be prudent prior to planning a major strike, both to prevent accidents and to assure maximum effectiveness. It may also be that the unknown nature of the hand wielding the weapon was part of the group's strategy, or perhaps a desire for optimum surprise and shock."

Olson made a prediction: "The person(s) responsible for Matsumoto certainly understand now that a significant quantity of nerve gas delivered into a warm, crowded urban area (such as a Ginza department store, or major subway station) could have catastrophic consequences. The Matsumoto incident has generally been referred to by authorities and the media in Japan as 'the accident.' There is compelling evidence that whatever the complete story of that deadly June night turns out to be, the events in that quiet city were anything but accidental. This case deserves further attention as the potential harbinger of the next phase of terrorist horror."

Olson circulated the document to friends in the government, including at the CIA and in the White House. Nothing happened. A few weeks later, he published the paper in a newsletter distributed by his think tank to several hundred government experts on intelligence and arms control. No one called to ask for more details. "Some of the people I gave it to said, 'Gee, that's interesting.' They'd pat me on the head and send me on my way," Olson said. "I got so little reaction, I put it on the shelf. I said, 'I've done what I can.' " But Sopko understood the power of Olson's story. It would make arresting testimony at the hearings Nunn's committee planned to hold in the fall.

On June 21, 1995, President Clinton signed a secret directive on counterterrorism, the results of a far-reaching review that had begun two years earlier after the World Trade Center bombing. Presidential Decision Directive 39 delineated the agencies that were to play the lead roles in handling terrorist incidents: the State Department overseas and the FBI inside the United States. Clinton ordered the Federal Emergency Management Agency to update its planning for "terrorism involving weapons of mass destruction."

Tacked on just before Clinton's signature was a sentence drafted by Dick Clarke: "The United States shall give highest priority to developing effective capabilities to detect, prevent, defeat and manage the consequences of a nuclear, biological, or chemical materials or weapons use by terrorists."

A few days later, nearly four hundred officials from the United States, Canada, Britain, and Japan met for a closed-door seminar on how those words might be turned into reality. Frank Young and Bill Clark, the Health and Human Services officials, had planned the three-day meeting long before the Tokyo subway gassing, but Aum Shinrikyo was weighing heavily on the minds of everyone who attended. "The copycat phenomenon, which we worry about a great deal in counterterrorism, is a real risk here," Philip C. Wilcox Jr., the State Department's coordinator for counterterrorism, told a packed auditorium in Bethesda, Maryland. "Once it has happened, others will take their cue and try it again. Once the barrier has been breached, what was originally unthinkable now becomes more likely."

Danzig spoke, as did Josh Lederberg. But it was Bill Patrick's presentation that provided some of the most chilling moments of the secret three-day meeting. Speaking in his usual relaxed, conversational tone, he detailed how a terrorist could mount a germ attack on the World Trade Center using a blender, cheesecloth, a garden sprayer, and some widely available hospital supplies.

Patrick gave several scenarios, including an attack with *Francisella tularensis,* a favorite agent of the defunct American offensive program and a germ that the Rajneeshees had ordered. In about thirty-six hours, he said, a terrorist who obtained a starter germ could grow enough bacteria on one thousand agar plates to produce five liters of material. Telling the story through the eyes of the terrorist, he said, "I am going to Waring-blend this mixture, and then I am going to filter it through cheesecloth." The concentration of germs in the solution would be nowhere near what a competent weapons scientist could achieve in a laboratory. But the home brew would have swarms of germs, about five hundred million cells per milliliter. An infectious dose of the bacteria, Patrick said, was about fifty cells per milliliter. A terrorist who aimed that garden sprayer at the building's air intakes would disseminate enough agent in a few minutes to infect half the people in the World Trade Center. "My conclusion today is not if terrorists will use a biological weapon but when and where."

There were few questions after the retired germ scientist finished his presentation. No one seemed to know what to say. One official asked if there was anything local authorities could do to protect Americans other than foiling terrorists before they could attack. "How do we react to something like this when and if an event occurs?"

"If it is an honest-to-goodness terrorist who knows what he is doing," Patrick replied, "your probability of defeating him is very slim."

Michael A. Jakub, an official with the State Department's counterterrorism office, closed the conference with a more uplifting assessment. For years, Jakub said, senior government officials had dismissed the dangers of chemical or biological attack. Those who dared raise the issue, he said, were told: "Here is a cookie, go back to bed, get out of my hair." Recent events had changed everything, propelling the issue to the top of the national agenda. Now, he said, those same officials were saying: "Tell me what needs to be done, and I will help you do it."

7

Evil Empire

IF American officials were worried about Aum and other terrorists in the mid-1990s, they were obsessed with Russia. The debriefings of defectors like Pasechnik and Alibek and Moscow's admissions about Sverdlovsk were disturbing enough. But not until Americans actually visited the premier Soviet germ factory did Washington fully grasp the enormity of what the Soviet Union had done.

The plant was in Stepnogorsk, a city in Kazakhstan, a newly independent republic that had broken away from Moscow in 1991. Stepnogorsk had always been closed to outsiders, even to most Soviet citizens. But the Kazakh government, eager for closer ties to the West, had decided to open it. Since Russia had barred visits by Westerners to comparable facilities, American officials had eagerly accepted the Kazakh offer. The first American mission to the facility, in June 1995, was led by Andy Weber, a young diplomat who was first secretary at the American embassy in Kazakhstan.

The Scientific Experimental and Production Base, as Stepnogorsk was formally known, was built in 1982 to develop and manufacture a new, more lethal variant of anthrax. The factory—the most advanced Soviet germ warfare production plant and the only one outside the Russian heartland—offered Washington its first look at the inner workings of the Soviet Union's germ empire.

Stepnogorsk was listed on no map. Its address was a post office box,

No. 2076. The remote site was only whispered about, almost reverentially, by military strategists in Moscow over a thousand miles away. But by mid-1995, when Weber's small team of Americans was permitted to visit the plant, the Soviet Union was no more, and the germ complex of Stepnogorsk was a decaying wreck.

Standing inside Building 221, the main production site, Weber stared at the ten, twenty-ton fermentation vats that towered four stories above him like a row of midwestern grain silos. Each vessel could hold twenty thousand liters of fluid. Together, they filled a building two football fields long.

Weber had read the intelligence reports about the mysterious facility before his trip. In the 1980s, the analysts had correctly deduced from the configuration of more than fifty buildings that the complex had been designed to produce anthrax or some other bacteriological agent. They had also accurately described the facility as large, even by Soviet standards, where *bolshoi,* or "big," was invariably a compliment.

But until the arrival in the West of the two Soviet defectors, spy agencies had little idea what kind of research was being conducted at Stepnogorsk, what weapons the plant was making, or how serious a threat it posed. Even the intelligence reports filed after the Soviet defections failed to convey the scale of the operation.

Weber quickly did the math: the vessels working at full capacity could brew three hundred tons of anthrax spores in a production cycle of 220 days, enough to fill many ICBMs. Iraq's entire production could just about fit into a single one of these gigantic vats. Since one hundred grams of dried anthrax was theoretically enough to wipe out a small city, the product of this plant alone was more than enough to have killed America's entire population. And the Stepnogorsk complex was only one of at least six Soviet production facilities.

Weber shuddered at the kill ratio. Destructive capacity on this scale was unimaginable to most Americans, even to him, who had few illusions about the Soviets.

Scientists had long struggled to overcome disease. Yet here at Stepnogorsk, the Soviet Union had spent billions of dollars mobilizing disease for war. It had devoted some of its best brains, virtually unlimited resources, and Russian science itself to the mass production of epidemics—a strategy as chilling as it was perverse. Stepnogorsk was a standby production facility, on orders to produce mass quantities of germs in the event of an international crisis. Disease by the ton was its industry.

Pasechnik and Alibek had already described the vast germ-warfare empire to Western intelligence analysts. American satellites had long focused on what were believed to be germ laboratories, production and testing facilities run by the Fifteenth Directorate of the Soviet Ministry of Defense. Stepnogorsk was an integral part of Biopreparat, the laboratories and plants that supposedly manufactured vaccines and other civilian pharmaceutical products. Known to the Soviets as "the Concern," Biopreparat was in fact a hub of Moscow's germ effort, a vast network of secret cities, production plants, and centers that studied and perfected germs as weapons.

Created in 1973, when the ink on the treaty banning germ weapons was barely dry, this vast Soviet infrastructure had, at its peak in the late 1980s, employed more than thirty thousand people, about half of the sixty thousand Soviets engaged in biowarfare, at more than one hundred facilities throughout the Soviet Union. Secretly run by the military, with an annual budget of close to $1 billion, the biological-weapons program had stockpiled plague, smallpox, anthrax, and other agents for the intercontinental ballistic missiles and bombers aimed at New York, Washington, D.C., Los Angeles, Chicago, Seattle, and other American cities. The Concern had studied some eighty agents and prepared a dozen or more for war. Almost every Soviet ministry—from the Ministry of Agriculture to the Ministry of Health, and including even the notorious secret police—had contributed to the germ-warfare program. The United States may have had a military-industrial complex, but the Soviet Union *was* a military-industrial complex.

Weber was given to understatement, but for him the only word to describe what he was seeing was *evil,* pure evil. And the United States was doing virtually nothing about it. He promised himself that day that if he did nothing else in government, he would try to understand what had happened at this plant and at dozens of other germ-warfare labs, institutes, and production centers. What had the Soviets produced here? What had happened to the "product," the Soviets' anodyne term for their germ stockpiles? And where were the hundreds of scientists and technicians who had once worked here? At its peak, this complex had employed 700 scientists and senior technicians. Now there were 180. Where were the others? Had they taken their deadly expertise to Iran, Iraq, or the other rogue states eager to recruit them? Were they driving taxis? Or working for terrorist groups? And what had the Soviets been thinking when they built these germ factories? How and when had they planned to use Stepnogorsk's "product"?

There were few answers to such questions in 1995 as Weber walked, dumbfounded, through the ruins of the secret facility. But he knew that hereafter he would strive to illuminate what the Russians call the "white spots" of their history, especially this grim chapter on the production of weapons of mass destruction. He would also try to ensure that Russia would finally, belatedly adhere to the treaty banning germ weapons that it had signed in 1972 and then flagrantly violated for the next twenty years. If he did nothing else in his life, he told himself, he would do this.

At first glance Andy Weber seemed an unlikely missionary. Polite and soft-spoken, he rarely raised his voice. At the Pentagon he was known for his long silences and unflappable disposition. His calm, pale face and gray eyes revealed none of the emotions that consumed him—the awe, fear, disgust, and obsessive curiosity he felt about the Soviet Union's most shameful, and still secret, military program.

Weber blended into almost any crowd. Slender and of average height, he looked older than his thirty-five years, with a receding hairline and the shapeless dark blue or gray suits, white shirts, and plain-colored ties he usually wore, the costume of the diplomatic corps he had joined right out of Georgetown University's Graduate School of Foreign Service. But his bland demeanor masked a passion for living on the edge. And he was a keen observer, a fluent writer, and a rigorous editor, talents that were evident in the cables he filed to Washington from his first consular post in Jeddah, Saudi Arabia, during the 1991 Gulf War, and from the American embassy in Bonn.

Weber came into his own in oil-blessed Kazakhstan, a former Soviet republic that was the largest and potentially the richest state in Central Asia. While serving at the embassy in the capital, then Almaty, between 1993 and 1995, he had skillfully managed political-military relations with a proud state that was still poor. Together with neighboring Uzbekistan, another even poorer former Soviet republic, Kazakhstan had renounced nuclear, biological, and chemical weapons soon after becoming independent in 1991.

The post was thrilling for a diplomat fluent in Russian and in search of political adventure. A year after his arrival, Weber helped plan and carry out Operation Sapphire, in which, for less than $30 million, Washington secretly transferred more than six hundred kilograms of highly enriched uranium—enough for more than fifty atomic bombs—from Kazakhstan to the United States for secure storage. Next came Operation Pivot Sail, in which he secretly negotiated the $40 million purchase of Moldova's twenty-one MiG-29 jet fighters, which could carry nuclear weapons, and more than five hundred air-to-air missiles. In another secret operation, this one code-named Auburn Endeavor, he quietly helped remove nuclear material from Georgia, a volatile Caucasus state, at a cost of $4 million. Through such missions, Weber developed a reputation within the arms-control bureaucracy for what was known in the jargon of nonproliferation as "preemptive acquisition."

Avoiding the diplomatic circuit of Almaty, Weber had rented a traditional wooden house with a large fireplace and sauna in the snowy mountains outside the capital. There he served home-cooked dinners for Kazakh military officers and national-security officials. Accompanied by his three-legged pit bull, Maggie, Weber and his Kazakh friends fished in summer and skied in winter and hunted pheasant, moose, and elk in the thick snow that blanketed the rugged Tien Shan mountains.

It was on one such trip that a Kazakh official told Weber of his visit to the eerie city of germs and death in the remote northeastern part of Kazakhstan's Central Asian steppe. Would he like to visit Stepnogorsk and some of the other places where the Soviets had tested their most lethal weapons?

Weber was more than eager, knowing all too well the failure of previous efforts to gain access to Soviet facilities. The 1992 "trilateral" pact between Britain, Russia, and the United States, which called for reciprocal visits and candid, classified disclosures of all past germ warfare programs, had never panned out, at least as far as London and Washington were concerned. American and British experts who had visited institutes known to have been deeply involved in the Soviet program were given limited access and repeatedly told lies—"legends," as the Soviet deceptions were known. The research in Communist times, the Russian officials insisted, was defensive, limited to vaccines and other antidotes.

American and British experts made a curious discovery when they visited Vector, the Siberian institute that Alibek had identified as the biggest, most sophisticated facility ever built to study and refine viruses for weapons. Russia had secretly moved its smallpox samples there from the Institute of Viral Preparations in Moscow—which the

World Health Organization had designated as the only acceptable storage site in Russia. The move violated WHO regulations, which required the Russians to inform the international organization immediately of any change in the status of its smallpox repository.

The Russians' tour of American facilities had also been a farce. The team insisted on visiting facilities owned by Pfizer—twice. A member of the Russian team had been found wandering the halls of the company's lab, far away from the designated tour, igniting suspicions that the Russians were using these visits for commercial espionage. During a trip to the Pine Bluff Arsenal in Arkansas, the germ factory the United States had shut down in 1969, a Russian colonel had stopped the bus and demanded to inspect a tall structure on a hill. As the Americans watched with amusement, the colonel had climbed to the top of what turned out to be a water tower. Even Alibek was shamed by the Russian team's conduct. He concluded that perhaps the Americans were not hiding an offensive program after all, a realization that helped prompt his decision to defect. Though the Russians had found no proof of an American biowarfare program during their tour, a Russian official later told *Izvestiya* that the team had found that the United States was violating the treaty.

The Americans were livid. Some officials concluded that the Russians were using the visits as a pretext to delay implementing Yeltsin's order to dismantle the program. The final breakdown of the 1992 agreement came in 1995, when Russia refused to allow Americans to visit labs controlled by the military and both sides exchanged secret briefings on their previous germ warfare efforts. American and British officials found Moscow's account hopelessly incomplete. But the Clinton administration chose not to challenge the Russians over their lapses. There were other, more pressing issues in the U.S.-Russian relationship. The monitoring regime fell apart. American officials hoping to find out more about the Russian program would have to find another way.

Weber knew that the opportunity to visit Stepnogorsk, take samples from there, and interview scientists who worked in what was now an independent country that had renounced unconventional weapons would do more than help refute Russia's claims; it would provide invaluable information about the germ weapons that the Soviets had made and other critical data about the still largely secret program. It might also reveal whether Russia was continuing such illegal work, as many CIA and DIA analysts believed.

Almost a year of intensive discussions ensued before Weber secured permission from President Nursultan Nazarbayev of Kazakhstan to visit Stepnogorsk and several other facilities. Weber then sent word to the four American experts waiting in Frankfurt to accompany him.

He knew that the Kazakh president, a charismatic autocrat and former member of the Soviet Politburo, had made a risky decision: Moscow would not be pleased when it learned that American scientists and intelligence officials had been visiting one of its former biological "holy of holies." But Nazarbayev and other senior officials were furious about the environmental and public-health disasters they had inherited from the Soviet Union. The Russians had used Kazakhstan and neighboring Uzbekistan not only to make and test the world's most advanced germ weapons but also to develop nuclear and chemical weapons. The Soviets had conducted 456 nuclear tests on the flat steppe of Semipalatinsk in eastern Kazakhstan, 116 of them open-air tests that had poisoned the soil and air and left many Kazakhs with radiation-related cancers and debilitating diseases.

At a facility in Pavlodar and at the Chemical Research Institute in Nukus, Uzbekistan, the Soviets had made, stored, and tested a staggering array of chemical weapons. In the midst of the shrinking Aral Sea, on Vozrozhdeniye Island, whose name means "Rebirth" or "Renaissance," the Soviet Union had killed thousands of mice, hamsters, guinea pigs, monkeys, baboons, horses, mules, sheep, and other animals in open-air tests of their newest, most lethal germ weapons. It had also accidentally killed an estimated thirty thousand Saiga antelope on the Usturt plateau in western Uzbekistan when, during one chemical weapons experiment, the winds unexpectedly shifted. As for the human inhabitants of this blighted region, no one seemed to know what had happened to them.

Kazakh and Uzbek officials knew that the Russians had viewed their countries and people as expendable, but until the Russians left, in the early 1990s, they had had little idea of the magnitude of the ecological disasters confronting them. After the Kremlin's new management rejected their appeals to assess and repair the damage inflicted by the weapons programs, both countries quietly turned to Washington for help.

THE reception of Weber's team in Stepnogorsk was icy. An emissary from the production plant greeted the Americans as they emerged from their chartered YAK-40 jet at the dirt-strip airport with a curt message: they were to leave the city immediately. Weber held his ground, insisting that Kazakhstan's president had authorized the visit. But he had no documentation to prove it. After being escorted to a small guest house in town, Weber and his team waited anxiously for permission to proceed. An hour later, a short, well-built man seemingly accustomed to giving orders entered the room.

"Who's in charge here?" snapped Gennady L. Lepyoshkin, a former Soviet colonel who had run the germ plant since Ken Alibek's transfer to Moscow in late 1987.

Weber identified himself, explaining that President Nazarbayev had invited his team to visit the plant. Trying to size up the mild-mannered young American who spoke such fluent Russian, Lepyoshkin declared that no one could enter Stepnogorsk, let alone his institute, without written authorization. "You are not welcome here," he said firmly.

Weber said nothing but called the Kazakh capital; several hours later, the fax authorizing the visit arrived. The next morning, the Americans traveled to Colonel Lepyoshkin's facility, some twenty minutes away from the city, where more testy encounters ensued. Though Kazakh leaders had told their American colleagues about the Stepnogorsk complex's work and their desire either to close the facility or redirect it toward peaceful endeavors, Lepyoshkin and his colleagues stuck to their Soviet-era script. The plant had been built for wartime production of vaccines, they insisted. Ken Alibek, the defector, had lied; he was a traitor. "Let's not argue," Lepyoshkin said. "I'll show you everything here and you draw your own conclusions."

Weber and his colleagues spent the next two days visiting every part of the enormous compound. What they saw horrified them. Everything Alibek had told the CIA was true. But even Alibek might not have recognized his former facility in its present state of decay. Weeds sprouted from the cracked sidewalks that linked the complex's twenty-five main buildings; there were more potholes than roads. The wire security fences surrounding the compound were torn and no longer electrified. The motion sensors outside the complex were gone. Winter winds that rip across the Central Asian steppe had blown out windows in most of the main buildings. Rain, snow, and ice storms had eroded the gray concrete walls. Air locks that had once kept deadly microbes from escaping now hung open. The plasticized lab floors were torn and buckled from neglect and the weather. Many buildings appeared to be abandoned.

Still, remnants of the plant's deadly past were unmistakable. In the infamous Building 600, the American team found the room that had housed the fifty-foot-high aerosol test chamber, the largest indoor testing facility the Soviets had ever built, in which Lepyoshkin's scientists had tried out their strains of anthrax, Marburg, Ebola, and other agents on animals. The walls were still crisscrossed with the pipes that had delivered fresh, breathable air to the space-suit-like garments worn by the scientists who worked with the deadly germs. Unmistakable, too, were the grounding strips designed to prevent the buildup of static electricity, which could cause accidental explosions in rooms where germ bombs and bomblets were tested, not to mention the glossy floors of epoxy to ease decontamination. The waste pipes that carried the contaminated water from the buildings were now dry and cracked, but there was no hiding their function.

Weber and the other Americans saw remnants of lab equipment that had been used to manufacture anthrax: the now abandoned refrigerated vaults in which hundreds of tiny vials of freeze-dried bacteria had been stored on metal trays in neat rows. In labs that lined the long, half-lit corridors, they found tiny pipettes that scientists had used to draw the nutrient medium and anthrax seed stocks out of small vials into larger ones. They counted the dozens of hot boxes, each about the size of a microwave oven, in which the brew had incubated before it was siphoned off into larger flasks, which were then taken to another room in Building 221, where air-bubbling machines turned the liquid into a rich, coffee-colored slurry.

The Americans roamed the underground bunkers, which had seven-foot-thick walls, and the rooms in which the finished product and raw materials had been stored. They photographed the facility from the top of giant mounds of earth intended to protect the agent during an aerial attack. They stood on the abandoned helicopter pad that had once received dignitaries from Moscow, and they toured the underground filling rooms, in which anthrax had been filtered into warheads, cluster bombs, and other delivery systems.

At night, the Americans returned to the rundown institute guest house to compare notes, badly shaken by what they had seen. Though crumbling and impoverished, the complex remained a danger. The plant, its equipment, and its scientists could still make germ weapons.

Somehow the threat had to be reduced. There was little alternative work available in town. The desolate city, like the institute, was decrepit—a ghost town with people. The weather was mild, but there were few people on the streets lined with rows of drab concrete apartment blocks or on the tree-lined walkways that connected one monotone block to another. The parks and playgrounds were also largely deserted. The movie theater and one hotel had been closed for years.

Institute officials estimated that there were still about forty thousand people in Stepnogorsk—just over half the city's peak population of seventy thousand in the mid-1980s. Many Russians who had abandoned Stepnogorsk for Russia, or for wherever they could find work after the Soviet Union's collapse, had been replaced by even more desperate souls from the countryside, where there was even less work, food, or shelter.

The city still had a few remnants of its proud if secretive past, including a giant metal-and-brick monument erected in 1964 to commemorate the military-industrial city's founding. Though Stepnogorsk's ball-bearing plant, which made parts for trucks and tanks, now operated only part-time, the red-lettered slogan on its walls, MY FACTORY, MY PRIDE!, recalled better times for people who had seen their standard of living collapse in only a decade. The germ-weapons plant had been built on land adjacent to a giant civilian biotechnological plant called simply Progress, which had once produced herbicides, animal-feed supplements, and ethanol. Now its fermenters produced mainly vodka.

People survived largely on the vegetables and potatoes they grew in summer, fish caught from the local river, and hope. "Hope dies last," Sergei Antonovitch, a former germ-weapons scientist who owned a poorly stocked pharmacy in town, told the Americans.

Hopelessness was perhaps the most difficult burden for the once envied enclave of scientists, technicians, and privileged workers. Stepnogorsk's giant sports stadium still stood. So did a large hospital, though it had no medicine, and a department store stocked with imported goods from Korea, Turkey, China, Russia, and even Italy. But with 50 percent unemployment, few residents could afford the costly goods; most people shopped at the outdoor bazaar, where poorer Kazakhs sold their own clothing, buttons, homemade pickles, and Russian army watches. (Weber paid twelve dollars for one with a week's guarantee. It stopped after two days.) There were fur coats and hats—essential during the fierce winters—as well as lanterns for when the power inevitably failed and gaudy Chinese bedroom slippers to wear on ice-cold floors—all at negotiable prices.

On the last day of the visit, Lepyoshkin invited Weber and his team to the institute's riverside dacha for a vodka and shish kebab farewell. Lepyoshkin brought his dog, a beige cocker spaniel named Chase who howled along in tune whenever his master played guitar and sang Russian folk songs. The official program was over, Lepyoshkin told his guests, filling their large glasses with vodka. Now they could speak honestly.

As the vodka flowed and the sun set over the Seleti River, the official lies that Lepyoshkin and his colleagues had told the Americans were stripped away. Of course the plant had never produced vaccines or any other product of possible benefit to mankind, Lepyoshkin eventually confided to the astonished Americans. He and his colleagues were only seeing whether they could still get away with the old cover stories—what in Russian is known as "hanging noodles on the ears" of the Americans. With embarrassed pride tinged with what seemed like real remorse, he confirmed that Stepnogorsk had not only made and tested anthrax but had also produced the world's most powerful strains, which were resistant to heat and cold, and even variants that would defeat antibiotics.

Stepnogorsk's pride was No. 836, the most virulent of all. The strain had been developed by Ken Alibek, who had found a more efficient way to produce and disseminate a pathogen created by the Soviet Union's first anthrax factory, which was in Kirov. Alibekov anthrax, as it was known, needed fewer spores to be effective in an attack and hence delivered "more bang for the buck," as Alibek himself had put it. This strain was three times as lethal in both dry and liquid form as the anthrax that was produced at Sverdlovsk, Russia, the site of the devastating accidental leak in 1979. In fact, Lepyoshkin confirmed, the Stepnogorsk complex owed its existence to that accident. The Kazakh plant had been built to fill the potential production gap after the Soviets were forced to stop producing anthrax at Sverdlovsk. Fearful of further contamination and growing Western demands to inspect that facility, the Soviets had built the Stepnogorsk plant in record time and, like other secret Soviet facilities, with prison labor.

Lepyoshkin and his colleagues did not try to justify what they had done. They did not say that they were only protecting their country and people from America's bioweapons program, though several of them did not believe that the United States had actually ended its own offensive germ-weapons effort. They regretted the past, they said. They had always known that what they were doing was wrong.

Raising his glass, Lepyoshkin toasted the Americans and his hope that trust, friendship, and partnership would develop between them. Scientists at Stepnogorsk were desperate, he said. Iran had tried recruiting them, but they had refused such offers. Still, people could not go on like this forever, he warned. Stepnogorsk needed help. Yes, it would be a gamble, but would Washington consider helping him salvage what was left of the facility and redirect its research toward peaceful pursuits?

The Americans were stunned by the contrast between the scientists' candor in this informal setting and their earlier behavior. Clearly, Weber concluded, if the desperation in Stepnogorsk was shared by scientists at Russia's own facilities, official démarches, angry protests, demands for inspections, and ritualistic visits by Americans would continue to fail. Recriminations against former Soviet scientists and military officers from the germ-weapons program would remain counterproductive. But this informal "dacha diplomacy" might do some good, he suspected.

WEBER was surprised but delighted when Lepyoshkin accepted his invitation to accompany his team to other germ sites in Kazakhstan. American U-2 planes and spy satellites had taken hundreds of pictures of Vozrozhdeniye, or Renaissance, Island, the Soviet Union's largest open-air testing site. No American had ever visited this forbidding place, and Lepyoshkin knew every inch of it.

In the 1980s, Lepyoshkin had spent weeks on the island testing the anthrax developed at the Stepnogorsk complex as well as agents for other diseases: tularemia, Q fever, brucellosis, glanders, plague, and, according to Kazakh and American sources, even smallpox. Despite its isolation and inhospitable climate—temperatures routinely plunged to zero in winter and soared to 140 degrees Fahrenheit in summer—Lepyoshkin had an odd affinity for the island. Between tests, he had passed the time reading, playing volleyball, drinking vodka, and painting watercolors of the island's stark, flat landscape and gray-blue sea. Some of those sketches still decorated a wall of his tiny apartment in Stepnogorsk.

Perhaps it was their visit to Renaissance Island that solidified the unlikely bond between Weber and Lepyoshkin. The day after he was told that the team's fixed-wing aircraft would not be able to land on the island's dilapidated airstrip, Weber had plunked down $8,000 in hundred-dollar bills to rent a Soviet-era Kazakh helicopter that could make the flight. When Lepyoshkin learned that Weber had not cleared either the new plan or the expenditure with Washington, he exploded with laughter.

"Andy," Lepyoshkin said, patting the young American on the back, "you are a real cowboy!"

To Lepyoshkin, whose knowledge of America was based largely on the pirated Western movies he had watched on snowy nights with friends, a cowboy was a free spirit who continued fighting losing battles against powerful government bureaucrats. In an ideal society, the righteous cowboys would win, thought Lepyoshkin, who had always dreamed of living in such a world.

"You're the cowboy, Gennady," said Weber, returning the compliment. Lepyoshkin had also broken the rules by agreeing to accompany the Americans on their tour after having been denied official blessing.

Weber would never forget their trip to "Voz" Island, as the American team called their destination. As their MI-8 helicopter sped toward the island in the middle of the Aral Sea, signs of life rapidly diminished. Fishermen in their wooden boats had vanished; what had once been a living sea was now marshland. Scraggly trees gave way to patches of sagebrush until, finally, there was nothing left to see but salt-covered, cement-colored sand that had once been seabed, now cracked and dry. There were no birds. Nothing seemed to live here.

After the helicopter landed, Weber and his colleagues put on the white germ-warfare suits, masks, and respirators that the Pentagon had given them for the mission. The suits were added insurance in case a year's worth of vaccinations against a host of gruesome diseases failed to protect them. While Cowboy Weber was willing to incur Washington's anger over the unauthorized rental of a chopper, he was taking no chances with the germs.

He had first learned about Renaissance Island from intelligence reports based on Ken Alibek's debriefing. Alibek, it turned out, had brought with him one of the most closely guarded secrets of the Cold War: Vozrozhdeniye, he had told the Americans, was not only the former Soviet Union's major germ-testing range; it was also the world's largest burial ground of weapons-grade anthrax. For the United States, the visit was a potential intelligence gold mine.

Weber recalled Alibek's account of the island's sad history as he double-checked his suit, mask, and connecting air filter for holes or leaks. In the spring of 1988, scientists from Sverdlovsk had been ordered to dispose of tons of anthrax bacteria that the Soviet factory had produced and stored at Zima, near Irkutsk. Though the scientists were not told why, they had deduced that President Gorbachev was getting nervous about pressing *glasnost* at home and *perestroika* with the West while the evidence that Moscow had blatantly violated the germ treaty was being stored throughout the country. In the new climate, the Soviet Union's anthrax stockpile was becoming a potential liability. What would Moscow do if Britain or the United States demanded an inspection and found the telltale product—a clear violation of the 1972 Biological and Toxin Weapons Convention, which the Soviets had been among the first to sign and ratify? The anthrax had to be destroyed. But where? And how? Anthrax spores were incredibly hardy, and Soviet science had made them even more so.

The scientists were ordered to dispose of the stockpile on Renaissance Island, their former test range, over a thousand miles away. Working in haste and total secrecy, they transferred the tons of bacteria—enough to destroy the world's people many times over—into stainless-steel canisters. Over the next several days, the canisters were packed onto a train two dozen cars long and sent on a lonely journey across the Russian and Kazakh republics to Vozrozhdeniye.

At the edge of the old test range, Soviet soldiers poured bleach into the canisters to decontaminate the deadly pink powder. Then they dug huge pits and poured the sludge into the ground, burying the decontaminated spores and, Moscow hoped, a serious political threat.

Now Weber was determined to dig up the anthrax and the past. As he and his team scrambled out of the helicopter, he noticed that Lepyoshkin declined the space suit they had offered him. If the anthrax hadn't killed him by now, Lepyoshkin joked, he was probably safe.

With Lepyoshkin at their side, the Americans explored the decaying laboratory and high-containment unit where Soviet scientists had handled the deadliest of agents. Though summer had just begun, the heat was a sledgehammer—over 100 degrees, one scientist's thermometer showed. In their nonporous protective suits, the Americans sweated profusely. But the adrenaline rush of the mission—knowing that they were the first Westerners ever to explore the island—made them indifferent to their physical discomfort.

To Weber's surprise, not only was the lab deserted—except for the occasional lizard, snake, or fly—but it also had been stripped bare. What little the Soviets had not burned or buried when they had abandoned the site in 1992, local scavengers had stolen to sell at bazaars in Nukus and faraway Tashkent—contaminated equipment, copper pipes, and even the floor and wall tiles, some of which they had left behind, shiny, lime-green tiles decorated with a fish motif. The vivaria that once housed thousands of smaller animals for tests—rabbits, guinea pigs, white mice, and hamsters, as well as larger animals like horses, sheep, donkeys, monkeys, baboons, and apes—were empty, their windows smashed or missing, their roofs collapsed.

In one bungalow, obviously a storage room, hundreds of small cages were stacked together; in another room stood a human-sized cage—for large "nonhuman primates," man-sized monkeys, Lepyoshkin assured them. Back in the 1980s, Lepyoshkin had joked with Alibek, his boss at the time, about envying the doomed monkeys: the animals, at least, had been given bananas and other fruit for dinner, a luxury even for privileged Soviet germ scientists. It was gallows humor; the thousands of animals sent here had died hideous deaths, hundreds of them in a single experiment. The animals were long gone, but a stench familiar to veterans of the world of germ warfare clung to the ruins—a blend of bleach, dust, animal waste, and death.

Just beyond the compound lay the vast desert test range—delineated only by telephone poles, evenly spaced a kilometer apart, to which the animals had been tethered during open-air testing. Air-sampling devices fastened to the poles measured the concentration and particle size of the agent and atmospheric conditions during the tests.

After the American team toured the compound, scooping up surface samples from the lab walls and test-range soil, Weber and Lepyoshkin inspected the facility's housing compound on the northern part of the island, less than a mile away from the test range. The three-story buildings that had served as barracks, homes, kindergarten, and cafeteria were deserted, but at the peak of the testing, Russian scientists, technicians, and their families had lived here—as many as a thousand people in all.

The children's playground, with its swings and a Russian version of a jungle gym, were an eerie reminder of those who had once called this forbidding place home. Lepyoshkin told Weber that most of the children had not been vaccinated against the agents that, until the 1970s, had been tested here, just a mile downwind—tularemia, Q fever, brucellosis, glanders, typhus, VEE, and plague. In some cases, no vaccines existed for the strains that had been disseminated from planes by spray tanks and from aerial cluster bombs.

"We didn't test unless the wind was blowing south, away from the living quarters," Lepyoshkin said. Still, there had been accidents. A plain wooden cross marked the grave of a young woman scientist who had accidentally become infected with glanders. Moscow had never disclosed her death, or those of several local fishermen who had ventured too close to the island during an ill-fated test in 1972 and were subsequently found in their drifting boat, dead from plague.

Lepyoshkin, of course, had been vaccinated against many of the lethal pathogens he had tested—several strains of which were apparently unknown to the Americans. Weber and his team had endured a year of painful shots to protect them against anthrax and other more exotic agents, but Lepyoshkin had been protected simply by stepping into a test chamber at the Stepnogorsk complex. As he had listened to soothing classical music, an aerosol vaccine had wafted across his body and into his lungs. The Soviet Union had pioneered mass immunization through aerosol, jet-injector, and oral vaccines, Lepyoshkin noted. In 1960, the Soviets had used an oral vaccine to inoculate some 6.6 million people, during a three-week period, against an outbreak of smallpox, which normally kills a third of unvaccinated victims. The United States would not begin testing aerosol vaccination techniques until the late 1990s, but aerosol protection had been routine in the Soviet Union. This was yet another area in which Soviet germ scientists had excelled; the United States had much to learn from them, Weber thought, as he listened to Lepyoshkin.

"The island was smaller and beautiful then," Lepyoshkin recalled wistfully. "From the lab you could almost touch the blue sea sparkling in the sun."

Because of wrongheaded Soviet irrigation policies, the Aral Sea was shrinking. Once the world's fourth-largest inland sea, it now ranked tenth. Its surface had shrunk by half between 1960 and 1990, its volume reduced by 75 percent. As a result, the dusty, tear-shaped speck of an island had grown from 77 to 770 square miles. At low tide on some hot summer days, it was already connected to the mainland. Soon this once isolated island would be permanently linked to the Uzbek mainland by a natural land bridge. Then the buried anthrax germs—and who knows what else—might escape their sandy tomb. Carried by gophers and other rodents, lizards, and birds, the spores could be brought to the mainland. Since anthrax bacteria spread from animals to human beings by direct contact, the potential for widespread contamination of people whose health was already abysmal would only increase—if the anthrax spores were still alive.

The local population already suffered from a variety of mysterious illnesses—rare forms of cancer—that may have been attributable not just to poverty and environmental degradation but also to the region's biological and chemical legacy. No one knew for sure. There had been no census in the region since 1989, and health statistics were either nonexistent or unreliable.

Weber's team had been told of the region's plight during their stay at Aralsk, Vozrozhdeniye Island's military support headquarters, where the Soviet army had kept a four-hundred-man battalion of soldiers skilled in germ warfare and some six hundred other troops. Before traveling to the island, the American team had spent a long night at the town's only hotel, a sweltering, bug-infested ruin, where they had talked to the former Soviet port's residents.

According to Doctors Without Borders, the volunteer physicians' group, people of the region, especially those in Karakalpakstan, a semiautonomous republic in Uzbekistan that borders the sea on the southern side, were among the former Soviet Union's most chronically sick people. Some 90 percent of the women were anemic. Infant mortality rates were comparable to those of sub-Saharan Africa. There was an alarming increase in kidney disease and various cancers. Two-thirds of the population suffered from some chronic illness, many from tuberculosis, whose incidence soared after the collapse of the Soviet Union.

At Aralsk, on the Kazakh side of the sea, the local government had turned one of the deserted Soviet army barracks into a tuberculosis colony where patients could be relatively isolated. But there were no drugs for them. Aralsk's population, along with the sea that had once supported a local fishing industry, had shrunk; there were now fewer than ten thousand people. There was no work. The children on the hot, dusty streets seemed listless and pale. Even the dogs and camels looked sick.

What would happen to them, Weber wondered, if the island's legacy of still-deadly anthrax came ashore as the sea continued to shrink? On the Uzbek side of the island, which included most of the testing complex, an Uzbek oil company had started exploratory drilling. Some oil riggers had even set up temporary headquarters in the barracks that had once housed the Soviet germ scientists. Was the buried anthrax still alive? Could there be still other deadly remnants of Vozrozhdeniye Island's past?

WHEN he returned to Almaty, the Kazakh capital, Weber briefed Ashton B. Carter, an assistant defense secretary who was touring the region. Weber's mission to Stepnogorsk had proved that what Pasechnik and Alibek had told the West about the former Soviet germ-warfare program was true. Biopreparat was, as Ken Alibek had called it, the Soviet Union's "Manhattan Project."

While Biopreparat institutes inside Russia were still closed to foreigners and even to most Russians, Weber's visit to some of the largest, most lethal sites in now independent Kazakhstan had created a unique opportunity, Weber and Carter agreed. If the Clinton administration could help Kazakhstan salvage and convert its facilities, especially the complex at Stepnogorsk, it might prevent the scientists there from returning to germ-warfare work and discourage the sale of their expertise to rogue states or terrorists. It could also create a precedent. A flourishing, prosperous Stepnogorsk would be a showcase to other former Soviet republics and even to Russia, which strongly resisted biological cooperation with the West. America, Weber urged Carter, had to make Stepnogorsk a "poster child" for the benefits of transparency, international cooperation, and nonproliferation.

8

Breakthrough

As Washington grappled with the implications of the Soviet collapse, the UNSCOM inspectors in the Middle East were bearing down on Iraqi officials, who were squirming to explain why they had bought forty tons of microbial food in the years leading up to the Gulf War. Iraq said that its records proved that the material had been used by laboratories and hospitals or had been lost. "How?" the inspectors demanded. Well, some containers of the germ food had fallen off the back of a truck. Other records had been destroyed in an unfortunate fire that had consumed only a single file cabinet. Iraqi officials added that the discrepancies could be explained by postwar riots in which citizens had attacked local research facilities and demolished their scientific supplies.

Rolf Ekeus and the UNSCOM team traveled to Iraq in late March 1995 to confront Rihab Taha, the Iraqi scientist assigned to escort the inspectors. Why had Iraq used far more microbial food than could be explained by civilian programs? Taha, who had trained in Britain and spoke excellent English, burst into tears, begging Ekeus to impose order on his unruly inspectors. Iraq could not possibly have imported forty tons of growth media, she insisted. The records were in error. She said her aides had done a calculation: if every fermenter in Iraq grew germs nonstop, it would take twenty-six years to use forty tons of microbial food.

Richard Spertzel, the chief bioinspector, sensed that Ekeus was wavering. The calculations sounded impressive. But he was sure that they were incorrect. He sent a U.N. inspector back to Al Hakam to count the fermenters again. An Iraqi technician acknowledged that several pieces of equipment identified on Iraq's previous declarations as "inactivating tanks" were, in fact, fermenters. Spertzel returned the next day and showed Taha UNSCOM's revised estimates. The fermenters at Al Hakam alone were big enough to have used much of the growth media Iraq had imported, he said. Taha began shouting. "Who told you this? These are not fermenters, they're inactivating tanks," she said. "*You* told us," Spertzel replied. "Here's the inventory we made yesterday with the Al Hakam people."

At an evening meeting on July 1, 1995, Iraq finally conceded. After four years of denials, Taha acknowledged that Baghdad had produced thousands of gallons of anthrax and botulinum at Al Hakam. It was an emotional moment and Spertzel sensed the Iraqi scientist was again close to tears. The strain of the past few months had been crushing as the UNSCOM inspectors tore apart Iraq's cover stories. "She knew that we knew she was lying." Spertzel recalled.

Iraq's admissions were summarized in a three-page document that Taha read aloud to the inspectors. She made no eye contact, staring at the ground or the papers before her. Iraq began its biological weapons program in 1988, she said, just as the Iran-Iraq war was winding down. When the fighting ended, "instructions" came to keep the project going, but "without rush." Production of germ agents began in 1989. She said the anthrax and botulinum were stored in stainless-steel tanks in a warehouse. According to Taha, the stockpiles were destroyed in September or October of 1990 at the order of "superior authorities" who feared allied forces might bomb Iraq's scientific laboratories.

The new account raised as many questions as it answered. It gave no hint as to who in the Iraqi leadership had directed the germ effort. And the weapons inspectors seriously doubted the claims that the anthrax and botulinum had never been poured into bombs or warheads. UNSCOM had told Iraq five months earlier that it believed the germs had been weaponized, and nothing in Taha's stiff, halting performance that night dissuaded them. Spertzel thought it was highly implausible that Saddam Hussein would go to the trouble of producing biological weapons only to destroy them on the eve of the mother of all battles.

A month later, on August 4, Iraq submitted a "Full, Final and Complete Disclosure" on its biological-weapons programs, the latest of a long series. Four days later, on August 8, the man who had run Iraq's biological-weapons program for most of the last decade defected to Jordan after a family quarrel. Lieutenant General Hussein Kamel, a son-in-law of Saddam Hussein's, was by far the highest-level Iraqi official to abandon the inner circle. The U.N. detectives finally had their own germ defector.

Iraq did not wait for Kamel to expose the inaccuracies in the August 4 declaration. Hours after he fled the country, Iraqi officials informed the U.N. that they had discovered a cache of documents hidden on a chicken farm owned by Kamel.

Iraqi officials assumed, incorrectly as it turned out, that Kamel was confessing all to UNSCOM. They immediately came forward with a new story: Iraq, they acknowledged, had weaponized its germs. The program had been run by Kamel, who they said had directed a wide-ranging conspiracy after the war to conceal the germ effort from the United Nations. Iraqi officials said the president's powerful relative had barred them from telling the truth.

Iraq invited Ekeus to the farm where the documents had been "discovered" to examine the evidence. Officials steered him to a small shack which contained a pile of wooden and metal boxes.

A bright orange photo album was resting on top of one of the boxes. Although the room was coated with grime, there was not a speck of dust on the album, which contained detailed records of Iraq's field tests of germ weapons. The album appeared to be a briefing book for top officials. The box of documents—a trove of valuable material on Iraq's germ program—included a sign-out sheet for the album. It bore the signature of Taha next to the date September 13, 1990, six weeks after Iraq had seized Kuwait.

A flood of disclosures followed the stage-managed discovery at the chicken farm. Iraqi officials admitted that they had produced an extensive array of germ weapons for use on the battlefield, from missiles to bombs to jet-mounted aerosol tanks. They said they had produced the germs over several years but had started a crash program to put them into weapons after August 1990, when Saddam Hussein had invaded Kuwait. General Schwarzkopf had been right to worry.

The Iraqis told UNSCOM that the fermenters at Al Hakam had produced their first botulinum in January 1989 and by August 1990 had made a staggering 13,600 liters (3,600 gallons), enough to fill a large swimming pool. The assembly lines had begun making anthrax in June 1990, several months before Saddam Hussein had first menaced Kuwait, and by December had turned out 8,350 liters (2,200 gallons). A second production line for botulinum was begun at the Dawrah veterinary vaccine plant near Baghdad in November, just as allied forces were massing in the desert. By December, it had produced an additional 5,400 liters (1,400 gallons) of the toxin. All told, the Iraqis, too, had made enough germs and toxins to kill every human being on the planet. Neither Al Hakam nor Dawrah was bombed by the allies, who had been fooled by the Iraqi shell game.

Iraq acknowledged that its weapons specialists had tested and produced artillery shells, bombs, and missile warheads that could deliver the germs. The work had begun in 1988, when scientists had tried out an airborne bomb on a test range, killing animals with a cloud of botulinum. A few months later, in July 1988, the scientists had performed several tests with an aerosol generator hung from a Soviet-made military helicopter, using bacteria that simulate the characteristics of anthrax.

The invasion of Kuwait, Iraq admitted, gave the weapons work added urgency. In the fall of 1989, Iraqi technicians had developed a new bomb, a biological version of an airplane-delivered munition that floated to earth with a parachute. The Iraqis had coated the insides of an R-400 bomb and filled some with anthrax and others with botulinum. Iraq said it had made one hundred bombs with botulinum, fifty with anthrax.

The Iraqi technicians also converted some chemical warheads designed for the long-range Scud missiles to carry biological agents. Iraqi officials said they had filled twenty-five warheads: sixteen with botulinum, five with anthrax, and four with aflatoxin, a poison from a fungus that causes long-term liver damage. The germ team also made a 2,000-liter tank that could spray anthrax or botulinum from a jet aircraft. American experts considered this one of the most threatening of the Iraqi systems. The bombs and missiles were crude, said one official, "but this would have worked." Iraq also said its scientists had begun experimenting with viruses in July 1990, shortly before the invasion of Kuwait. These included camel pox, a relative of smallpox; hemorrhagic conjunctivitis, which causes temporary blindness; and a rotavirus, which causes severe diarrhea. The Iraqis said none of these had been produced in large quantities.

Shortly before the allied bombing began, Iraq said, its military officers buried ten of the biological warheads in an abandoned railway tunnel and hid fifteen more in pits next to the Tigris River. The aerial bombs, the R-400s, had been distributed around the country. Iraq disclosed that 157 of them were buried near the Aziziyah firing range and a military landing strip the Iraqis called Airfield 37. On the eve of the Gulf War, in December 1990 and January 1991, Iraqi pilots tested the spray tanks in their fighter planes at the Abu Obeydi air base. Iraq claimed that it had destroyed all of its bombs and warheads in the summer of 1991, but provided no evidence.

Iraqi officials offered conflicting explanations of why they had produced a germ arsenal. Chemical and biological weapons, they said, could be used to counter a "numerically superior force." Iraqi field commanders also had the authority during the Gulf War to launch a retaliatory germ strike in the event of a nuclear attack that destroyed Baghdad. UNSCOM noted dryly that this "does not exclude" other possible doctrines for using biological weapons.

The Iraqis' account of their program and where they hid their biological weapons matched the information obtained by the CIA in the summer of 1991. The CIA's spy or spies had correctly identified three of the four locales in which Baghdad had buried its germ arms. The CIA reports had pointed to Al Hakam and Fudaliyah, two of the places where Iraq operated production facilities. Even the account of the bomb sizes appeared to match the Iraqi disclosure. Iraq said it had poured its germ agent into R-400 bombs, which had a capacity of about 90 liters, and into warheads for the Scud missile, which held 145 liters, just as the CIA reports had said.

How much of a threat were the Iraqi germ weapons? Years later, Spertzel and American intelligence officials still struggled to answer the question. Baghdad told UNSCOM that the anthrax and botulinum produced at Al Hakam were liquid, the most primitive form of biological weapons. Iraq had rushed biological bombs to several military airfields just before the war began. But the Iraqi R-400 was a highly inefficient way to deliver death; most of the pathogens were killed when its conventional explosive detonated. Spertzel and the UNSCOM team were more impressed with the aerosol generators and spray tanks. If the Iraqi pilots had been able to evade the allied air defenses, they could have carried out a significant attack. UNSCOM had no evidence that Iraq had ever made the freeze-dried, powdered form of anthrax invented by Patrick and the American offensive program. But the sample of biopesticide taken from Al Hakam was worrisome. It showed that Iraqi technicians had the skills to produce dried agent of exactly the right size. If Iraq ever resumed making germ weapons, they would surely be much more deadly.

IN Washington, Stephen Joseph, the Pentagon's top civilian health official, was still trying to convince Admiral Owens of the need for anthrax vaccinations. On July 25, he sent the vice chief of staff a memo assuring him the vaccine was both safe and effective. The files at Fort Detrick and the FDA, he said, contained no reports of serious injury caused by the vaccine. Some people had suffered mild to moderate side effects, but they were temporary, comparable to those seen with children's vaccines. Joseph assured Owens that the Michigan vaccine worked well against the aerosolized form of anthrax. Data from animal studies, he said, "is quite convincing that the anthrax vaccine is an efficacious product."

The memo left some things unsaid. An unvarnished accounting of the vaccine's drawbacks and selling points would have taken up many more pages and would have given Owens a lot more ammunition to fight the program. In fact, there were some questions about whether the Michigan vaccine was effective against the full array of anthrax strains available to germ warriors. While the vaccine had performed well on rhesus monkeys in the fall of 1990, a number of other studies on guinea pigs and mice had found less impressive survival rates, with some strains as low as 10 percent. In early 1995, as the debate in the Pentagon over the program was heating up, army officers circulated a proposal to start work on a new, genetically engineered anthrax vaccine. "Recent reports," it said, suggest that the existing vaccine "may not provide universal protection against all strains."

No one understood exactly why the vaccine appeared to work better against some strains of anthrax than others. Dave Franz, director of the army lab at Fort Detrick, believed the results with mice and guinea pigs reflected some quirk in their immune system that did not exist in primates. The army lab did an experiment in which immunized monkeys were exposed to a strain that had defeated the vaccine in rodent tests. All survived. The main component of the Michigan vaccine, protective antigen, prevents the disease by blocking the *Bacillus anthracis* toxins from penetrating the cells of its host. If all anthrax strains used similar mechanisms to infect animals, as Franz believed, the vaccine should be effective against all of them. Additional experiments in which primates were tested against more of the strains would help settle the question. But Franz had little money for such research in the mid-1990s and rhesus monkeys were expensive. Franz was confident the vaccine was effective, but he phrased his briefings carefully, pointing out the limits on what could be concluded from animal models. Franz would tell his colleagues: you're protected—if you're a rhesus monkey.

The original trials of the anthrax vaccine in the late 1950s had never resolved the issue of whether the vaccine protected people against inhaled spores. The tests involved wool workers who were routinely exposed to anthrax spores as they processed animal hides and goat hair imported from overseas. A significant number of unvaccinated workers contracted anthrax through their skin, establishing that the vaccine worked against that form of the disease. But only five of the unprotected workers fell ill with inhalation anthrax, not enough to draw conclusions. There was also one bit of anecdotal evidence. A secretary at one mill who had not received the inoculations walked onto the factory floor and contracted the respiratory form of the disease. Her vaccinated coworkers remained healthy.

Those tests were conducted with an earlier version of the vaccine developed by Merck & Company. By the time the Michigan lab had completed work on its own similar product in the 1960s, the wool industry was dying and there were hardly any unvaccinated workers left to study. The FDA allowed Michigan to use the Merck results, which drew no distinctions between inhaled and cutaneous forms of anthrax, to support its application for a license. The label accompanying the vaccine specified only the occupations that might involve exposure to anthrax.

In the fall of 1995, just as they were reassuring Admiral Owens, army officials asked the Michigan lab and a contractor, Science Applications International Corporation, to draw up plans for a new round of animal research. The goal was to amend the vaccine's license so that it was explicitly approved as a defense against germ weapons. Five years after the Pentagon gave anthrax shots to 150,000 Persian Gulf soldiers, SAIC offered a startling assessment: "This vaccine is not licensed for the aerosol exposure expected in a biological warfare environment." That view was not widely accepted by Pentagon officials, who felt the ambiguously worded license did permit its use as a protection against germ attack.

SAIC said it was also planning animal studies that would examine whether full immunity could be attained with fewer shots. If the experiment was successful, and the FDA could be persuaded, the logistics of administering the drug would become much easier. Several recent experiments with primates had suggested that the vaccine was effective after only two injections. The FDA-approved regimen of six shots was based on guesswork, not science. Researchers testing the vaccine decades earlier doubled the number of shots, from three to six, after some of the vaccinated wool workers had contracted anthrax.

Questions of vaccine efficacy are inherently difficult to settle. The lethal nature of anthrax precluded testing on people, and no reliable animal model had been found. Immune systems of primates, mice, and guinea pigs resembled those of human beings, but researchers at Fort Detrick were still looking for ways to achieve a more precise correlation.

Admiral Owens's qualms about the safety of the anthrax vaccine came at a time when Persian Gulf War veterans were growing increasingly concerned about their health. Thousands of servicemen were reporting systemic illnesses with a host of mysterious symptoms, from memory loss to muscle and joint pain. Doctors could find no immediate cause, and Joseph suspected that much of the sickness stemmed from combat stress. The veterans insisted that their illnesses were caused by the medicines they were given, the air they breathed, or, perhaps, exposure to low levels of Iraqi biological or chemical weapons. Some blamed the anthrax inoculation. There was no evidence that the Michigan vaccine had ever caused such long-term problems. But if the military were challenged on this, it would have difficulty defending its position. Because such poor records had been kept of who received the anthrax and botulinum shots during the war, it would be difficult, perhaps impossible, to conduct a scientifically valid, long-term study of their effects.

ON the morning of October 31, 1995, the Senate's Permanent Subcommittee on Investigations opened its hearings on Aum Shinrikyo.

John Sopko, Senator Nunn's aide, had obtained a wealth of information about the Japanese group, and reporters were handed a report prepared by the panel's Democratic staffers that described Aum's global quest for chemical and biological weapons. In his opening statement, Nunn sounded a favorite theme: the dangers posed by the fall of the Soviet Union. Never before, he said, had the collapse of an empire left behind tens of thousands of nuclear, chemical, and biological weapons and the scientists who knew how to make them.

The Tokyo attack deepened the threat. "The scenario of a terrorist group either obtaining or manufacturing and using a weapon of mass destruction is no longer the stuff of science fiction or even adventure movies," Nunn declared. "It is a reality which has come to pass." Senate hearings are political theater. Witnesses are paraded before spotlights to recite canned testimony that has been closely reviewed by their bosses. The serious work of government does not take place in such public forums.

But the Permanent Subcommittee's hearings on Aum were different, pointing out in vivid detail how little America's vast intelligence apparatus had known about the Japanese cult before the attack. On the second day of the sessions, after hours of testimony on Aum's procurement efforts in the United States and elsewhere, Senator Nunn asked top officials from the FBI, the Pentagon, and the CIA what seemed a straightforward question: Had anyone detected the threat posed by the cult before the subway attack?

John O'Neill, the FBI's chief counterterrorism official, said the bureau had only a "small office" in Tokyo, which acts as a liaison to the Japanese authorities, whom he accused of being tight-lipped. "We received no information from the Japanese National Police," he said. Gordon C. Oehler, chief of the CIA's Nonproliferation Center, said it was up to the local authorities to detect cults like Aum. "The world is full of very crazy organizations that have designs against the U.S.," he said. "You are certainly welcome to argue that, quite frankly, we have not followed religious cults around the world and we do not have right now the resources to be able to do that." The CIA, Oehler said, had taken note of the earlier gas attack in Matsumoto, which turned out to have been a dry run. But he said the agency had believed the initial Japanese reports that the case was closed with the arrest of the local man.

Senator William S. Cohen reminded Oehler that Kyle Olson, the

Washington consultant hired by Japanese television, had given a report to government officials months before the Tokyo subway attack suggesting that the Matsumoto gassings were the work of terrorists. "From what I can detect," said Cohen, a former vice chairman of the Senate Intelligence Committee, "the report was ignored."

Cohen and his colleagues pressed their attack.

What about the threat against President Clinton in an Aum magazine article?

It was, Oehler said, "the kind of thing we see routinely around the world."

"This is not routine," Cohen snapped back. "This is not one of those routine sightings here. We already had the attack at Matsumoto. The Japanese press was carrying stories on a regular basis on the threat this particular cult group posed."

After the hearings ended, the picture of Aum's activities became clearer as Japanese authorities hauled the group's members into court and elicited confessions. The testimony, which trickled out over the next several years, showed that the cult had repeatedly tried—and failed—to carry out biological attacks. The guru's chauffeur told a Japanese court that the group, using a truck with a crudely improvised aerosol device, had tried to spray botulinum toxin in downtown Tokyo and at the nearby Narita airport. Three years later, he said, the group had tried again, this time aiming anthrax spores at the Imperial Palace and the Parliament in Tokyo. Finally, he said, the cult had also staged unsuccessful attacks on the American navy bases at Yokohama and Yokosuka.

These attempted strikes against American facilities had gone unnoticed by American intelligence agencies and the military, who learned of them when the *New York Times* published an article quoting the courtroom testimony. To this day, it is not clear that Aum ever succeeded in making and spreading germs potent enough to harm anyone. Aum's failures were instructive: they showed that making a biological weapon was harder than some experts had claimed. Experts frequently suggested that anyone who had studied science in college could make a biological bomb. Aum had invested lavishly in laboratory gear, had even dreamed of doing recombinant experiments. The group's scientists had taken graduate courses in modern universities. Yet they couldn't master the engineering challenges of producing a terror weapon from pathogens.

Although the cult's weapons makers had some advanced training in science, they did not appear to have understood the basics of germ dissemination. They attempted some of their anthrax attacks from the roof of a building during the day, which guaranteed that sunlight would kill off many of the pathogens. The cult also had difficulty obtaining the starter germs from which to make a biological weapon. Both the Iraqis and the Rajneeshees of Oregon had ordered pathogens from the American Type Culture Collection, the American germ bank. But Aum members had tried to culture botulinum from soil samples obtained in the wild, trial testimony shows. The group also had trouble finding a virulent strain of *Bacillus anthracis.* There is no indication they managed to produce the dried anthrax spores that Bill Patrick favored for their efficient distribution.

A team of Japanese and American scientists in time reported that a sample taken from the site of a July 1993 attack by the cult matched the Sterne strain, a weakened form of anthrax used to make animal vaccines. It would have been virtually impossible to turn those germs into a biological weapon, the researchers concluded. Aum may also have fouled up the process of growing germs and refining them into a slurry that can be sprayed as an aerosol. For all their money and determination, Aum lacked the expertise of a Bill Patrick, a Vladimir Pasechnik, or a Ken Alibek. But American officials tracking international terrorism were not comforted by the group's ineptitude. The significance of the Aum case, they felt, was that a line had been crossed. A terrorist group, however wacky, had decided to use chemical and biological weapons against a modern city full of defenseless people.

In January 1996, in the aftermath of one of the biggest blizzards in the history of Washington, D.C., Richard Danzig, his aide Pamela Berkowsky, and Stephen Joseph met for lunch at a Thai restaurant. The subway was not running that morning and Joseph, who lived in northern Virginia, skied across the Key Bridge into downtown Washington for the lunch. The Pentagon officials were joined by Richard Preston, a regular contributor to *The New Yorker* whose flair for storytelling had brought the experts' fears about the spread of infectious diseases to a much wider audience. Four years earlier, Preston had been searching for a way to write a popular book about the dangers posed by exotic viruses. The science was dense, and Preston told Lederberg he needed a story that would draw in ordinary readers. Lederberg casually mentioned an outbreak of Ebola virus among monkeys at an animal-supply company in suburban Washington. How bad was it? Preston asked. Pretty bad, Lederberg replied. Army officers dressed in space suits had killed every single monkey that might harbor the disease. That conversation led to *The Hot Zone,* a nonfiction book published in 1994 that read like a thriller and helped put Lederberg's concerns about emerging diseases on the public agenda. Some of its strongest passages described the panic of health officials as they realized what the virus might do to the population of Washington if it escaped and began infecting people.

As he was completing *The Hot Zone,* Preston learned a fascinating fact: Ebola virus was on the military's list of possible biological warfare agents. As he began thinking about his next book, he toyed with the idea of writing a novel structured around a germ attack by terrorists. Preston returned to Lederberg and said he was thinking about a book on biological weapons. The Nobel laureate was discouraging, saying it was best not to say too much publicly about the potential of germ warfare. Preston was not dissuaded and began researching the novel as if it were a work of nonfiction. He interviewed government officials and discovered that few of them knew anything about the biological threat. Those who did, however, were deeply worried.

Preston sketched out a plot for his novel in which a terrorist dispersed anthrax in a metropolis. He discussed the outline with an FBI source, who asked him to reconsider. The threat from anthrax was too real. It would not serve the public to have a book appear that vividly described its powers. And so the author revised his story line to focus on a less plausible germ: a biologically engineered superbug.

Preston's novel was well underway in January 1996 when he was invited to lunch with Danzig by Commander James Burans, a medical doctor who had been tutoring the navy undersecretary on the fine points of germ weapons. Burans directed the navy's biological research programs and his unit had developed a portable laboratory that could perform on-the-spot tests of biological samples. It was exactly the sort of device the army lacked when it accompanied the FBI on the covert break-in at the Aum Shinrikyo offices in New York.

The conversation was freewheeling. Was the biological threat real? What could the United States do to improve its defenses? Danzig asked

Preston to whom he should talk to deepen his understanding of germ weapons, and Preston suggested Bill Patrick. What are you writing your next book about? Joseph asked. Preston was deliberately vague about the format but said the topic was germ terrorism. That would be a great plot for a novel, Joseph told him. Danzig encouraged him to stick to nonfiction, but Preston demurred. Such a book would have to rely too heavily on anonymous sources. There was another problem. Preston said he wanted to describe how public-health authorities responded to biological attack on a city, but none had yet happened.

Shortly after the lunch, Danzig enlisted Preston in his long-running effort to change the minds of the nation's military leaders. He organized a dinner party for General Ronald Griffith, the Gulf War commander who was now vice chief of the army, at a fashionable Washington restaurant. The real target was Griffith's wife, Hurdis M. Griffith, who was dean of the College of Nursing at Rutgers University. Danzig had Preston give her a copy of *The Hot Zone,* hoping this would sell her on the germ threat and establish a beachhead within the Griffith household.

The gambit failed. General Griffith continued to oppose the anthrax vaccinations. Inoculating American soldiers, he felt, would make it more difficult to build coalitions in the future. Why would other nations with unvaccinated troops take more risks to fight alongside Americans? Preston acknowledged over dinner that he was writing a novel, and in the months that followed, Danzig encouraged him, opening doors inside the government and offering literary suggestions. Preston, a first-time novelist, was buoyed. "The principal help I got from Richard Danzig," Preston recalled, "was that he said he believed in me as a writer."

THE events of 1995—the Aum attack, the Oklahoma City bombing, the Iraqi admissions—should have served as exclamation points in Danzig's campaign. But the message was not taking hold where it counted: in the federal budget. The number of people working at Fort Detrick, the nation's premier germ-defense laboratory, had dropped about 30 percent since the Gulf War. Colonel David Franz, the commander at Fort Detrick, had begun by laying off technicians, but in 1996 he was forced to fire scientists. Visiting Russia in the early 1990s, Franz learned that Moscow was years ahead of the United States in using recombinant techniques to create an improved vaccine against botulinum. A new vaccine would cause fewer reactions and could be effective against a broader array of botulinum strains. Franz wanted to push his own researchers in the same direction, but he couldn't come up with the several hundred thousand dollars needed to pay for it. Money was also tight for the research on the next-generation anthrax vaccine.

At the Pentagon, Admiral Owens was targeting the budgets for chemical and biological defense for a $1 billion cut over five years, a reduction of nearly one third. Owens felt the chem and bio programs were poorly managed, even wasteful. They were not spending enough money on developing better vaccines. He was particularly aggravated by an army project to mount biodetectors on helicopters. The navy admiral considered it a sneaky way for the army to buy nine more helicopters.

Owens recruited an important ally in his campaign against the anthrax vaccinations: his boss, General John Shalikashvili, chairman of the Joint Chiefs. In the "chairman's program assessment," a Pentagon planning document, Shalikashvili endorsed Owens's key recommendations, calling on the Pentagon to stockpile vaccine and invest in developing better antidotes. "No immunizations," a terse summary of the document said. Shalikashvili endorsed Owens's idea of brandishing the nuclear deterrent against potential biological foes, suggesting that the United States warn its adversaries that "use of BW weapons against US personnel will result in a response involving the full force of our capability." Shalikashvili supported the idea of inoculating the troops, but he felt the questions about the vaccine's efficacy remained unresolved. He also believed the Pentagon should do more to show there was no connection between Gulf War syndrome and the anthrax shots. "We were suffering very low credibility with many groups on this subject," he recalled. "Not just the vets, but Congress and others."

A new GAO report summarized the low standing of chemical and biological defense in the Pentagon. American soldiers, the auditors wrote, "face many of the same problems they confronted during the Persian Gulf conflict in 1990 and 1991." Local commanders were using money designated for chemical and biological training to pay for operating expenses. Three of the army's five divisions were missing half their required protective suits and other items "critical" for defense against a germ or chemical attack. U.S. forces "continue to experience significant weaknesses in (1) donning protective masks, (2) deploying detection equipment, (3) providing medical care, (4) planning for evacuation of casualties, and (5) including medical and biological issues in operational plans." The report blamed the military's senior officers, the Joint Chiefs, and the regional commanders around the world for failing to give the issue a high priority.

Successful leadership in government agencies is about picking your battles, about knowing when to give up. Danzig had plenty of other issues to worry about as undersecretary of the navy. Still, he refused to accept defeat. If the Joint Chiefs did not see the need to inoculate the troops or build biodefenses, it was because the case had not been made effectively. Owens retired at the end of February. Danzig redoubled his efforts.

IN early 1996, Congress moved to tighten control over laboratories and companies selling pathogens to scientists and medical researchers. The previous year, a right-wing army veteran in Ohio named Larry Wayne Harris had bought three vials of plague bacteria for about $300 from American Type Culture Collection. Harris had placed the order by using the state laboratory license number of the scientific company where he worked as a microbiologist. He eventually pleaded guilty to one count of mail fraud—the most serious charge possible under existing law. The case, which broke soon after the Aum attack, received widespread publicity.

A March 6, 1996, hearing before the Senate Judiciary Committee was dominated by a discussion of Harris, a survivalist who said he had bought the germs to make his own plague antidote. Even the American Type Culture Collection called for tighter rules that would go beyond the Commerce Department's 1989 ban on exports of germs to countries such as Iran and Iraq. "It is apparent," said the committee's chairman, Orrin G. Hatch of Utah, "that there has been kind of an ignoring of the potentials for harm." Weeks later, Congress passed a law that imposed tough rules on the domestic transfer of pathogens and made it a crime to threaten a biological attack. The estimated two hundred U. S. laboratories that maintain germ collections would have to register and submit to federal inspections.

Elsewhere on Capitol Hill, Senator Nunn and the investigations subcommittee were preparing for a new round of hearings on the implications of the Aum Shrinrikyo case. The Georgia Democrat was planning to leave Congress at the end of the 1996 session, and his name was already attached to the program that provided American money to Russian military scientists pursuing domestic projects. He wanted to do something for domestic defenses against germ or chemical attack.

The image of the Tokyo emergency personnel choking in the sarin fumes continued to haunt Nunn and John Sopko, his investigator. Bill Richardson, the recently retired Pentagon expert who was advising the committee, told Sopko that a modest program, $35 million a year at most, could train emergency personnel in key cities. Sopko responded, "Nunn's going to retire. We may never get another shot. Let's make it $100 million."

The subcommittee focused on civil defense at its hearings, which opened on March 27, 1996. A staff report written by Sopko and his colleagues asserted that terrorist attacks with germs were all but inevitable. The threat of a terrorist group using a nuclear, biological or chemical weapon is real, the report said. "It is not a matter of if but rather when." Senators Nunn, the New Mexican Republican; Pete V. Domenici, the Indiana Republican; and Richard Lugar closed ranks behind a plan to train "first responders" in 120 cities.

At the Pentagon, Richard Danzig began courting the vice chiefs, organizing a new round of seminars that took on the arguments against anthrax vaccinations one by one. To the oft-repeated view that nuclear weapons were a deterrent, Danzig responded that rogue nations could attack the United States through surrogate terrorist groups. American presidents were reluctant to order punishment unless they had undeniable proof of culpability. Intelligence officials, he reminded the vice chiefs, had argued for years over whether the 1989 bombing of Pan Am flight 103 could be linked to Libya, Iran, or Syria. Danzig brought in Josh Lederberg to address the scientific objections. Some of the vice chiefs had said anthrax vaccinations could be delayed because researchers would soon produce a single inoculation that would protect against multiple diseases. Not so, Lederberg told them. Such a vaccine was still years, if not decades, away. What of the view that genetic engineering rendered vaccinations useless? Lederberg patiently explained why making superbugs that could survive outside a test tube was harder than commonly understood. "You'll have substantial gains if you inoculate," he assured the officers.

In the spring of 1996, a serious obstacle to the plans for vaccinating the troops surfaced. Two teams of consultants who had visited the aging Michigan lab warned that it stood a substantial chance of flunking its next FDA inspection. The consultants said Michigan had failed to carry out the promises for improvements it had made to address the findings of the 1993 and 1994 inspections. Federal rules require manufacturers of anthrax vaccine to do the most dangerous work in a lab equipped to meet the strict rules of biosafety level 3. Michigan was well short of that standard, one expert said, and it was "dubious" whether it could even be classified as biosafety level 2.

One of the senior Pentagon officials overseeing the lab, General Walter Busbee, sent an e-mail to his colleagues urging them to begin considering what they would do if the FDA shut Michigan down. Busbee wrote that he had tried to plant the "seeds of doubt" about the facility at Danzig's most recent meeting with top officials. "No one really responded," he wrote. Busbee's aides scrambled to figure out whether anthrax vaccine could be bought anywhere else. The only manufacturer of anthrax vaccine outside Russia was Britain's Porton Vaccines, which would need six months to gear up for producing fifty thousand doses a week. The U.K. vaccine, made from a different anthrax strain than Michigan's, was not licensed for use in the United States. Without the Michigan lab, there would be no program of anthrax vaccination anytime soon.

On June 21, 1996, the Pentagon made an announcement that shattered the military's credibility on medical issues and changed the politics of the debate on vaccinations. For nearly five years, military officials had insisted that no one serving in the Persian Gulf War had been exposed to chemical or biological weapons. Now they were forced to recant. Tens of thousands of allied soldiers might have been exposed to nerve gas, they said, after American soldiers blew up an Iraqi ammunition depot containing chemical weapons. The detectors picked up no sign of a chemical release that day, but officials said the detection equipment appeared to have been faulty. There were reliable intelligence reports dating from 1991 that said the Iraqis had stored chemical weapons at the depot. But the Pentagon said the information had been overlooked for years through a series of bureaucratic mishaps. Veterans suffering from Persian Gulf symptoms were stunned. The military, they felt, had lied to them about the cause of their illnesses, just as it had lied to previous generations of soldiers about the effects of radiation and of Agent Orange. A team of government experts pored over the records and discovered a second incident in which allied soldiers had blown up chemical weapons, deepening the veterans' suspicions. The Internet buzzed with theories of cover-up.

On June 25, four days after the Pentagon's first disclosures about the release of nerve gas during the Gulf War, Islamic militants in Saudi Arabia detonated a truck bomb near an apartment building that housed hundreds of American servicemen. The attack on Khobar Towers killed nineteen and wounded as many as five hundred. To Danzig and his colleagues, it added another powerful argument in favor of improving what the military calls "force protection." The Pentagon's experts did some calculations. If the truck had been packed with chemicals or germs, the effect on the troops in Saudi Arabia would have changed from tragic to apocalyptic. The soldiers at Khobar Towers were not prepared for attack. None had their gas masks at hand when the bomb went off.

Danzig and Joseph pressed their case for a relatively modest program that would implement William Perry's 1993 directive: vaccinations for those servicemen in high-threat areas or assigned to early-deploying units. But the commanders did not see how they could operate with a force that was only partially protected. If the assumption was that germ weapons would be used against rear areas, logistics bases, one would also need to inoculate the national guardsmen who worked in support positions.

A turning point in Danzig's meetings with the vice chiefs came when General Richard Hearney, the vice commandant of the Marine Corps, declared that anthrax vaccinations should be given in boot camp, along with the other routine inoculations administered to new soldiers. Danzig had wanted to begin modestly, vaccinating the early-deploying forces and those in high-threat regions like the Middle East. But the consensus had swung. If inoculations were needed, the officers concluded, they should be given to everybody. Hearney said Lederberg "drove home" the case for vaccinating before hostilities begin. "People would argue: 'We don't want to do it'" and insist that the vaccinations should instead be given on the battlefield, said Hearney. "I didn't want to take that chance."

Danzig had an ally in General Dennis J. Reimer, the army's chief of staff. Reimer had served in the Pentagon during the Gulf War and had directed the effort to improvise biological detectors. He was part of the debate over how to allocate the scarce vaccine supplies, an experience that framed his thinking on biological defense. "You can't play God on this," Reimer said later. "You've got to find a solution and the only solution is to vaccinate everybody." The army vice chief, Ronald Griffith, continued to question the policy. He had heard from colleagues in the army that the Michigan facility was in trouble with the Food and Drug Administration and might not be able to make enough vaccine. The issue did not make much of a dent. It was assumed that the problems were technical. They could be overcome.

By the fall of 1996, the chiefs had turned around and endorsed the recommendation to vaccinate the entire force. Owens's recommendation to cut $1 billion from the biological- and chemical-defense budget had been reversed as well, transformed into a $1 billion increase over five years. That decision, a $2 billion shift from January, reflected a radical change in priorities. Danzig was not one to gloat over victory. On October 2, 1996, the *Washington Post* published a story on its front page under the headline "Military Chiefs Back Anthrax Inoculations." Reversing earlier opposition, the story began, "the nation's military chiefs have endorsed a plan to vaccinate all U.S. forces against anthrax." The names Richard Danzig, Stephen Joseph, and Josh Lederberg did not appear.

9

Taking Charge

THE proposal to vaccinate the entire U.S. military against anthrax came before John P. White, the deputy secretary of defense, in January 1997. An economist whose son had been a platoon leader in the Gulf War, White took the recommendation seriously. He respected Danzig, and it was not every day that the chiefs and the Pentagon's civilian officials could unanimously agree on such a contentious issue. But White thought that the proposal to improve the military's biodefenses was spectacularly ill timed. The uproar over Gulf War syndrome was escalating. Congress was holding hearings. First Lady Hillary Clinton was pressing the Pentagon to do more for the veterans. This was no time to be proposing new vaccinations. It was politically tone-deaf. Are you guys out of your minds? White asked the Pentagon's health experts.

The program, they told him, was a straightforward effort to protect the force, insurance against a clear threat. White was unmoved. Ordering 2.4 million people to take six injections in the arm was far more intrusive than issuing soldiers a new type of flak jacket or helmet. White thought the immunizations should be deferred. Several years might be a good start. He ordered a new review of the vaccine's safety and efficacy. He also asked for a detailed briefing on how, precisely, the military planned to persuade servicemen to accept the program.

At about this time, President Clinton named William S. Cohen, the

Republican senator, to replace Bill Perry as defense secretary. Cohen had taken a keen interest in Aum Shinrikyo during the 1995 hearings. He had served as vice chairman of the Senate Intelligence Committee, and he came to his new job already persuaded that more should be done about the threat of biological weapons.

On March 4, Stephen Joseph, the Pentagon health official, asked the Food and Drug Administration to settle a crucial question: Did the license for the Michigan vaccine approve its use against the aerosolized form of the disease? Although the label was "non-specific as to route of exposure," Joseph wrote, the Pentagon had long interpreted this to encompass inhaled anthrax. Days later, Michael A. Friedman, the acting FDA commissioner, offered a reply that was sufficient, barely, to give the Pentagon the legal authority it needed. There was a "paucity of data" on the effectiveness of the vaccine against inhaled anthrax. But he said animal studies "indicated" that the vaccine was effective against spores in the lungs. The 1970 license "did not preclude this use." "Your interpretation is not inconsistent with the current label," wrote Friedman.

Just as the last obstacle was cleared, disaster struck. On March 11, 1997, the FDA warned the Michigan lab it could lose its federal license if it did not immediately make improvements. The problems, the inspectors wrote, "represent a failure to comply with the regulations that safeguard the drug and pharmaceutical industry." Pentagon officials worried that the lab might not survive. Michigan's legislature had voted the previous year to sell the facility. And many key production officials were taking early retirement, encouraged by a state program that gave state employees generous incentives to do so.

The FDA report on the Michigan factory detailed a litany of failings. It did not follow its own manufacturing procedures. Its equipment was rusting. It had failed to "clean, maintain and sanitize equipment and utensils at appropriate intervals." The FDA inspectors found "dead insects" in one room and "a live insect was observed in the capping room" where workers prepared to enter the plant's sterile areas. After previous inspections in 1993 and 1995 that warned of "serious deficiencies," the regulators had run out of patience.

Joseph and his aides demanded that the Pentagon test the millions of doses of vaccine in the stockpile for sterility, potency, and safety, which some Pentagon experts resisted. It was almost inevitable, they knew, that some lots already cleared for use would fail a second review.

Potency tests, in particular, were notoriously variable. "Are you sure you folks really want to go down this path?" one Pentagon official wrote. Joseph and White insisted that the military hire an outside firm to oversee the tests and assure servicemen that the stocks conformed to federal rules.

The Pentagon sent an "assistance/assessment team" to study the Michigan plant. The name summed up its conflicting missions. Military officials were desperate to preserve the Michigan lab. State officials wanted to complete its sale by February 1998, and the problems would have to be addressed before anyone would buy it. A memo prepared for White by Joseph and his aides outlined the dilemma. If the military limited its purchases to "stockpile quantities" of vaccine, there would not be enough sales to keep the production lines moving. Without a program of peacetime vaccinations, "it is likely that the existing industrial base will be lost." The medical officials reminded White of Operation Desert Storm in the Gulf War, of the frantic attempts to produce more vaccine, and of the fears of biological attack against unprotected troops. "There was a six month lead time for ODS and there may be little, if any lead time, for the next crisis," the memo said. Had Iraq used its germ weapons, the casualties among those not immunized "would have been extensive and disastrous."

In June 1997, Andy Weber went to Russia for the first time. After touring Moscow and its countless monuments to Russia's war dead, he concluded that no one who had not visited this country could appreciate either the vastness of a nation that still covered eleven time zones or its determination never to be invaded or overrun again.

Indeed, Russia had been forged by a thousand years of warfare with the Tatars, Mongols, French, English, Poles, and, of course, Germans. The Great Patriotic War, the Soviet Union's valiant struggle against the Nazis that had made the Allied triumph over fascism possible, was a fundamental touchstone of the post-revolutionary state and the ever-suspicious national psyche. The Soviet Union had paid a harrowing price for the victory: an officially recognized 20 to 25 million deaths, 10 million more wounded, 25 million left homeless, a third of the national wealth destroyed. Soviet losses were forty times those of Britain, seventy times those of the United States. The Soviets had used the war to justify, and even temporarily obscure, the memory of

Stalin's horrific purges against putative internal enemies, the ruthless collectivizations, the gulags, and the state-induced famines in which millions had died.

In Moscow, Weber stood before the eternal flame honoring Russia's unknown, often unacknowledged, victims. As he contemplated the country's unimaginable suffering, Russia's germ-warfare program suddenly became comprehensible, if not defensible. He knew that many Russian officials still believed that the United States had never abandoned its offensive germ-warfare program, that its research and even development of agents as weapons had continued, President Nixon's dramatic renunciation of such weapons in 1969 notwithstanding. And given their history, who could blame them for thinking so?

After Moscow, Weber went to Kirov, a fourteen-hour train ride east of the capital, for a meeting sponsored by the International Science and Technology Center, a multinational organization formed by the United States, the European Union, and Japan. Since March 1994, the Moscow-based ISTC had been financing peaceful research by former Soviet nuclear-, chemical-, and germ-weapons scientists to dissuade them from fleeing to rogue states or, as the State Department would eventually prefer to call them, "countries of special concern." But Weber and other American officials thought that the organization and America's National Academy of Sciences had been too cautious, avoiding scientific collaboration that might be diverted to the purpose of making weapons even if such research would enable the West to learn how much the Russians knew about the pathogens they had turned into weapons.

One evening after the end of a conference session, Weber went to the *banya,* the large cedar-paneled sauna as popular as vodka with Russians of every class and status, perhaps because such overheated places were hard to bug. There he encountered two Russian scientists from the Obolensk State Research Center of Applied Microbiology. At the still-secret research complex known simply as Obolensk, a two-hour drive from Moscow, the Soviets had perfected dozens of strains of deadly bacteria for military use. Making sure that they were alone in the dimly lit room, the scientists began telling Weber what they were prohibited from discussing at the official proceedings. They were worried about their institutes and their country, they told him. A delegation of Iranians had recently visited Obolensk and Vector, another leading former germ-warfare center, which studied viruses. The Iranians had tried to recruit them, dangling salaries of up to $5,000 a month before former Soviet germ warriors who earned one-fifth of that a year, when they were paid at all.

The offers, though tempting, were troubling. First of all, some of the Iranian "scientists" seemed to know almost nothing about science; second, the Iranians said they were interested in substances useful in biological assaults not only against people but also against crops and livestock, among America's largest exports; third, they were especially interested in Russian expertise in genetic engineering. In germ weaponry, Weber knew, attacking human genes was the most chilling possibility of all. In theory, genetically engineered agents could kill or cripple selectively by race or gender.

Although the former Soviet Union was not known ever to have shared such research with any other country, let alone with a radical Islamic regime that had tried to undermine the secular Soviet empire, the Russian government seemed to be encouraging Russian-Iranian cooperation. That May, the Russian Ministry of Science and Technology had sponsored a biotechnology trade fair in Teheran. Participants included more than one hundred leading biologists from Russian laboratories that had been part of the germ-warfare program, including scientists from Obolensk and Vector. Several impoverished Russians from institutes in Moscow had already accepted posts in Iran or agreed to provide information to Teheran by computer, the scientists in the *banya* told Weber. In their empty, unheated labs, without pay, Russian scientists in thick sweaters were trying to conduct research. Obolensk alone had lost 54 percent of its staff between 1990 and 1996, including 28 percent of its top scientists. With Russia's growing economic and political chaos, who could blame scientists for trying to survive? Russians might not trust or like the Iranians, they told Weber, but these were neither love matches nor even marriages of convenience. They reflected post–Cold War necessity.

While the Iranians were courting Russia's best biologists and offering them real money, the Americans so far had offered their former foes nothing comparable. Congress had generously funded several programs to stop the spread of unconventional-weapons expertise under the legislation sponsored by former senator Nunn and Republican Senator Lugar of Indiana. But almost all of the money had gone to the former Soviet nuclear cities, to isolated enclaves where once prosperous scientists were also impoverished and vulnerable. In mid-1997, as Iran was courting Russian germ scientists, Washington was spending well under $1 million a year on biology assistance programs, a pittance given that more than fifteen thousand Russian biologists were badly underpaid and vulnerable to rogue-state recruitment. Nor was Iran the only hunter. Intelligence agencies had detected similar overtures in Russia from Iraq and North Korea.

The disagreements among American intelligence officials over Russia's germ program had reemerged. Some believed that Moscow was cheating—still making weapons at military institutes that were closed to Western visitors. The sort of scientific collaboration Weber was promoting, they argued, should be limited to relatively benign germs that had not been, and could never be used to make weapons.

Weber had just the opposite view. He believed that the United States had to be bolder. The limited joint projects attempted thus far involved relatively innocuous pathogens, those that caused diseases such as liver fluke infection, tuberculosis, and hepatitis. The Americans were learning little about the former Soviet program or Moscow's military capabilities, scientists at Fort Detrick complained. Few in the U.S. government wanted to risk sanctioning joint research that involved highly dangerous pathogens. But Weber believed that only through such projects could the United States gain access to the key labs or learn how far Russian science had taken its offensive program.

Weber's hunch was reinforced the next night of the Kirov conference, again in the *banya,* when the two Obolensk scientists, now joined by a few trusted colleagues from other institutes, discussed the kind of collaboration they were seeking with the Americans and what kind of help they needed. Yes, they were grateful for the money to study liver flukes, they told Weber and a small group of U.S. Army scientists. They appreciated any money that kept their institutes alive and their families fed. But while the projects launched under the auspices of the U.S. National Academy of Sciences had helped open once closed Biopreparat institutes, the Russians said, they wanted to work directly with their American military colleagues on more scientifically challenging work. Only American military scientists were familiar with the pathogens they had perfected, the Russians told the Americans. For the safety of all mankind, they argued, leading military scientists from both countries had to collaborate on developing vaccines and antidotes for such germs to protect their populations against an attack by a rogue state or terrorist group. Later in the conference, the Obolensk scientists openly discussed their desire to undertake joint research to decode the genetic structure of one of the pathogens they had weaponized: they suggested anthrax.

Weber was shocked. Few agents interested American intelligence as much as anthrax, the hardy spores that Ken Alibek had enhanced and that the Soviets had hastily buried on Vozrozhdeniye Island in the late 1980s. While anthrax had been a weapon of choice for both the Soviet and the American offensive germ-warfare programs—not to mention Iraq's—the Soviets had developed the world's toughest, most virulent strains, as the tragic 1979 accident at Sverdlovsk had suggested.

American intelligence agents had been watching Obolensk especially closely since 1995, when several of its leading scientists had disclosed in a poster display at a scientific conference in Winchester, England, that they had genetically engineered a more virulent strain of anthrax. Though the poster contained few details of the research, veterans of America's germ-warfare program had found it ominous. Andrei Pomerantsev and Nikolai A. Staritsin announced that they had created a new super germ by inserting virulence genes from *Bacillus cereus,* a bacterium that attacks blood cells but normally does not cause deadly disease in human beings, into anthrax microbes. In the experiment virtually all the hamsters that were infected with the newly created pathogen had died, even those that were given Russia's anthrax vaccine. Russia's first line of defense against this genetically engineered new microbe was apparently useless. Was America's own vaccine ineffective against the new pathogen?

Weber had a chance to find out more about the Obolensk research when he met young Staritsin at the Kirov conference. Staritsin told him that although the Russian team had done the research in 1993, when money from Russia's germ-warfare program was still trickling in, they had not performed the genetic manipulation for military purposes. They had not intentionally tried to develop a modified disease that was impervious to vaccines or antibiotics. Their motives had been benign, he insisted. Because anthrax and *B. cereus* were closely related and often found in the same soil, they might naturally exchange genes one day without any external intervention. If that occurred, the Russians wanted to understand what the result might be.

While Weber rejected this explanation, his concerns were allayed somewhat by the anthrax-monitoring proposal from the Obolensk scientists as the conference was closing. Obolensk had some three hundred different strains of anthrax in its collection, the scientists volunteered. Why not work together, using the latest gene-sequencing techniques America had pioneered, to study various strains and determine which genes were responsible for important properties, such as virulence and antibiotic resistance? As part of the monitoring program, Russian and American scientists could exchange strains—eventually, even the Pomerantsev-Staritsin strain that was of such concern to Washington.

Hard-liners in the Clinton administration were skeptical of both Staritsin's explanation and the proposed monitoring project. Of course Russian scientists would have an ostensibly peaceful cover for their military research, they argued; Alibek and others had repeatedly warned that Soviet military scientists had been ordered to develop such stories, or "legends," to conceal the true nature of their work. Staritsin's concern about "natural gene exchange" smacked of such duplicity, American intelligence hawks maintained. "Guys, don't be so gullible," Alibek had often scolded Weber and his friends. "They're lying to you." Obolensk's scientists, skeptics argued, simply wanted access to the latest American expertise and technology, which their country could no longer afford to develop or buy.

For Andy Weber, the weeks after his "*banya* diplomacy" in Kirov were a blur of meetings, conferences, and flights between the United States, Europe, and Russia. At almost every stop, Weber found that Iranian germ hunters had left their business cards, sometimes only days before American officials had arrived. He was now persuaded that scientist-to-scientist exchanges and research collaborations were the only way to learn how far the Soviet germ-weapons program had advanced, to pry open Russia's still largely closed key facilities, and to stop the spread of lethal germ expertise to rogue states. In the summer of 1997, he and a handful of allies at the departments of State and Energy, the intelligence agencies, and the Pentagon worked to persuade skeptics in their own administration and in Congress that some risks were worth taking. Washington had to increase funding for scientific collaborations, even in advanced fields, and end America's indifference to the plight of Russia's biologists—not for humanitarian reasons but to protect its own national security. Senior officials, however, were preoccupied with other issues. Even within the arcane arms-control community, Clinton officials were far more worried about "loose nukes" than the possibility of "loose bugs."

Meanwhile, U.S. government scientists were having trouble securing full access to Vector, Obolensk, and other key institutes where American intelligence officials claimed secret military research was still being conducted. After visiting Stepnogorsk and hearing scientists discuss the desperate conditions inside Russia's labs, Weber doubted that the Russians could afford to be doing such work. But neither he nor the intelligence hawks knew for sure.

In July 1997 in Washington, D.C., Weber met Lev Sandakhchiev, Vector's energetic director who had surprised Josh Lederberg back in 1993 by calling for greater disclosure of the Soviet germ warfare program. An eminent scientist whose discreet lapel pin identified him as a member of the prestigious Russian Academy of Sciences, he had clout in Russian scientific and political circles. Weber was aware that Sandakhchiev, known for his fierce devotion to his staff, had privately been promoting the conversion of Vector into an open, international center for research into viral vaccines and defenses against the spread of emerging infectious diseases. Weber was intrigued. Could America do business with this small but formidable man?

American hard-liners despised Sandakhchiev. A member of the Russian team that had accused the United States of continuing its germ-weapons program after visiting American drug companies and government labs in late 1991, Sandakhchiev had joined his comrade Alibek in signing a report assailing the Americans. Nonetheless, Washington was growing more alarmed each day about intelligence reports indicating that Iran and other countries were targeting Vector's scientists in its relentless hunt for germ-weapons expertise and deadly microbes.

Sharing a ride to a meeting at Fort Detrick, an hour's drive from the capital, Weber tried explaining to Sandakhchiev why Washington was so alarmed by Iran's quest for lethal pathogens. The 444-day-long hostage episode—the kidnapping and detention of American diplomats at the U.S. embassy in Teheran during the Carter administration—had scarred America's psyche, Weber said. Sandakhchiev looked puzzled. "The Iranians did that?" he asked tentatively. "When?"

Weber was stunned. Sandakhchiev, who knew everything about

lethal viruses, apparently knew nothing about an episode that had gripped America for over a year and had helped elect Ronald Reagan president. For the first time, Weber grasped how deeply isolated Vector and its scientists really were.

For his part, Sandakhchiev bleakly described life in Koltsovo, the small Siberian town near his giant complex whose still secret work into deadly viruses worried American intelligence analysts. Deprived of most Russian government funds since the Soviet Union's collapse, Vector was struggling to pay its scientists and technicians—less than half the number at the lab during its heyday. Many of its installations were now empty, unfinished, or deteriorating. Since military funding ended in 1991, Vector had had no money even to buy monkeys for tests. Salaries were paid intermittently. The Russians and Americans had been talking for months about an assistance program, but nothing significant had materialized.

While some Americans were gleeful about Vector's desperation, Weber knew that the institute was vulnerable to offers of assistance from North Korea, Iraq, and especially Iran. Sandakhchiev, whose forebears came from Armenia, feared the implications of the spread of Islamic militancy for his beloved Russia. By the year 2015, he said, Russia would be 60 percent Muslim. The insurgency in Chechnya was a harbinger, he feared. Russia, he conceded, had moved smallpox strains from the designated repository in Moscow to his remote Siberian complex without notice, violating World Health Organization rules. But the aim was not to do secret research, he told Weber, but to protect the strains against specific terrorist threats from Chechnya. Would the Americans like to see "Emir" Khattab, the ruthless militant Arab hero of the Chechen insurgency, acquire smallpox samples? Vector provided a more secure site for Russia's 120 smallpox isolates—the world's largest collection, twice the size of America's—as well as its bank of some ten thousand isolates of the world's most exotic viruses and strains of other lethal pathogens: Marburg, Ebola, Lassa fever, and a variety of encephalitis strains.

Just as Sandakhchiev knew every strain in his collection, he also knew the location of every senior scientist who had left his institute, he said. There had been no thefts of dangerous agents from Vector. And none of his senior people had gone to work for Teheran—yet.

Weber, encouraged by his conversation with Sandakhchiev, went to see his boss, Frank Miller, the acting assistant secretary of defense for international security policy. Miller, a longtime champion of the small threat-reduction program and a seasoned bureaucrat, liked Weber's proposal for how to open up Russia's closed military labs. He approved of Weber's suggestion that the U.S. army's surgeon general to invite his Russian counterpart to visit American facilities as part of an exchange.

In addition, Weber wanted money to help Sandakhchiev open Vector to the world and deflect Iranian overtures. How much money would it take? Miller asked. About $3 million, Weber replied.

Miller looked up quizzically over his wire-rimmed glasses. Was that all? he asked his protégé. In Pentagon dollars, that was chump change.

But getting authority to spend even chump change was complicated, especially since the intelligence agencies believed, and had persuaded Congress, that military-related germ research was still being conducted at Vector. He would get the money, Miller told Weber, but it would take time.

Weber knew that Vector was running out of time. Scientists had to be paid. Something had to be done.

In September 1997, Andy Weber and Anne Harrington, his State Department counterpart and a key ally, stared anxiously at the e-mail letter they had written to Sandakhchiev on her computer. They shared his frustration, Weber and Harrington wrote, with the pace of U.S.-Russian exchanges and the scope of assistance to Vector. Could a U.S. team visit Vector to tour the facility and discuss expanded, accelerated cooperation?

No one had authorized such a letter, never mind a visit to Vector or a significant increase in foreign aid for it. Normally such initiatives required weeks of bureaucratic clearances and endless red tape. Weber and Harrington knew they were taking a risk. But Weber was desperate to stop the departures of Russian scientists. And Harrington, an energetic woman who had become obsessed with Russia while living in Moscow as a diplomat's wife in the early 1990s, was equally committed to overcoming the bureaucracy's lethargic response to what she also considered a pressing threat. "Should I push the 'send' button, Andy?" she asked nervously. "What can we lose?"

Not much, Weber thought. If worse came to worst, he could always get a job assigning parking slots at the Pentagon, and Anne might enjoy stamping visas in the consular section of some obscure embassy in Africa. Who needed this aggravation anyway?

At the Pentagon, momentum was building to begin the anthrax vaccinations. In a dramatic decision, Secretary Cohen denied a promotion to Air Force General Terry Schwalier, the commander of the base in Saudi Arabia bombed by terrorists. Schwalier retired soon after. Protecting the troops, Cohen made clear, was a top priority and military officers would be held personally accountable for how they performed. White, the deputy defense secretary who had hoped to delay the inoculations indefinitely, left his job that month. He was replaced by John J. Hamre, the Pentagon's comptroller. Hamre had been skeptical of the vaccine initiative when it involved hundreds of millions of dollars of spending on a government-owned factory. Judas Iscariot, he later joked, was the patron saint of Pentagon comptrollers. "We'll do anything to save a few shekels." But Hamre quickly came to favor the anthrax proposal, which was strongly supported both by civilian health officials and by General Henry H. Shelton, the new chairman of the Joint Chiefs of Staff.

The Pentagon team sent to Michigan concluded that the laboratory "has shown significant improvements" and could meet FDA standards if it was renovated as planned. The joint program office, which was set up to manage biological defense, assured Cohen that the Michigan lab would be producing new vaccine within about a year. In the meantime, there was more than enough stockpiled to begin the shots. General Anthony Zinni, the American commander for the Persian Gulf region, had told the Pentagon that his troops in Kuwait needed the immunizations immediately. If military action was ordered against Iraq, Zinni did not want to face the same vulnerabilities that had frustrated Schwarzkopf. The revelations about Iraq's program, which accelerated two years earlier with the defection of Hussein Kamel, had seeped through the American government. No one doubted Baghdad's ability to turn germs into weapons. After nearly three years of debate, the program had solid support from both civilians and military officers.

Senior Pentagon officials understood they might have some difficulty selling the program to the troops. Hamre told Cohen that the defense secretary, the chairman of the Joint Chiefs, and others in visible positions would have to be among the first to roll up their sleeves and take the shots. Cohen agreed.

The Pentagon's civilian leaders also worried about whether the military was able to manage a complex vaccine program that would involve more than 14 million shots. The failings in the Gulf War were now abundantly clear. The immunization records were a shambles. In addition, field commanders had repeatedly violated the Pentagon's agreement with the FDA, giving soldiers experimental drugs, including vaccine against botulinum and pyridostigmine bromide tablets, without informing them of the risks.

The peacekeeping mission to Bosnia which began in 1995 had only deepened doubts about the military's competence in handling medical issues. Health officials had decided before the deployment to immunize the troops against tick-borne encephalitis, an endemic disease in Europe that can be fatal. No one wanted any more outbreaks of mysterious illnesses among American soldiers. There was a widely used vaccine against the disease available in Europe, but it was not licensed by the FDA and was considered an experimental drug. The Pentagon nonetheless decided to give it to the troops sent to the Balkans.

The FDA was determined to avoid a repetition of the administrative problems in the Gulf War. It insisted the military keep meticulous records and inform soldiers of the risks of taking an experimental drug. At the Pentagon, medical officers spent hours briefing the commanders on what was needed. The result, amazingly enough, was another fiasco. Even though the mission involved only a few thousand soldiers, the records were once again incomplete. The FDA also discovered that the documents describing the vaccine omitted some key details, including the fact that it could cause life-threatening allergic reactions and, in rare cases, permanent paralysis. FDA officials sent the Pentagon a stinging letter about the lapses in July, just as the Pentagon was nearing a final decision on the anthrax shots. The Cohen team vowed the anthrax program would be better run.

It had been a tough day even by Pentagon standards. Pamela Berkowsky, Cohen's special assistant, and other top aides spent most of Saturday, November 15, 1997, preparing for the defense secretary's appearance the next morning on the ABC news show *This Week.* The Clinton administration was girding for a new confrontation with Iraq. After admitting that he had lied about his germ programs, Saddam Hussein had refused to permit United Nations inspectors back into his country. Cohen's immediate goal that weekend was to build public support for the administration's plans to bomb Iraq if Baghdad did not relent. He also wanted to prepare the public for the anthrax vaccinations, which he planned to announce shortly. In their conversations that Saturday, Cohen's aides were searching for a way to dramatize the biological threat.

The continuing uncertainties about Iraq's biological program gave Cohen some potent material. Despite six years of the most aggressive inspection in arms-control history, the fate of more than 150 Iraqi bombs and warheads built before the Gulf War to disperse germs remained unknown. So did the status of a dozen special nozzles that Iraq had made to spray germs from helicopters. Baghdad claimed that it had destroyed its twenty-five germ warheads made for missiles with a range of four hundred miles, but had offered no proof. These facts would impress the cognoscenti, but Cohen's aides wanted something more visual to drive the point home on national television.

Berkowsky brought several possible props to the "murder board," as Pentagon officials called the mock grilling before an important appearance.

A young, savvy Princeton graduate, Berkowsky had served as a senior aide to Danzig during his three years as navy undersecretary. Germ weapons were a crucial part of her portfolio. She had sat with Danzig through dozens of briefings and had attended many of his meetings with the senior military officers. She knew the subject as well as any civilian in the building. On that Saturday, Berkowsky and her colleagues debated how Cohen could best translate the potential threat—the "kill ratios" of germ warfare—into terms that civilians could understand. Exposure to no more than ten thousand anthrax germs—all of which would fit comfortably into the period at the end of this sentence—could kill a human being.

To help viewers grasp the threat, she suggested, Cohen could hold up one of the tiny vials of simulated germ agents that she had brought to the meeting. One vial contained tiny castor beans plucked from a plant that friends had given her as a joke. The plant itself was harmless, but its beans contained ricin, a protein toxin that when correctly processed was two hundred times more potent than cyanide and caused vomiting, high fever, weakness, and ultimately death. There was no known antidote.

The ricin seemed misguided, one official argued. Anthrax, not ricin, was the most likely threat from Iraq. The group decided that Secretary Cohen should hold up a bag of flour and give an estimate of how many people would die if an equivalent amount of dried anthrax bacillus were spread over a city. At home that night, Berkowsky told her husband, a surgeon, about the choice of visual aids.

"Why not a bag of sugar?" he said. Not everyone used flour, but everyone was familiar with the heft of a five-pound bag of sugar. The yellow paper sacks were so familiar that he usually instructed postsurgical patients to lift nothing heavier.

On Sunday morning, Bill Cohen unveiled his simulated weapon of mass destruction—a five-pound bag of sugar. "Anthrax," he said, stunning the normally loquacious Sam Donaldson and Cokie Roberts into momentary silence. If Saddam Hussein spread this amount of anthrax over a city the size of, say, Washington, D.C., "it would destroy at least half the population of that city.

"One breath," he continued, "and you are likely to face death within five days." Still worse was VX, a potent nerve agent that Iraq had also developed. If properly dispersed through aerosols, VX could kill "millions," he said. Cohen reminded viewers that Iraq had initially denied making both nerve gas and germ weapons. In fact, he said, Iraq had made four tons of VX and at least 2,100 gallons of anthrax. Hoisting his bag of sugar ever higher, he repeated the number. Roberts, tired of being upstaged by a bag of sugar, interrupted to ask: Would the secretary please put that bag of sugar down?

Thus began the great sugar debate. Some military experts and scientists were outraged by Cohen's performance. Yes, a five-pound bag of anthrax bacillus spores could theoretically kill half of Washington's population—or about 300,000 people—but only if the atmospheric conditions for such an attack were perfect, the germs very potent, and the dispersal highly efficient. If an attack occurred during a rainstorm or at midday in hot, sunny August, the germs might not harm anyone. Some experts privately grumbled that Cohen had presented a worst-case casualty estimate to terrify Americans.

The man who had helped prepare those numbers, Bill Patrick, watched the news show that morning. A day or two before Cohen's appearance, Dave Franz, then the chief of the army's biological research arm at Fort Detrick and a member of Berkowsky's murder board, had asked Patrick to help with the calculations. The "old fossil" was the logical choice for such a job, having routinely made such estimates when he worked to develop germs for America's arsenal in the 1950s and 1960s.

Patrick told Franz that a terrorist hoping to kill half the population of the greater Washington metropolitan area would have to spray seventy-five to eighty pounds of dried anthrax agent from an airplane or helicopter flown from Rockville, Maryland, to northern Virginia. But, Franz complained, Cohen couldn't hold up an eighty-pound bag of sugar. How much anthrax would be needed to kill half the population of, say, just the District of Columbia? That would require the equivalent of a fifty-pound bag, Patrick replied, still quite a prop to hold up on television. "They finally decided they would kill off only a part of the population of a part of D.C.," Patrick said. "But by then it was almost airtime and we were still discussing the numbers." It would never be entirely clear how Cohen ended up misstating the casualty estimate. Somewhere in the human chain between Patrick, Franz, Pentagon officials, and Cohen, an error was introduced which made its way into the national debate on germ weapons.

Although Patrick said he regretted Cohen's inaccurate estimate of the casualties, he approved of the presentation. The secretary might have overstated the lethality of anthrax, but he had publicized a threat that had been largely ignored. Perhaps the sugar bag was melodramatic. Perhaps Cohen should have made it clearer that American intelligence analysts viewed Saddam Hussein as highly unlikely to attack either an American city or the seventeen thousand troops still stationed in the Persian Gulf. But Patrick felt that the right points had been made. Finally someone in the political leadership had said out loud what he, Josh Lederberg, and others had been saying for years in secret government sessions.

In late November 1997, culminating a drive that began after the Gulf War, the Pentagon awarded a $322 million, ten-year contract to DynPort, a British-American venture, to develop and obtain licenses for smallpox and seventeen other vaccines for the military, including a new recombinant version of the one for anthrax. The plan focused on building up a stockpile of advanced vaccines for the next century.

On December 15, the Pentagon announced that it would vaccinate the military against anthrax with the old vaccine—an unprecedented step to counter the threat of germ weapons. "We owe it to our people to move ahead with this immunization plan," Cohen said in a written statement.

At a news conference, Pentagon officials estimated that the program would take six years and cost $130 million. They also said that the vaccine appeared to cause fewer side effects than its counterpart for typhoid and, hence, did not threaten the health of soldiers. Even so, officials noted that the Pentagon would not require pregnant women to take the shots.

The Pentagon's briefing that day played down the problems at the Michigan lab. The FDA's March 1997 threat to revoke its license was depicted as a bland warning about "some production problems" that had nothing to do with the anthrax vaccine. Officials reassured reporters that there was a plan in place to modernize the facility. Cohen said the program would begin in about six months, provided four conditions were met. The military would have to test the vaccine stockpile for potency, sterility, and purity. It would have to commission an independent medical review, in effect, a second opinion. It would have to set up an accurate system to track vaccination records, and it would have to draw up an acceptable plan to explain the policy.

The day after Cohen made his announcement, a reporter asked President Clinton at a news conference whether he, as commander-in-chief, would take the shots. "Secretary Cohen made a quite vivid demonstration not long ago on TV that a primary threat of anthrax would be a terrorist attack against a civilian population," the reporter said. "Should civilians be vaccinated against anthrax?"

Clinton defended the Pentagon's program but said he had no plans to take the shots himself. "I know of no expert opinion," he said, "that would say that those of us that are essentially in the civilian population of the United States should be vaccinated."

As the Pentagon prepared to vaccinate soldiers, new evidence of the biological threat—and even graver questions about the wisdom of the U.S. vaccination program—emerged in the December 1997 issue of *Vaccine.* The London-based scientific journal's publication of the Russian research on genetically modified anthrax that Weber had first learned of in 1995 now alarmed military scientists and civilian germ-warfare experts alike. In government offices and labs throughout the country, intelligence and defense analysts dissected the dry, difficult prose of the Russian scientists, searching not only for information about the superbug that the Russian team at Obolensk claimed to have created but also for clues about why the Russians had published their findings.

The Obolensk institute and several of the article's authors—Pomerantsev and Staritsin—were well known to Western intelligence agencies for their involvement in the former Soviet germ-warfare program. But the article confirmed what some Western military scientists and intelligence analysts had long argued: Russia was far more advanced in some areas of recombinant research than Washington had assumed. Pomerantsev and his colleagues must have known that publishing their research would astonish and concern microbiologists throughout the world.

In an interview with the *New York Times,* Staritsin disclosed some of what he had told Andy Weber privately in 1997 when they'd first met in Russia. He insisted that the research had been done in 1993 and was motivated by public-health concerns. The Russian strains of *Bacillus anthracis* and *Bacillus cereus,* he said, were "closely related and often found in soil in close proximity." If the two organisms naturally exchanged genes without any external intervention, Russian scientists needed to know "what the result might be." In any event, he added, repeating what he had told Weber in confidence, the new strain was too unstable to be useful in weapons.

Still, the experiment raised disturbing questions. Could it be made sufficiently stable to work as a germ weapon? Would the Americans be able to secure a sample of the new superbug from Russia? Should the Pentagon force millions of American soldiers to take the shots if America's anthrax vaccine was not effective against the new genetically modified pathogen?

Lederberg was among those deeply concerned about the experiment, which he considered a scientific watershed—the first public example of a new pathogen with military implications created in a laboratory. When it came to recombinant research, he said, the cat was now "out of the bag." Biowarfare had entered a new phase.

The article soon helped fuel the debate that had been raging among intelligence analysts since the 1992 defection of Ken Alibek. In secret debriefings, Alibek had charged that the Russians were continuing their military germ research, an allegation he now bolstered with the *Vaccine* article and other published work. The evidence suggested that post-Soviet Russia was still making bioweapons, he insisted, including new "chimera" pathogens that combined several types of microbes. Russia, he said, had not abandoned its effort to blend Ebola with smallpox, mankind's greatest scourge.

Because Alibek had not emphasized such research in his initial debriefings, his charges polarized the intelligence community. Some analysts argued that Alibek was undoubtedly right, even though his assessment was drawn largely from a review of recent scientific literature and from secondhand information he had gathered in Moscow before his defection.

Other analysts remained unconvinced that Russia, or any other country, had been able to create such chimera weapons, despite the claims of Alibek and Pasechnik. Though such breakthroughs were theoretically possible, they argued, the requisite genetic technology was still years, if not decades away. Besides, the analysts reasoned, why would anyone bother to make such diabolical new agents? The "oldie moldies" were heinous enough. Why would anyone need to improve on nature?

ONE of the men who had led the Soviet Union's effort to make deadly pathogens even more so was General Yuri T. Kalinin, the long-serving director of Biopreparat, the civilian cover for the Soviet biological-weapons program. After the Soviet Union's collapse, Biopreparat had supposedly been reborn as a civilian drug company, 51 percent of which was owned by the Russian government. But it was still led by Kalinin and several others who had helped spearhead and conceal the Soviet Union's germ-weapons programs. President Yeltsin had promised the Americans in 1992 that he would fire Kalinin and other Cold War holdovers from the Soviet biowarfare program, but he had not done so. Because Biopreparat's operations were still shrouded in secrecy, Kalinin and his colleagues were the focus of intense concern.

Andy Weber had been encountering resistance from Kalinin since their first meeting in September 1997. From his gloomy, Soviet-style offices on Samokatnaya Street in downtown Moscow, Kalinin had tried to impose ground rules on the U.S.-Russian biological relationship. Biopreparat had no objection in principle to American assistance to Russian labs and institutes it had once rigidly controlled, he had informed Weber and a small delegation of American officials. But he disliked the American tactic of circumventing Biopreparat in favor of building direct relationships with his former institutes. Wouldn't it be more efficient, he had suggested, to channel all assistance and research proposals for Russian biological institutes through Biopreparat?

Neither Kalinin nor Weber had discussed the past during their brief meeting. And Weber had avoided making any commitments to him. Now the Biopreparat chief seemed determined to prevent his organization from being sidelined in the flurry of new Russian-American partnerships and grants. Kalinin's organization was surely far less powerful than it had been when it ran dozens of institutes across the Soviet Union, employing more than thirty thousand scientists and technicians. But it still claimed to oversee nearly twenty institutes and other enterprises that made about a thousand different vaccines, medicines, and other biological products. Weber had been eager to avoid alienating the general, whose aristocratic demeanor hid the iron determination that had helped him rise in the Soviet system. Weber's fledgling program did not need another enemy, especially the well-connected, ubiquitous Kalinin.

Since that 1997 session, Kalinin had increasingly become an obstacle, American officials felt. Russian scientists complained that Kalinin had undermined their independence by using his bureaucratic powers to deny them the right to travel abroad to conferences and training programs. He had confiscated their passports, delayed the issuing of visas, and pressured institute directors not to develop independent ties to Western labs and companies. He had dismissed a prominent institute director from his post after the scientist had accused Kalinin of "illegal practices," including pocketing Western assistance money. Kalinin denied those charges.

Although he did not agree with the most hawkish intelligence assessments circulating in Washington, Weber knew firsthand of other troubling indications that Russia's military remained committed to biological weaponry. Russia's new military doctrine reserved the right to use nuclear or other unspecified weapons of mass destruction against an enemy. And the Russians were continuing to deny aspects of Soviet history, including the accidental release of deadly anthrax at its lab in Sverdlovsk. Perhaps the most troubling aspect of Russian germ-warrior culture, Weber concluded, was that so many former Soviet scientists continued to tell lies about their previous work, while others resolutely maintained an utter silence about the past. Soon after he had ordered an end to the clandestine germ effort, President Yeltsin had issued a decree forbidding Russian scientists and officials to discuss any aspect of their secret work. This edict, which Russian scientists confirmed was still in force, made it dangerous for them to speak out.

As a result, Weber and the American analysts had to work intensively to piece together solid, accurate information about the scale, scope, and state of technological advancement of the former Soviet germ program. While the defectors Pasechnik, in 1989, and Alibek, in 1992, had given the West a detailed outline of Soviet germ activities, even they knew little about some areas and institutes. The program had been a deliberate maze of false fronts, secret projects, and parallel organizations that often conducted both military and peaceful research. The structure was designed to enhance secrecy.

Weber could not afford to contemplate such obstacles; there was so much to be done. For instance, several former Soviet institutes in Kazakhstan that had been dedicated to developing germs to destroy plants and animals could not afford even rudimentary protection—electrified fences, guards, locks, and computerized access—for their hundreds of strains of highly virulent pathogens. At the Anti-Plague Institute in Almaty, Weber had just visited a collection of more than one thousand lethal germ strains that was protected only by a rickety wooden door and a single skeleton-key lock. Asked if he would like to see a sample of the institute's plague, Weber found himself holding enough plague in a rusty pea can to depopulate the entire region.

10

The President

In late December 1997, President Clinton traveled to the resort island of Hilton Head, South Carolina, for the Renaissance Weekend, an invitation-only gathering of officials and private citizens who liked to ring in the New Year with readings from Kierkegaard and half-day seminars devoted to such topics as "Moral Compasses for Modern Leaders."

One night, Clinton arranged to have dinner with J. Craig Venter, a molecular biologist who was pioneering the effort to map the genes of germs and people. Venter and his wife, Claire, also a scientist, had been invited to dine with the Clintons at the previous year's Renaissance Weekend and were thrilled and slightly surprised that scientists would be asked again. The dinner party included only four couples, and Clinton and Venter spent much of the evening locked in conversation. Their talk turned to what the strides in genetic mapping would mean for germ warfare, an issue on which Venter had strong views. As head of the Maryland-based Institute for Genomic Research, a nonprofit group that deciphered microbes, he had recently helped map the smallpox virus.

Venter warned Clinton that such work could be extraordinarily dangerous in the wrong hands, allowing a foe to reconstruct the rare microscopic killer. "Most of the discussion was about smallpox and synthetic microorganisms and how to come up with effective defenses," Venter recalled. "The president was extremely knowledgeable already. He said throughout history the biggest changes have come when there were new offensive weapons without defensive ones, and he was worried that we were entering such a period."

Clinton asked if smallpox could be spliced with another bug to make it more harmful. Venter replied that it could and that a new novel—*The Cobra Event,* the thriller that Preston had written with Danzig's encouragement—presented just such a scenario. He urged the president to read it.

Venter described smallpox as more dangerous than anthrax, even in its unaltered state. It was contagious wildfire, he told the president, a killer that leapt from person to person. By contrast, anthrax was incommunicable. It killed only where sprayed. Venter's comparison made a deep impression on Thomas Schneider, a Washington businessman and friend of Clinton's who had organized and attended the two dinners. The smallpox warning, he recalled, "was *really* talking about the impact on humans," especially civilians. And the superbug issue, Schneider said, made smallpox and other bugs seem even more of "a poor man's weapon."

Offering quick reassurance, Venter told Clinton that the gene mapping done at such places as his institute offered a new way to strengthen the nation's germ defenses. It could help create innovative cures and identify pathogens used in an attack by revealing their genetic signatures.

The next day, Clinton sent an assistant to see Venter. The aide said the president would be grateful if the biologist would prepare a list of experts who could advise him on the topics they had discussed. Venter was delighted. In the months that followed, he talked repeatedly with the White House about possible candidates for the panel.

Clinton had long been fascinated by the promise and peril of the new biology. His concern arose in part from the collapse of the Soviet Union, which had put tens of thousands of scientists privy to the technological secrets of mass destruction on the job market. The threat, Clinton would tell his aides, was obvious. He had also come to see the danger of germ weapons in the context of terrorist incidents that had marked his presidency.

The attacks on the World Trade Center and the Murrah Federal Building in Oklahoma City had been done with simple fertilizer bombs, the subway gassings in Tokyo with home-brewed sarin. Clinton saw germs as the next logical step for terrorists. "The president started inserting warnings in all his speeches," said Dick Clarke, the

National Security Council adviser on counterterrorism. "Even if the subject was not relevant, he would start talking about bioterrorism as the 'dark' side of globalization. Again and again he would say, What if the World Trade Center and Oklahoma City had been a biological or chemical event?" Clinton was a hands-on editor of his own speeches, and Clarke said the germ references were his idea. "At first we thought, well, this will pass. The president is interested in lots of things," Clarke said. "But it didn't pass. He kept after us. He wouldn't let it go. It wasn't just a whim; he was obsessed with it."

Clinton's fascination with germ weapons was deepened by his reading. He devoured histories, newspaper and magazine articles, and especially fiction. Tom Clancy's *Rainbow Six,* a thriller about a counterterrorist team's efforts to prevent Armageddon, made a big impression. Another favorite was a Patricia Cornwell novel that focused on a female medical examiner's battle against a shadowy figure intent on using mutant smallpox for mass murder.

But nothing caught the president's attention as much as *The Cobra Event,* the novel Venter had recommended and that Clinton read in early 1998. It depicted a mad scientist's determination to thin the world's population by infecting New York City with a designer pathogen. By combining smallpox, a virus similar to that of the common cold, and an insect virus that destroys nerves, the scientist invented an ideal doomsday germ—a "brainpox" that spread quickly and melted the brain.

The book opened with seventeen-year-old Kate Moran heading off to her private high school in upper Manhattan. She had a bad cold. By art class, her teeth were chattering and her nose was gushing. Disoriented, she soon found herself seized by convulsions and, bizarrely, biting her own mouth. She collapsed, her body lashing back and forth, her face twitching uncontrollably. With classmates and a teacher standing by helplessly, she died a violent death, her spine cracking under the strain of her contracting muscles.

The psychopath behind the killings also eventually fell victim to the bug he had created. It was a plot twist Lederberg had suggested to Preston at the outset of his research to make germ weapons seem less attractive to a potential terrorist.

Clinton was impressed by the book's grim narrative and apparent authenticity. Preston's acknowledgments included more than one hundred experts—military officers, intelligence analysts, doctors, scientists, and officials in Clinton's own administration, including

Danzig. The novel's science, Preston wrote in a preface, "is real or based on what is possible."

Clinton began asking his friends, cabinet members, even House Speaker Newt Gingrich whether they had read the book and what they thought about it. After a White House meeting on another topic, the president suddenly turned to John Hamre, the deputy secretary of defense, and asked whether he could speak to him privately for a few moments. As the two men walked into the Oval Office, Clinton said he had recently read *The Cobra Event* and asked Hamre whether he thought the novel's scenario was plausible. Could a terrorist unleash an unstoppable plague with designer pathogens?

Hamre was somewhat taken aback. He had not read the book and was no expert on superbugs or recombinant technology, he told the president. "But I'll have one hundred colonels reading the book at dawn," he promised. Returning to the Pentagon, Hamre searched in vain for the book. He finally borrowed the only one he could find—a copy from Secretary Cohen's private library.

The next day, Hamre delivered a preliminary assessment to the White House. While the novel was not based on secret government data and contained no classified information, the scenario was theoretically plausible.

In February 1998, the arcane intelligence debates about Russia's biological program exploded into public view. Ken Alibek emerged from anonymity to charge publicly what he had said in secret government debriefings: the Russians were still making sophisticated bioweapons under the guise of developing vaccines for defense.

In interviews with the *New York Times* and ABC's *Prime Time Live,* Alibek said that the Soviet Union's plans for World War III had included "hundreds of tons" of anthrax bacteria and scores of tons of smallpox and plague viruses. And he repeated in public his controversial charge that the Soviet labs had made hybrid designer germs from Ebola and smallpox that were impervious to vaccines and antibiotics.

As Alibek stunned legislators on Capitol Hill who had not known of the defector's existence, Andy Weber was making real progress in Russia.

The gamble that Weber and his State Department colleague Anne Harrington had taken the previous September had paid off. Less than twenty-four hours after receiving the e-mail they had sent him, Lev Sandakhchiev, Vector's director, had invited the Americans to visit his Siberian institute. He did more than just welcome his former Cold War rivals to his deteriorating facility; he gradually permitted them to photograph the most sensitive labs in the sprawling one-hundred-building complex that previously had been off limits to foreigners and even most Russians.

Before the visits, an interagency group at the White House had carefully weighed the invitation as well as the potential risks and benefits of scientific cooperation with Vector and other former Soviet institutes. Senior administration officials recognized that Russia might try to use American money or expertise for its own purposes, perhaps even in an ongoing germ-weapons program. But they sensed that cooperation might also provide valuable insights into Soviet biowarfare activities and help thwart Iranian attempts to recruit Russians. After several tense meetings in the fall of 1997, the interagency group endorsed the Vector visit and broader efforts to engage former Soviet bioweaponeers through cooperation with their institutes. It insisted that military officials, intelligence analysts, and other government experts review each proposed project for its risks and scientific merit, adding enormously to the analysts' workload. The group also ensured that money for such projects would be paid directly to the Russian scientists and that the projects would be closely audited. The group's recommendations were described to President Clinton in a memo from Samuel R. "Sandy" Berger, the national security adviser. Clinton blessed the approach by scribbling his assessment of the gamble in the memo's margin—difficult, he wrote, but necessary.

Subsequent American visits to Vector were a resounding success.

American experts were eventually permitted to tour Building 6, which housed the well-guarded smallpox laboratory, the huge aerosol test chamber, and the high-containment labs where research on other exotic agents was being conducted. They spoke at length to Vector scientists about their work and saw a video describing a controversial Vector expedition to a remote Siberian village. There scientists had unearthed the long-frozen bodies of turn-of-the-century smallpox victims and taken samples from them. American teams visited new projects where the scientists who had once turned viruses into weapons were struggling to produce such commercial products as yogurt and a new line of cosmetics. And they were told that American researchers would soon be invited to spend weeks, even months at Vector, working on joint projects with their Russian counterparts.

Now, on a bitterly cold February day, Weber emerged from touring Buildings 6 and 6A, two adjacent four-story brick buildings where the Soviets had done some of their most sensitive work on Marburg and other deadly viruses. The U.S. effort to engage the Russians was working. The American team, having visited every part of the compound that intelligence analysts had previously identified as suspicious, was now convinced that Vector was no longer engaged in offensive germ research and development.

Equally important, the American team and Sandakhchiev had quietly struck an agreement restricting Vector's contacts with Iran for at least a year. The so-called gentlemen's agreement, negotiated by a towel-clad Vector scientist and U.S. officials in the institute's *banya,* proscribed cooperation that had military uses with Teheran. While Vector could sell pharmaceutical products to Iranian companies, Sandakhchiev agreed not to exchange scientific visits, engage in collaborative research, or provide Iran with Russian research-and-development expertise and technology.

The agreement was unprecedented. It was repeated in subsequent verbal compacts with other leading, cash-strapped germ institutes, including the one at Obolensk, whose collection of several thousand bacterial pathogens had long worried Western analysts. General Nikolai N. Urakov, Obolensk's long-serving director, pledged not only to forgo commercial and scientific contacts with Iran but also to lift what he called the "curtain of secrecy" that enshrouded the twenty-five-year-old complex and spurred doubts about its supposed conversion from military to peaceful research.

Meanwhile, Weber received encouraging news from Washington. American military scientists had analyzed soil samples that American teams had brought back from anthrax burial pits on Vozrozhdeniye Island in 1997. Samples from six of eleven burial pits showed that some of the bleach-soaked anthrax spores the Russians had buried there almost a decade earlier were still alive and potentially dangerous. Preliminary analysis suggested that the anthrax shots being given American soldiers seemed to be effective against the strain found on the island. Whether America's vaccine would work against the "superanthrax" Alibek had created or other Soviet strains was still unknown. But now there was one less thing to worry about.

Not everyone shared Weber's enthusiasm for what administration officials were calling the "Vector model." Weber and Harrington had dispensed with some of the Kabuki dance of diplomacy—formalistic inspections and letters of protest. They were working on the theory that cooperative projects, paid for with American cash, could build trust, foster partnerships between Russian and American scientists, prevent the spread of bioweapons expertise, and persuade the Russians to open up previously closed facilities. The Russians would reject Iranian overtures and reveal the remaining secrets of the Soviet biowarfare program, Weber believed, if they had some incentive to do so.

While many American intelligence analysts remained skeptical about the growing Russian-American biological cooperation, they acknowledged that they now lacked compelling evidence that either Vector or Obolensk was continuing its germ-warfare research and development. In early 1998, the institutions were removed from the administration's list of suspect sites. But the analysts soon focused their suspicions on Russia's four leading military labs: the facilities in Zagorsk (Sergiyev Posad), Kirov, Sverdlovsk (Yekaterinburg), and Strizhi. All remained closed to Western government officials, and little was known about what was going on inside them. So there was no way to ensure that American money and biotechnological expertise were not being shared with Russian germ warriors at these closed military labs to make new generations of deadlier germ weapons. Weber and his colleagues were well-intentioned fools, a former intelligence officer complained. The Russians were taking the American handouts and "laughing all the way to their secret labs."

GIVEN what he was learning about the former Soviet program, Weber couldn't discount such concerns. The breadth and depth of the program were shocking, even to supporters of the new cooperative effort. Just as Alibek had asserted, almost every Soviet ministry and important institution had played some role in the top-secret program: the Ministry of Defense; the Ministries of Agriculture, Health, and Science; the Communist Party Central Committee; and even the eminent national academy of science. The Ministry of Internal Affairs had provided prison labor to build the vast biowarfare complexes, as well as additional guards to supplement the military in protecting them. The Ministry of External Trade had secretly helped purchase equipment and animals for experiments. A special office in the Soviet Ministry of Justice had provided legal services and special courts for the weapons institutes' personnel. The KGB had its own germ centers for developing methods of assassination.

The murkiest parts of the Soviet effort were two closely related programs that were aimed at using genetically engineered germs and toxins to cause psychological and physiological changes. The first, called Bonfire, was overseen by a part of Biopreparat that even Alibek knew little about. That program had apparently created a new toxic weapon that manipulates peptides, the short chains of amino acids that send signals to the central nervous system. Such weapons could in theory alter moods, sleep patterns, and heart rhythms—all without detection. They could even kill, which explained why the KGB was particularly interested in them. The second program, whose code name was Fleyta, which translates to "Flute," was run by the Ministry of Health's Third Directorate. Its main goal was to create psychotropic and neurotropic germ agents for the KGB, again using peptides and other "bioregulators."

Another focus of concern was the Research Center of Molecular Diagnostics and Therapy which would soon start work on enzymes and vaccines to combat a variety of diseases, thanks to International Science and Technology Center grants largely from the United States. During the Cold War, the institute had developed psychotropic agents and peptides to alter mood and behavior as part of the Flute program. Situated near a hospital for alcoholics, drug addicts, and other social misfits, the relatively plush institute was in the heart of downtown Moscow.

American officials had been told by what they called an "impeccable source" that the facility had tested its peptides and mood-altering agents on the patients confined to the hospital next to the lab. Though its director and staff vehemently denied the allegations, other Russian scientists insisted that the lab, and other institutes as well, had violated a profound moral taboo—the ban on human experimentation. Russian officials had adamantly denied the charge, which had been raised before. But Americans felt certain their sources were in a position to know about the testing.

American officials were unable to learn more about the fate of the patients who had allegedly been subjected to Flute testing. Few Russian scientists knew which particular peptides or agents patients had been given or how many such experiments had been conducted. No Russian official would confirm, even in private, that such experimentation had occurred.

Washington also remained uneasy about Vozrozhdeniye Island, the Soviet Union's major open-air germ-testing site in the Aral Sea. Since the early 1990s, American intelligence agencies had collected unconfirmed reports that the Soviet military had tested smallpox on part of the island that it controlled. According to scientists and residents of the region, the Soviets had buried not only anthrax from Sverdlovsk but also debris associated with the smallpox testing in as many as thirty separate pits on the island. Biopreparat scientists had told the Americans they knew nothing of such testing. But that proved nothing, given the intense rivalry between the Soviet military labs and the civilian cover organization. Similar reports continued to flow into the CIA. They were marked top secret and shared with only a handful of specially cleared analysts.

The director of a former Soviet lab said in an interview that he had information suggesting that the Soviets *had* tested smallpox on the island. Bakyt B. Atshabar, the director of the Anti-Plague Institute of now independent Kazakhstan, said that the testing had probably caused an outbreak of the disease in neighboring Aralsk in 1971. According to detailed records filed by physicians from a branch of his institute who had struggled to contain the epidemic, he said, nine people, including children, had developed smallpox, three of whom died. Though the physicians' reports did not specifically mention Vozrozhdeniye Island, he and other scientists at his institute, which was once part of the Soviet germ-warfare empire, were convinced that the epidemic had its roots in tests on the island. There had been no other outbreaks in that region for several years, he said. And the illnesses had never been reported to the World Health Organization.

In early 1998, Dick Clarke, the White House counterterrorism coordinator, set out to build the case for strengthening the nation's defenses against germ attacks. In his view, not enough had been done since the 1995 gas attack in the Tokyo subways. The money appropriated by

Congress had gone mostly to hire FBI agents and train "first responders": police, firefighters, and other emergency-management officials. A tiny fraction of the funding was used to subsidize Russian scientists. But few steps had been taken to protect the civilian population. The United States had no stockpiles of vaccines or antibiotics, and planning for an outbreak was at an impasse, an awkward standoff among the Federal Emergency Management Agency, the departments of Justice and Defense, and the FBI.

The training for such "first responders" as firefighters and other emergency personnel would be useful in a chemical attack but would be worthless against anthrax or smallpox. In a germ assault, there would be no "scene" at which experts could converge. In germ terrorism, the "first responders" would be doctors and nurses, and the first signs of attack would be the arrival of sick people at an emergency room or clinic. The bulk of the government's spending on biodefense—billions annually—went to the Pentagon for studies and for battlefield equipment: detectors, suits, vaccines, and masks. Little money had been allocated to strengthening the local health offices and clinics that would have to contain an infectious outbreak.

In March 1998, more than forty senior members of the Clinton administration gathered at Blair House to rehearse how they might handle a biological attack by terrorists. It was what the government calls a tabletop exercise. Clarke had convened the officials to give them a hands-on sense of the decisions they might face as an infection rocketed through a panicked population. For help in drawing up a scenario, Clarke's staff had turned to William A. Haseltine, an expert on genetic engineering and the founder of Human Genome Sciences, a company that identified and patented genes for drug development.

In Clarke's scenario, terrorists spread a genetically engineered virus in California and the Southwest along the Mexican-American border. After doctors diagnosed the epidemic as smallpox, public-health officials rushed in with vaccine to immunize the population. But what was initially diagnosed as smallpox turned out to be the sort of hybrid that Alibek said the Soviets had been developing—a combination of smallpox and the Marburg virus. Though not highly contagious, Marburg causes a hemorrhagic fever for which there is no known vaccine or cure.

The terrorists' trick, according to its designers, was to present policy makers with a recombinant virus that would express itself in distinct phases. The law-enforcement and public-health officials who had raced to the affected region to contain and treat the smallpox outbreak would themselves become the unwitting vectors of a secondary epidemic of deadly Marburg after they returned to their own communities.

The scenario, Haseltine assured the White House staff members, was scientifically plausible. "You could make such a virus today," he said. In fact, any "trained molecular virologist with a really good lab could do it." Other experts later disagreed, dismissing the scenario as science fiction. John W. Huggins, the head of viral therapies at the army lab at Fort Detrick, thought that using an exotic virus made sense, since the exercise was intended to stretch the nation's biological-warfare preparedness system to its limits. But he cautioned that scientists were not even close to being able to manufacture such a diabolical weapon. The hybrid weapon envisioned in the exercise was many years away.

Whatever its scientific plausibility, the results of the drill were not encouraging. While officials playing the role of state and local officials eventually managed, with great difficulty and after thousands of casualties, to bring the smallpox epidemic under control, they were flabbergasted when their ostensible victory over smallpox evaporated. They soon realized that they lacked the knowledge, resources, and stamina to contain or treat a secondary outbreak of Marburg or any deadly disease.

After only a couple of hours of role-playing, the administration officials pretending to be state and local officials were overwhelmed by the demands of thousands of hypothetically sick and dying people. Local medical offices rapidly exhausted their stocks of antibiotics and vaccines. Federal quarantine laws turned out to be too antiquated to deal with the rapidly spreading epidemic, and no state had adequate plans to take care of the people it had isolated. Officials did not know where to store and bury the still-contaminated dead. What began as a domestic disaster became an international crisis as the epidemic threatened to spread into Mexico. Discovering huge gaps in logistics, legal authority, and medical care, officials began quarreling among themselves and with Washington over how to stem the epidemic. No one seemed to be in charge.

The hush of defeat fell over the room.

IN March, Secretary Cohen announced that the Pentagon would assume a much more direct role in domestic defense. Speaking to reporters at the National Press Club, he said the National Guard was creating ten rapid-reaction teams that would rush to an American city, town, or village attacked by chemical or biological weapons. The units—RAID teams, for Rapid Assessment and Initial Detection—would be trained to detect chemicals or germs and organize the initial flow of federal and military assistance. Guard officers would provide crucial early help to local authorities overwhelmed by the task of treating the sick and wounded.

The March 17, 1998, speech was vintage Cohen, quoting the poets T. S. Eliot, Robert Frost, and Seamus Heaney, as well as Sun-Tzu, the ancient Chinese military theorist. Cohen paraphrased Winston Churchill as he summed up the dangers of the new biology, saying: "We can return to the Stone Age on the gleaming wings of science just as quickly as we can glide into the twenty-first century." The defense secretary did make one error of scientific fact, asserting that a person could die from inhaling a single spore of anthrax. In fact, the fatal dose is several thousand spores. Afterward, a reporter asked whether the Pentagon truly saw an imminent danger of germ attack against civilians. Training the reserves for biological defense was a serious step, "almost a code red," the reporter said.

Twenty-five nations had or were developing chemical and biological weapons, Cohen replied, and the expertise was spreading rapidly through the Internet and other means. Terrorist groups were also likely to acquire germ or chemical weapons in the coming years. We have the finest military in the world, he said, the best equipped, best educated, and most capable in American history. Few, if any, countries dared challenge the United States directly. Foes would be searching for vulnerabilities to exploit. "We have to be prepared. This is not a scare tactic. It's a reality. And we're likely to see more of it in the future."

The story behind Cohen's announcement that day was more politics than poetry. Whatever the ultimate truth of Cohen's assessment—and it was not universally embraced—germ defense had become yet another prize in the quest for federal dollars by government agencies and beltway consultants.

The National Guard could call on some influential political allies both in Washington and in state capitals. It had the support of the nation's fifty governors and a network of members in every congressional district. In the mid-1990s, it was fighting to protect its turf as the post–Cold War military shrank. The guard's relationship with the army was acrimonious; guard officials believed they had been given short shrift when the military services drew up the long-term planning document, the Quadrennial Defense Review. Legislation was pending to give the guard a seat on the Joint Chiefs, a proposal that stirred anxiety among the other branches of the military. Cohen and his aides were eager to integrate the reserves more closely with the active duty force.

In the fall of 1997, the guard's supporters on Capitol Hill added $10 million to the defense spending bill that would define the guard's role in the fight against chemical or biological terrorism. Soon after, the guard awarded $7.5 million of that money to Science Applications International Corporation, which had helped lobby Congress for the appropriation. SAIC was already working for the government on biodefense issues and was, among other things, advising the Pentagon on managing the Michigan vaccine facility.

Cohen believed that the guard should have a place in domestic defense. The citizen-soldiers reported to the governors and were trained to work with local authorities on disaster relief. They were, he would often say, "forward-deployed in the United States." Cohen's aides were already reviewing how the guard and reserves would fit into the military's domestic defenses against chemical or biological attack and did not want to get corralled into a huge new spending program by Congress. In January, just as SAIC was beginning its study, a team of Pentagon officials recommended creating fifty-four RAID units, one for each state and territory. Cohen cut the number to ten. The guard, he told aides, should learn to walk before it ran.

Lawmakers lavished money on the new program. When the Pentagon asked for five more RAID teams, Congress provided money for an additional twelve. The SAIC study was eventually delivered to lawmakers. Crammed with colorful charts and graphs, it listed forty-seven potential new roles for the guard. The study was quietly shelved.

ON April 10, 1998, Dick Clarke's staff summoned seven of the nation's most distinguished scientists and public health experts to make their case for biodefense directly to President Clinton. It was the culmination of the planning that Clinton spurred after his dinner with Venter. One of the first experts Venter had recruited for the advisory group was Josh Lederberg, who doubted that a meeting with the president would do any good. "Be careful when you dance with elephants," Lederberg had warned him. The other invited experts were Lucille Shapiro, a Stanford University microbiologist; Barbara H. Rosenberg, an authority on the biological-weapons treaty; Thomas P. Monath, a vaccine researcher and former head of the army's biodefense lab; and Jerome M. Hauer, New York City's head of emergency management. The delegation was led by Frank Young, who had retired from his emergency preparedness post at HHS and become a pastor.

Originally, the list of invitees had also included three other experts, among them Brad Roberts, a research staff member of the Institute for Defense Analysis. Roberts, a seasoned Washington military analyst, was rather more skeptical than some of the other invitees about whether states or terrorists would be able or willing to use germ weapons to kill masses of people anytime soon. He planned to highlight the inherent obstacles to the use of such germ agents and to put what he termed some "balance" into the debate about how serious a threat bioweapons posed and how much the United States ought to spend to counter it. But days before the meeting, a member of Clarke's staff called him and two other invitees to say their invitations were being rescinded. The White House's tabletop exercise the previous month had generated so much cabinet-level interest in the experts' meeting that there was not enough room for all of the outsiders. Since Clinton had requested a small session, some experts had to be taken off the list. They were very sorry, the official added. Maybe Roberts could attend a future meeting with the president.

Though they differed on the likelihood of a germ attack in the near future, the invitees all agreed that the United States had dangerously neglected its public-health system. An audience with Clinton, they felt, was the chance of a lifetime. No president had ever been more predisposed to understand their pleas. The scientists saw America's vulnerability to germ attack as an extreme example of the country's pervasive failure to deal with infectious diseases. Persuading Clinton to increase spending on public health would help millions of Americans, even if no one ever tried to attack America with germs.

Compared with health care in other economically developed states, the U.S. public-health system was a disaster. Almost half of all local health departments did not have the use of e-mail; at least one thousand of them had no access to any on-line or Internet service; 20 percent of them still had rotary phones. Ralph D. Morris, the president of the National Association of County and City Health Officials, had warned Congress in 1998 that 70 percent of health directors had little or no expertise in using computer services. Some local health departments could not even afford a microscope, much less sophisticated computer technology.

The Centers for Disease Control and Prevention's epidemic-surveillance system, created a half century earlier to detect germ-warfare attacks, was in tatters. Most epidemiological investigators and even emergency-room physicians were unlikely to recognize a case of anthrax, a rare disease. Nor would they correctly diagnose a patient with smallpox, which had theoretically been "eradicated."

The meeting was scheduled for the Cabinet Room in the White House. President Clinton apologized in advance for seeming a little slow that morning. He had been up most of the night mediating between Northern Ireland and the Irish Republican Army to avoid the collapse of the peace process and had gotten only an hour or two of sleep. Behind the experts was a ring of chairs occupied by cabinet officials and their top aides—among them Defense Secretary Cohen, Attorney General Janet Reno, Health and Human Services Secretary Donna E. Shalala, Director of Central Intelligence George J. Tenet, National Security Adviser Sandy Berger, and Dick Clarke.

According to notes of the meeting provided by a participant and interviews with all the experts, Clinton told the group that he had become increasingly worried about the dangers of the new biological era. The ability to sequence and manipulate genes had profound implications for national security and for traditional strategies aimed at defending the nation against unconventional attacks. Because rogue states and terrorists could now manipulate genes, he said, "offense may be getting ahead of defense." Nuclear weapons were expensive and relatively easy to detect, he said. Biological weapons were neither. Germs were a "low-rent way to be a big player." Iraq's resistance to relinquishing its biological-weapons program, for instance, showed that Saddam Hussein obviously placed a huge value on "not giving it up." Iraq had forfeited tens of billions of dollars in oil revenues to retain its germ-warfare agents and expertise.

Recombinant biotechnology created the possibility not only of new treatments for age-old diseases but also of "new threats," Clinton said. What was their assessment of Richard Preston's book *The Cobra Event*? Was the plot a "forecast of what's in our future"? Was it scientifically plausible? Several of the experts said later that while they personally doubted that any scientist could conduct such a terrorist strike, they stayed silent at the meeting. Others told the president that yes, such an attack was possible. Steering the conversation away from the novel, Lederberg said that any terrorist attack involving an infectious agent, be it naturally derived or genetically modified, would be, in his deliberate understatement, a "very serious event."

Lederberg detailed the importance of scientific breakthroughs in the history of war: the dramatic results when offensive capabilities had outstripped defensive measures. The introduction of iron weapons, for instance, had made bronze weapons ineffective. Gunpowder had made the defensive armor of medieval knights obsolete. Horses and cavalries had enabled Asians to sweep through Europe. And now the offensive capability of germs as bioweapons seemed to be outstripping defenses against them. Germs, he said, were "strategic" weapons that could strike deep into the nation's heartland. Such attacks could be attempted by states, terrorist groups they supported, or lone actors. "Individuals can make war with these new weapons," Lederberg warned. To counter such attacks, antiviral drugs and therapeutics were needed. So were new vaccines based on new technologies.

Shapiro, a colleague of Lederberg's from Stanford, outlined the new possibilities offered by recombinant technology. To illustrate the awesome powers of new genetically manipulated germs, she described to the president how natural selection created new antibiotic-resistant microbes. "What bugs do naturally," she said, was worrisome enough. Federal money was desperately needed to help science combat emerging new diseases, whatever their origin. Shapiro reminded the administration officials of the Russian article on bioengineered anthrax. Much of the government's support for research over the past twenty years had been spent on fighting heart disease and cancer. The NIH, she told Clinton and the others, had not been given sufficient resources to study pathogens and understand how they operated, or to do other basic antibiotic and antiviral research. Microbiology in general had been underfinanced. "The bugs are smarter than we are," she declared, "and the bugs are winning."

Venter pushed hard for federal support for genetic sequencing and gene identification. The techniques were crucial, he argued, to detecting germs and designing drugs against new strains of ancient scourges—tuberculosis, among others. Scientists needed these genetic codes to protect the country. Recombinant technology was rapidly spreading, for good and evil, he warned. Even high school students were now genetically engineering bacteria and viruses.

Hauer, New York's emergency-management director, outlined measures that the city had adopted to respond more effectively to a germ-weapons attack. To detect sudden shifts in the health of its population, New York City now monitored not only daily admissions to hospital emergency rooms and the deaths of young or unlikely victims of mysterious ailments but also less obvious public-health indicators such as sales of Kaopectate and other over-the-counter drugs used to treat diarrhea. He had also designated close to three hundred facilities as "PODs," or points of distribution, for antibiotics and drugs in a crisis. He had selected Madison Square Garden and the Jacob K. Javits Convention Center for emergency-care facilities where people could be treated if the city's hospitals were filled. New York had taught thousands of doctors, nurses, police, firemen, and other emergency personnel what to do and what not to do in responding to a chemical or germ attack. His department was planning to negotiate contingency agreements to buy drugs in huge quantities at affordable prices should the need arise.

And yet, Hauer said, New York was still not prepared to respond to such an emergency. Even the city's army of forty thousand physicians and fifty-four thousand policemen, firemen, and emergency officials would be rapidly overwhelmed by an epidemic. New York and other cities would desperately need federal assistance quickly to control unrest and protect drugstores and hospitals that might be stormed as people sought antibiotics and vaccines for themselves and loved ones. And the federal government was still unable to provide timely help.

Tom Monath, a vaccine specialist who had joined a private vaccine company after a distinguished government career, stressed the need for new vaccines, antibiotics, and other antidotes to respond to infections from classical germs and genetically modified agents. He called Clinton's attention to a touchy issue: the limits on research spurred by an overly strict adherence to the biological-weapons treaty. Many scientists, he asserted, were reluctant to conduct important research because they feared that it could be seen as part of a biological-weapons program, and thus a violation of the treaty. Monath said the government had underfunded vaccine research. The nation needed a vaccine for Ebola, for instance, but even the military said it could not afford to develop one. The private sector would not do this work because it was not profitable.

Barbara Rosenberg, the group's specialist on the treaty banning biological weapons, told Clinton that he would have to place greater priority on negotiating a protocol to enforce the treaty if he wanted to stop the spread of germ weapons. American leadership was desperately needed if progress at the Geneva talks was to be made. After the Russian scientists' disastrous visits to Pfizer in 1994, the American pharmaceutical industry was understandably opposed to demands for routine inspections of its plants, though it accepted the idea of "challenge" inspections, spot checks that would be conducted if a country was charged with specific violations of the treaty. While the drugmakers were worried about the theft of trade secrets, their doubts could be assuaged by meaningful consultation with government officials, Rosenberg argued. Without such arrangements and adequate "verification," she warned, the treaty was toothless.

Clinton frequently interrupted the brief presentations, firing off questions to the experts and his cabinet. How long would it take for the nation to develop the technology needed to quickly identify and diagnose deadly pathogens? Who was doing such research and where? Was such work being coordinated by anyone in the government? In the private sector? If a terrorist "let something loose" inside an airport, how long would it take the government to figure out that the nation had been attacked, and how long after that to identify the agent? And finally, what about money—how much would an adequate defense cost? The cost would be high, the group agreed, adding that it would soon provide the president with specific proposals and cost estimates.

As the meeting ended, Lederberg gave Clinton a copy of the August 1997 issue of the *Journal of the American Medical Association,* which was devoted entirely to the threat of germ weapons and bioterrorism. Lederberg had edited the issue and written its lead editorial. The issue also featured an article by CDC scientists that described for the first time the full epidemiological story behind the Rajneeshees' salmonella attack in Oregon. Thirteen years after the event, in the wake of the sarin gas incident in Tokyo, the investigators had finally decided to give a fuller account of the incident. The benefits of wider public discussion, they felt, now outweighed the risks of inspiring a copycat. Clinton read the articles and circulated the scholarly journal to his staff a few days later with his handwritten notes scrawled in the margins. Told of Clinton's attention by a White House staffer, Lederberg smiled: it was surely the first time a president had ever read an entire issue of the medical journal.

WHILE President Clinton told his staff that he found the meeting very useful, the session was not without its critics. Within days, experts who had not been invited complained that several participants had used the gathering to promote personal agendas.

Venter's call for increased spending on gene sequencing drew some heat. His nonprofit institute was already receiving millions of dollars in federal support, and some viewed his presentation on the value of his work as special pleading. Venter dismissed the criticism, saying that his federal money came through grants that were competitively awarded by expert panels on the basis of scientific merit, not lobbying. His institute, for example, had recently won a grant to sequence the cholera microbe, a deadly pathogen.

Venter did have a broader, commercial interest in the techniques he was developing. When the experts gathered at the White House, he was already preparing a for-profit venture, Celera. The new company planned to beat a federally funded group in the race to decode the human genome. Venter wanted his company to be among the first to realize the commercial potential of that breakthrough.

Participants in the meeting and some White House officials said they were also troubled by Tom Monath's potential conflict of interest. Monath was widely admired within scientific and government circles for his work at both the army lab at Fort Detrick and the Centers for Disease Control and Prevention. As a tropical disease specialist, he had repeatedly risked his life fighting epidemics around the world. But most of those present said they had no idea that his struggling vaccine company, OraVax, was trying to win a Pentagon subcontract to make smallpox vaccine. After a *New York Times* story disclosed his financial interest, colleagues rushed to Monath's defense. Monath, they said, had devoted his life to public service. The White House meeting was about domestic defense and Monath's company was seeking a contract to make vaccine for the military, not civilians. Monath said that he had disclosed his company's financial interest to White House officials.

On May 6, Frank Young, the moderator of the experts group, sent the president a seventeen-page report summarizing its unanimous recommendations. The report, which was not made public, concluded that all levels of government were "woefully unprepared" for a bioweapons attack. It recommended focusing on two areas: the use of new genetic-engineering techniques to develop better antibiotics and vaccines, and a plan for reducing civilian casualties if a dangerous pathogen was deliberately released. The document included a separate section suggesting ways in which the proposed protocol to the treaty banning germ weapons should be strengthened. Finally, it urged the president to establish an "advisory committee" to set priorities and foster cooperation among the many government fiefdoms with sway over biodefense and public health.

The experts had been unsure how much money to request. At a gathering in a cloakroom just before the meeting, Young had said the group dared not ask for more than $100 million a year, no matter how grave a threat they painted. Lederberg had strongly disagreed. It was not worth even taking up the president's time for that amount of money, he had argued. Moreover, $100 million annually would not begin to address the problem. Lederberg said the group should propose spending about a half-billion dollars a year for the next four to five years. That was a minimum, he asserted.

Frank Young was staggered by the estimate. A veteran of many a federal budget battle, Young said the sum was too large. Others agreed. But Lederberg insisted on the larger figure. In Washington, serious business was defined by serious money.

The document Young submitted to the White House reflected Lederberg's thinking. It called for more than $1.9 billion in spending over the next five years, with a total of $420 million for national stockpiles of antibiotics and vaccines. This would be enough to provide 40 million doses of smallpox vaccine, which would reduce death and illness from a smallpox epidemic by "10 to 100 fold." The group called upon the government to invest $860 million in research to develop new antibiotics, antivirals, and vaccines; and to increase spending on the study and sequencing of genes, the molecular blueprints of microbes, and on diagnostic equipment to detect dangerous germs. Finally, the panel urged the president to spend $1.04 billion over five years to improve emergency responses to biological attack, with at least $170 million for the dilapidated public-health system.

One of the most politically delicate sections of the report had nothing to do with money. The document noted, for instance, that many of the experiments needed to study genetically manipulated organisms and to develop antibiotics and vaccines for them "would not be permitted by recombinant DNA review board and biosafety regulations in the United States." If such research were to be done anyway, and the results were published, the research might well "raise alarm on the part of other nations (as did the Russian report on anthrax resistant to vaccine)." It might also provide "recipes," the group warned, to those "interested in using genetic engineering for nefarious purposes."

The experts endorsed such "defensive research" and urged that it be conducted under the guidance of "an objective review board composed of civilian scientists and public health specialists." Such a panel could oversee Pentagon research and programs "that would otherwise not be approvable under current regulations." The report did not address the issue of how this could be done without tipping off possible adversaries to America's vulnerabilities. Without directly saying so, the panel seemed to be arguing that the existing restrictions on genetic research should be sidestepped in the interests of national security. Dangerous new recombinant organisms had to be studied, made, tested—even if such work pushed to the limits of what the germ warfare treaty allowed in the name of defense.

The results of the presidential audience were somewhat less than the experts had hoped. A little more than 25 percent of the funds Young requested was initially approved for the first year. To be sure, Clinton's rhetoric changed measurably. In May 1998, the president traveled to the U.S. Naval Academy in Annapolis to present his plan for germ defense. "We must do more to protect our people," Clinton said. His speech touched all the bases. It urged the creation of a stockpile of vaccines and antibiotics for the first time in the nation's history. It said strengthening the treaty banning germ weapons was a "major priority." It committed money to support basic research into deadly pathogens. It reflected Venter's influence by stressing the potential importance of decoding the human genome in designing new antibiotics. And it promised money to upgrade the nation's public-health clinics "to aid our preparedness against terrorism and to help us cope with infectious diseases that arise in nature." "We must not cede the cutting edge of biotechnology to those who would do us harm," the president told the cheering cadets.

The president also signed a new executive order, Presidential Decision Directive 62, creating a White House office to coordinate the government's programs in "security, infrastructure protection, and counterterrorism." The job of counterterrorism czar went to Clarke, greatly expanding his responsibilities.

But the applause of the naval cadets had barely subsided when the resistance began—at home and abroad. Faced with unanticipated skepticism about the president's sweeping initiatives from both its own agencies and outsiders, the administration began trimming its pledges. In June, Clinton asked Congress for $294 million to finance all of the experts' proposals. While the group recommended spending $200 million to build a stockpile of vaccines and antibiotics, the White House initially requested $41 million. Less than a month later, the White House decided against spending any money at all to buy vaccines that year. More effective, cheaper vaccines might be available in two to three years, administration officials said. It was better to wait.

THE Pentagon began vaccinating the troops in the Middle East against anthrax in March 1998, before Cohen gave his final approval to the program. General Zinni, the regional commander, told Washington he saw an imminent threat from Iraq and could not wait any longer. Most servicemen took their shots, but more than two dozen sailors aboard the navy carriers the USS *John C. Stennis* and USS *Independence* refused, fearing the immunizations could harm their health. The navy moved to discipline the sailors for disobeying a direct order, and several asked for help from Mark S. Zaid, a Washington-based lawyer who had made a name for himself challenging the government in national-security cases. Zaid's clients already included Patrick G. Eddington and Robin Eddington, a husband-and-wife team who worked for the CIA and contended that their agency had covered up evidence about Gulf War syndrome. Taking advantage of the new technology available aboard ships, several of the sailors had e-mailed Eddington with their concerns about the vaccine program. He passed their messages to Zaid.

A slender, kinetic young lawyer with a knack for obtaining documents through the Freedom of Information Act, Zaid was a government bureaucrat's worst nightmare. His solo practice was eclectic. He was one of the lawyers suing Libya for the bombing of Pan Am flight 103 in 1988, and he had a continuing interest in the assassination of John F. Kennedy. He had amassed a thick file of reporters' phone numbers and had learned how to frame a story to give it broad appeal. When the first sailors began to call, Zaid, barely thirty, encouraged their mothers to be interviewed by the press. "The generals can yell at me, but they can't yell at them," he later explained.

At first Zaid saw little grounds for a legal challenge. The anthrax vaccine appeared to be fully approved by the FDA and the navy's orders to take it seemed proper. Still, Zaid felt he needed to learn a great deal more. He filed a lawsuit under the Freedom of Information Act for every document relating to the program. Government programs, he knew, usually looked different from the inside.

Pentagon officials were optimistic about the prospects for the program and for the Michigan vaccine facility. It had finally been sold to BioPort, a new company whose partners included Admiral William J. Crowe Jr., the former chairman of the Joint Chiefs of Staff. BioPort did have some experience in the vaccine field. Its chief executive, Fuad El-Hibri, had already converted a British government vaccine plant into a successful private venture. Still, there were some worrying signs about the Michigan lab's prospects. A February 1998 inspection by the FDA reported that while some of the deficiencies cited in the scathing 1996 review had been corrected, numerous problems remained. The inspectors found shortcomings in almost every phase of the production process, suggesting that BioPort would have to come a long way before the regulators would be satisfied.

In May 1998, two months after the military began giving the shots to the troops in the Middle East, Cohen gave his final blessing. The military began inoculating active duty troops around the world in August. Reservists were included in the initial round of vaccinations but only if they spent more than thirty days in a "high threat" area. Air national guard pilots, who handle a significant number of the military's cargo flights into the Middle East while on weekend duty, were largely exempt from the shots since they only spend a few days in the region.

By the summer of 1998, Pentagon officials were dealing with a new potentially controversial question: How to structure the command for the guard and other military units that would respond to a chemical or biological attack within the United States. Hamre told NATO officials in late June that the Pentagon was considering whether to appoint a new regional commander to direct "homeland defense" within America's borders. "We don't believe we have primary responsibility, but within minutes of an event, people are going to turn to us,"

Hamre said. "If there is a biological attack, you can easily see regional governors calling out the national guard to quarantine the highways. It could get crazy very fast." Pentagon officials saw a host of valuable roles for the guard or reserve forces in the aftermath of a germ or chemical attack, from pitching tents to setting up field hospitals to burying the dead.

The proposal alarmed civil liberties experts. Gregory T. Nojeim, legislative counsel on national security for the American Civil Liberties Union in Washington, felt the idea had been poorly thought through and posed a potentially serious danger to American freedom. A germ or chemical attack by terrorists would unleash chaos, and military units working with local authorities would be quickly pressed to maintain order, a mission for which they were not trained. Soon enough, soldiers would be arresting civilians. The United States has a long, proud history of restraining military operations on American soil. The Posse Comitatus Act, passed after the Civil War, bars federal troops from doing police work within U.S. borders. The *New York Times* editorialized against the plan for establishing a new command, saying it "would erode the long-established legal principle that America's armed forces should not be involved in domestic law enforcement."

Pentagon officials said they did not want to supplant local police, and they pointed to the federal disaster-response plans drawn up in 1995 which called for the Federal Emergency Management Agency to take the lead in managing the consequences of a terrorist attack. Hamre invited Nojeim to lunch at the Pentagon. The military, he said, would support FEMA and had no intention of overstepping bounds so deeply rooted in law and tradition.

Reacting to the criticism, the Pentagon scaled back the plan for homeland defense, reserving the command for a two-star general drawn from the National Guard. Hamre felt the change would make a big difference. Four-star generals, he had seen, were a breed apart, treated like demigods by their subordinates and accustomed to giving orders that would be obeyed without question. A citizen-soldier from the guard would be much better equipped to understand America's civilian, civil-liberties culture.

Nojeim was not mollified. Changing the rank of the commanding officer did not settle the question of how, precisely, the government would prevent the military from seizing control in the panicked hours after a terrorist assault. Popular fears about the issue were fanned by the movie *The Siege,* which came out in the fall of 1998. In the film, New York is attacked by a group of Arab terrorists and a particularly obnoxious military officer, played by Bruce Willis, imposes martial law, rounds up American citizens, and puts them in barbed-wire-fenced camps. Pentagon officials dismissed the movie as utter fantasy, but it had resonance.

COLD rain came down in torrents at the White House on January 21, 1999, as President Clinton discussed the threat that he said kept him awake at night. Not even nuclear or chemical weapons were as frightening as germ warfare, he told the *New York Times* reporters who had been invited to the Oval Office for the first interview he had given to a major newspaper in a year, since the Monica Lewinsky scandal had erupted. While a chemical attack would be horrible, he said, it would be "finite." But a biological attack would spread through the land, particularly if its perpetrators unleashed a contagious agent that was not quickly identified and properly treated. A germ attack would be, as he put it, "the gift that keeps on giving."

In deciding to speak publicly about what he considered a growing danger to the United States, Clinton said he was trying to strike a balance between educating people and throwing them into unnecessary panic. He wanted Americans "to know what they need to know and have a realistic view of this, not to be afraid or asleep. I think that's the trick." But he added that a serious attack or the threat of one "is highly likely" in the next few years, which was why the government was erecting defenses. "We'll just deliberately work on this, and do the very best we can."

If Clinton was preoccupied with the ongoing scandal that threatened his presidency, he showed no sign of it. Seated in a large wing chair in front of a fireplace, a portrait of George Washington staring down at him serenely, Clinton seemed relaxed but focused on the interview's agenda. The performance was impressive—perhaps even surreal—since at that very moment Clinton's lawyers were defending him at Senate impeachment proceedings on Capitol Hill. The White House itself was quieter than usual; most reporters and cameramen had flocked to the Hill for the Senate debate.

Clinton seemed eager to discuss something other than the Lewinsky affair, every detail of which was being dissected on television and in the halls of Congress. He had clearly studied the biological issue and, echoing the views of Venter and Lederberg, said he saw germ weapons as posing a unique threat to the nation's security. Defending against such dangers, he said, would tax the resources and creative powers of the military, law enforcement agencies, scientific research institutes, and the nation's ailing public-health system. But it could—and would—be done. He recalled that as a boy, during the Cold War, he had lived in a world where protective shelters against radioactive fallout were common "and every school had its drills." Those steps were sensible at the time, he said, and now new threats to the nation demanded a new kind of civil defense. "What I want us to do is everything within reason we can to minimize our exposure and risks here." For the American people, he said, this push "shouldn't be a cause for alarm" but reassurance. "They should want us to be well organized on these things."

Yes, he said, the collapse of the Soviet Union was a blessing, but it was also a challenge. During the Cold War, the Soviets and Americans had been giants in straitjackets, deterred from using their unconventional weapons for fear of retaliation. But new enemies could now strike the United States with bioweapons without identifying themselves. People who dreamed of being martyrs were "more likely than the Soviets" to strike out at Americans. And deterrence would be more complicated if Russia's impoverished scientists sold their expertise or lethal germs to people who were "more likely than the Soviets ever were to use it." Osama bin Laden, the renegade Saudi sponsor of anti-American terrorism, had tried hard to acquire chemical weapons and "may have" tried to get germ weapons as well.

The terrorism Clinton had already witnessed in his presidency, the growing number of hoaxes in which people claimed to have dispersed anthrax (there had been twelve the previous month in California alone), and the numerous intelligence reports he had read had convinced him that it was "highly likely" that a terrorist group would launch or threaten a germ or chemical attack on American soil within the "next few years." He said his administration would have to do more to bolster the nation's biodefenses.

At the same time, he said he worried about the threat to civil liberties inherent in expanding the government's authority through its antiterrorism programs. Noting that America was founded by people who had been abused not by someone else's government but by their own, he stressed his determination to safeguard individual freedom and civil liberties while strengthening the nation's defenses.

Ultimately, he added, America's best defenses against unconventional warfare, and bioterrorism in particular, were scientific advances in deciphering genetic material in microbes and people so that vaccines could be tailored for quick response to an attack. This would allow the defense to stay ahead of the offense, he said. The Human Genome Project, the costly federally financed effort to map out human genetic material, was extremely valuable, an important part of the defensive shield he was trying to build. Once the secrets of the human genome were unlocked, Clinton predicted that one day scientists would be able to "take a blood sample, and there would be a computer program which would show us if we had—let's say we had a variant of anthrax."

The president had clearly been impressed, perhaps too much, by Craig Venter's enthusiastic predictions about the antibiotic and biodefense wonders that would miraculously flow from deciphering the human genome. But he hoped that America would use "each new wave of technology to close the gap between offense and defense"—an optimism not widely shared by the government's military analysts and researchers.

Until such rapid detection and response were possible, the United States would have to concentrate on deterring attackers armed with biological weapons. Alluding vaguely to unheralded but successful actions his administration had taken to delay efforts by rogue states and terrorists to acquire germ weapons, he warned that the use of such a weapon against Americans, provided that Washington knew who had used it, would prompt "at least a proportionate, if not a disproportionate response." It was unclear what he had in mind, and Clinton declined to elaborate, seeing virtue, as had his predecessor, George Bush, in leaving such threats vague.

Clinton described elements of a proposal that he would announce the next day to strengthen defenses against terrorists. Although he did not say so directly, the president seemed to be suggesting that his initiatives to combat terrorism and strengthen America's biodefenses were more important—and ultimately should figure larger in his legacy—than an affair with a young White House intern. Granting the interview was surely part of that strategy.

When his aides told him that the thirty minutes scheduled for the meeting had elapsed, Clinton deflected a question about what kind of toll the Lewinsky scandal was taking on his family. He thought the worst was over, he said. Other than refusing to say whether he had been vaccinated against anthrax—he had not been, though the Secret Service had asked him not to discuss this—the scandal-related question was one of the few he sidestepped during the interview. But even as his White House handlers insisted that it was time to go and other aides, who had been discreetly taping the interview, switched off their recorders and slipped out of the room, Clinton kept on talking.

He dismissed those who argued that his administration had exaggerated the threat of bioterrorism. He said Anthony Lake, his former national security adviser and an early advocate of biodefense, had once told him only half jokingly that the administration's proudest accomplishments would be preventing bad things from happening—the "dog that didn't bark," he said. The $10 billion that he would seek from Congress to counter all forms of terrorism would be money well spent, despite what detractors and skeptics of the new, exotic twenty-first-century threats might say. And if there were no terrible incidents, "nobody would be happier than me twenty years from now," he said. His critics would probably say, " 'Oh, see, Clinton was a kook, nothing happened,' " he continued. "I would be the happiest man on earth."

THE following day, President Clinton announced his decision to ask Congress for $2.8 billion to thwart and prepare for attacks with biological and other exotic weapons and to combat computer-warfare threats. The money for fiscal year 2001, more than double what he had sought the year before, was part of an overall request of $10 billion to defend the nation against terrorism.

With Josh Lederberg at his side, Clinton outlined the major elements of his expanded antiterrorism program in a speech to the National Academy of Sciences, a decidedly friendly audience. Members of his cabinet, senior military and prominent civilian scientists, and scores of officials from emergency-preparedness agencies applauded and cheered as the president defended the program to the three hundred–some people who had packed the academy auditorium.

The initiatives included $683 million to train emergency workers in American cities to cope with a chemical or biological attack; $206 million to protect government sites; and $381 million for research on dangerous pathogens, the development of new vaccines and therapies, technology to detect and diagnose rare illnesses, and decontamination.

The White House also asked for $87 million, a 23 percent increase over the previous year's spending, to improve the nation's public-health surveillance system, and $52 million to continue building a national stockpile of antibiotics and medicines against anthrax, smallpox, and pneumonic plague. Clinton also wanted to triple current spending—to a total of $24 million—to create twenty-five urban medical-response teams in major American cities. Only four cities—Miami, Denver, Washington, and Charlotte, North Carolina—had been able to assemble and train such teams.

While the bulk of the $10 billion request would be spent strengthening security at U.S. embassies and at other American facilities, public-health officials were delighted. For the "first time in American history," said Donna Shalala, the secretary for health and human services, the public-health system was being integrated into national-security planning.

Lederberg, too, had gotten some of the money he wanted for basic research into dangerous pathogens. In remarks introducing the president and the scientist to the audience, Sandy Berger, the president's national security adviser, noted that Lederberg had secured his place in scientific history by winning the Nobel Prize when he was only thirty-three. But the president hadn't done badly: Clinton was already governor of Arkansas by that age. When Clinton rose to speak, he told the audience that it could infer from his decision to enter politics that he had been no good at chemistry and biology. The easy banter on the stage that day between the president and "Josh" sent a powerful signal to the federal bureaucracy: Lederberg and his fellow advocates of biodefense had influence at the White House that could no longer be ignored.

THE first major test of the biodefenders' clout came only a few months later, when the administration debated whether to destroy the world's last known remaining stocks of the smallpox virus.

The destruction of smallpox seemed an easy call. Eradication of smallpox had been one of the great achievements in the history of public health. The last known outbreak of the disease had occurred in

Somalia in 1977, and three years later the World Health Organization had formally declared the disease eradicated. In 1996, WHO's governing body had recommended the destruction of the last remaining stocks, which were under tight security at two places—the Centers for Disease Control and Prevention, in Atlanta, and Vector, Lev Sandakhchiev's institute in Siberia. The target date for destruction was June 2000.

But Russian and American intelligence analysts believed that other countries maintained hidden stocks of the virus. Sandakhchiev had said he was certain that North Korea, among other countries, was secretly keeping smallpox, and American analysts shared his view.

Pentagon officials argued that a public ceremony announcing the elimination of the smallpox virus would be a charade. First, WHO had had no means to determine whether all countries had carried out their pledges to destroy existing stocks or transfer the virus to WHO repositories. It had simply accepted national declarations. Second, Pentagon officials believed that several nations, including American allies, had also secretly decided to maintain clandestine stocks. Even if every country was smallpox free, the deadly virus might be recreated from related microbes if a country wanted to make new stocks. Smallpox, they argued, was simply too tempting a weapon to abandon.

Finally, they argued, destroying the Russian and American stocks would complicate efforts to develop a new vaccine and antiviral drugs against the disease. America's population, which had not been vaccinated against smallpox since 1972, was particularly vulnerable. If the contagious virus was used in a terrorist attack, one expert said, "we are all Indians," a reference to the devastating effects smallpox had on the Native American populations after European settlers had arrived.

President Clinton was under heavy political pressure from arms-control advocates and other largely Democratic constituencies to proceed with the destruction of the virus anyway, as a "moral" statement and a political example to the world. Respected scientists such as Donald A. Henderson, a former White House science adviser who had led the campaign to destroy the virus, argued that scientists did not need the actual virus to develop a new vaccine. His view was finding a sympathetic audience among many administration officials and in many developing countries.

Lederberg and other opponents of destruction realized that they would have to marshal compelling arguments in order to secure a stay of execution for the virus. Fortunately for their cause, they had an ally in the White House—an effective bureaucrat with impeccable scientific credentials who happened to head an office in the National Security Council that Clinton had established to evaluate issues of public health and national security.

For Kenneth Bernard, a physician who had served as assistant surgeon general before assuming the new White House post, the correct position was a "no-brainer," as he had told friends in the biodefense community. In addition to concerns about Iraq and North Korea, Bernard knew that Ken Alibek had warned that the Soviets had weaponized smallpox and that the Russians might still be keeping genetically modified stocks for use in weapons. Of course the destruction of the virus had to be postponed, Bernard believed. But building the case would not be easy. After all, the administration had initially favored the destruction in 1996 and now would be reversing itself.

Realizing that Clinton would want to rely on expert scientific opinion and knowing that one of the nation's leading smallpox experts, Donald Henderson, fervently favored destruction, Bernard asked the National Academy of Sciences to form a committee to study the issue. Rather than ask the panel whether the virus should be kept or destroyed, Bernard persuaded the academy to explore whether critical treatment advances would be lost if the virus were destroyed.

The framing of the issue proved decisive. In March 1999, the panel concluded that such opportunities might indeed be sacrificed if the virus were eliminated. Proponents of destruction were furious, knowing that they were surely likely to lose the argument, at least for now. A month later, Clinton announced that the United States would fight to delay destruction when the WHO assembly met in Geneva in May. With Russia and the United States in agreement, the world reluctantly followed. WHO decided to postpone destroying the stocks until at least 2002. Bernard, Lederberg, and their allies had won an important round.

EARLY in 1999, Bill Patrick traveled to Alabama to address officers at the Air Force Counterproliferation Center, located at Maxwell Air Force Base. He described how terrorists use germ weapons. Near the end, Patrick flashed on the screen two slides. They compared American and Soviet industrial capacities for germ production at their peak levels and revealed how Moscow's was extraordinarily large. Washington's germ agents were more refined, he noted, and its munitions were more efficient at dispersing germs. Still, what America had done was more modest.

Comparison of Dry Agent Production (metric tons per year)

	UNITED STATES	SOVIET UNION
staphylococcal enterotoxin B	1.9	0
F. tularensis (tularemia)	1.6	1,500
Coxiella burnetii (Q fever)	1.1	0
B. antracis (anthrax)	0.9	4,500
Venezuelan equine encephalitis virus	0.8	150
botulinum	0.2	0
Yersinia pestis (bubonic plague)	0	1,500
variola virus (smallpox)	0	100
Actinobacillus mallei (glanders)	0	2,000
Marburg virus	0	250

"Look down at the bottom of the chart," he said. "Look at these killers—plague, glanders, smallpox, Marburg—all very lethal." He expressed disbelief at the numbers, calling them unreal. "What in the name of God are they going to do with 4,500 metric tons of anthrax?" Patrick asked. "Look at smallpox. We didn't weaponize smallpox, but they had a hundred metric tons."

SMALLPOX was very much on Andy Weber's mind in the fall of 1999. He was attending a party in Koltsovo for the twenty-fifth anniversary of Vector, Russia's official smallpox repository, which had once specialized in turning viruses into weapons. The daylong festival was held on the once secret Siberian city's soccer field. The occasion was marked by tributes to and speeches commemorating the institute's great but unmentionable achievements, as well as by parades by Vector veterans, displays of ballroom dancing by the children of the institute's two thousand scientists and technicians, and demonstrations of self-defense skills by Russian special forces.

After lunch on a bitterly cold day, Weber and an American friend accompanied Lev Sandakhchiev, the institute's director, and some of his Russian colleagues to the town's cemetery. Sandakhchiev raised a plastic cup of vodka to the memory of Nikolai Ustinov, a scientist who had died in the spring of 1988 after accidentally infecting himself with the Marburg virus, a hemorrhagic killer that he and his colleagues had been trying to perfect as a weapon. Standing beside Ustinov's widow, Yevgenia, at the foot of her husband's still-contaminated grave, Sandakhchiev toasted their fallen colleague and friend, who had documented the progress of the disease in a diary stained with his own blood. Sandakhchiev then emptied the remaining contents of his cup on his friend's grave, whose zinc-lined coffin was now covered with flowers and grass. Vodka, Weber thought, as he stared at the ground. How many Russians, and germ-warfare scientists in particular, had drowned their misery and whatever moral compunctions they had about their work in vodka?

There was no plaque or public tribute to Ustinov, no acknowledgment of how he had died. But Ustinov had achieved a perverse, Soviet-style immortality. After his death, his colleagues at Vector had cultured the virus that killed him. The virus had mutated while passing through Ustinov's body, and the new variant, according to Ken Alibek, was particularly virulent and had been weaponized as a replacement for the original. Vector had called it "Variant U."

Sandakhchiev turned away, tears streaming down his face. "Thank God it is over," he said.

But was it really? Weber wondered.

11

Defenders

NEAR the end of a hot, dry summer, on August 23, 1999, Deborah S. Asnis, an infectious-disease specialist at Flushing Hospital Medical Center in Queens, called the New York City health department with some distressing news: two elderly patients had come in with symptoms that seemed to be caused by a neurological illness—fever, muscle weakness, and confusion. Four days later, she called again. Now there were four patients with similar symptoms. Though each of the patients had seen different doctors and one was at a different hospital, Asnis found their complaints disturbingly similar, suggesting the possibility of a dangerous outbreak. Marcelle Layton, who headed the New York City Health Department's Bureau of Communicable Diseases, told Asnis to send samples for possible testing to the State Department of Health. Layton also called the CDC in Atlanta.

By the week's end, eight cases of the mysterious illness were identified. Layton and her medical assistant who had interviewed the victims' families were struck by the fact that all of them lived within a sixteen-square-mile area of northern Queens and had spent time outdoors in the evening—prime mosquito time. While the symptoms were consistent with viral encephalitis, a disease that mosquitos can carry that causes inflammation of the brain, the severe muscle weakness afflicting the patients was unusual for that ailment.

New York City was considered among the best prepared in the

country to detect and deal with a germ outbreak. After the 1993 bombing of the World Trade Center, Mayor Rudolph W. Giuliani had invested heavily in counterterrorism. New York was among the first cities to receive federal funds for training emergency-response officials against attacks involving unconventional weapons. The city had spent more than a million dollars on two mobile vans equipped to identify dangerous chemicals, but not germs, in the event of an attack. While most cities could barely scrape together tens of thousands of dollars for such precautions, New York had already spent millions a year building emergency-response units, working out deals with regional hospitals for emergency care.

Jerome Hauer, recruited by the mayor in 1995 to head a new office aimed at strengthening the city's defenses, had worked closely with the city's public-health department to enhance the city's ability to respond to an unconventional attack. Hauer was especially concerned about the potential impact of a bioterrorism attack on New York, whose twenty-four thousand people per square mile made it twice as crowded as the country's second most densely populated city, Chicago. A fifty-year-old former emergency-preparedness chief in Indianapolis who held a master's degree in public health, Hauer was one of the experts who had briefed President Clinton in 1998 on what steps the government should take to defend against biological attack.

Soon after taking over the new Office of Emergency Management, Hauer had created a computerized system to track admissions to eleven city hospitals. New York had also begun training emergency-room personnel to look for clusters of symptoms that could be caused by dissemination of anthrax or other germ weapons.

In August 1999, Asnis of the Flushing Medical Center was performing exactly as Hauer had envisaged, aggressively investigating an unusual pattern of disease.

Serum samples from Queens were forwarded to the CDC's facility in Fort Collins, Colorado, for more sophisticated testing. Duane Gubler, an expert on disease spread by insects, and his scientists suspected that the cause of the illness might be Saint Louis encephalitis, or SLE, which is often found in the southeastern United States. After checking the samples for antibodies against six viruses transmitted by insects commonly found in the United States, their tests came back positive for SLE. The New York City Department of Health and the CDC announced on September 3 that their tests of blood and spinal fluid had supported the lab finding. The city's press release stated that the death of one elderly patient and the illness of two others in Queens were "confirmed to be associated with St. Louis encephalitis, a viral disease transmitted by mosquitoes."

Hauer announced the start of a $6 million pesticide-spraying campaign to eradicate the city's mosquitoes. While there were objections to spraying large parts of the city with malathion, a potent pesticide, Hauer had little choice. New York had not experienced a major epidemic of mosquito-borne diseases in the twentieth century, and the city, unlike neighboring New Jersey and Connecticut, had ended its mosquito surveillance and control program in the late 1980s. The city decided to spray large areas to kill the mosquito population quickly. There is no known vaccine or cure for SLE.

In addition, Hauer's office quietly cornered the nation's supply of insect repellent. Some 400,000 cans were distributed by an army of five hundred city employees at neighborhood firehouses and police precincts. His office and the city's health department wrote and printed 250,000 brochures about the disease in eight languages and put information about it on the city's public-health Web site. Health alerts and frequent updates were sent on a broadcast fax system to the city's hospitals. The mayor, Hauer, and public-health officials held daily press conferences.

City officials had reason to be pleased with their performance and relieved that the epidemic seemed relatively mild. By the time that fall's cooler temperatures ensured at least a temporary end to the epidemic, sixty-two people had become ill from the virus, most of them elderly, and only seven of them had died. Thanks to Asnis's rapid reporting and the health department's equally swift response, a diagnosis had been made in only eleven days—record time.

But there was, it turned out, a significant problem: the diagnosis was wrong. Clues pointing to the error came from an unlikely source. Since mid-August, wildlife veterinary pathologists at the Bronx Zoo had been worrying about a serious, but what seemed to be a separate, biological event—the death of large numbers of crows and other birds. Four days after the city's Saint Louis encephalitis announcement, several exotic birds at the zoo were dead. But not the emus. And that made Tracey S. McNamara nervous. McNamara, the head of the zoo's pathology department, knew that SLE, which is carried by birds and transmitted through mosquitoes, does not normally cause disease in the birds that are its hosts.

True, the dead birds had hemorrhages and inflammation in the brain, symptoms similar to those produced in people ill from SLE. But the birds also had badly damaged hearts and other affected organs. She considered whether influenza or another common virus might be the culprit, as they caused similar symptoms in birds, but the pathology did not fit. Having taken a course in recognizing foreign animal diseases at the U.S. Department of Agriculture's animal lab at Plum Island, New York, she knew that both those diseases would be affecting the zoo's domestic poultry, but the chickens were thriving. She also rejected the notion that mosquito-borne Eastern equine encephalitis could be responsible, since EEE caused a different kind of inflammation and lesions in other parts of the brain than what she was seeing in the birds piled high in her cooler. Moreover, the zoo's flock of emus, an Australian ostrich which is exquisitely sensitive to the EEE virus, was unaffected. So something more complicated had to be the cause.

Convinced that the avian and human deaths were connected, McNamara sought help. Public-health officials in New York later noted that she did not contact the public-health task force, some of whose members were also uneasy with the initial diagnosis. McNamara said she had decided not to call the task force because it had no veterinarians and because the group lacked a secure laboratory in which her dead zoo bird and crow samples could be tested. Instead, she sent off her samples to the U.S. Agriculture Department's National Veterinary Service Lab in Ames, Iowa. She also called the CDC to alert its scientists to the possibility that the birds' deaths and the human epidemic were linked. The CDC's lab at Fort Collins, overwhelmed with human samples from New York, dismissed her suggestion as unlikely and initially refused even to analyze her bird tissues. After all, birds were the natural hosts for flaviviruses, a group of pathogens known to cause encephalitis in people. Simultaneous outbreaks in humans and birds had never before been reported. Undeterred, she sent them out to Colorado anyway.

McNamara was not alone in her convictions. Ward Stone, the chief wildlife biologist for the State Department of Environmental Conservation, had alerted his veterinary counterparts in New Jersey and Connecticut to the strange illness that had struck the local bird population. But throughout the late summer, the investigations of the bird and people illnesses seemed to be proceeding along mostly independent tracks. While the CDC eventually agreed to analyze some of her samples, McNamara complained that agency workers had not returned her daily calls for almost a week. Struggling to diagnose the mysterious ailment that had appeared in New York, the doctors and scientists who focused on humans were slow to grasp the possible relevance of the birds' illness. In the federal government's planning for how a city might detect and handle a sudden outbreak of disease, officials had foreseen many different scenarios. No one, however, had anticipated an outbreak in which crucial evidence would be uncovered by a wildlife specialist.

When McNamara finally heard from the USDA lab in Ames, the news was alarming: the virologists could not identify the virus that was killing the birds. They had determined that the ailment was definitely some kind of flavivirus. But this was not a virus that veterinarians had ever seen before. Because no flavivirus had ever been known to cause disease in animals, neither the Ames lab nor any other veterinary lab in the country had the testing material needed to pin down a specific diagnosis. Only the CDC labs and other public-health facilities that did human testing had such reagents and chemicals. McNamara sent more samples of animal tissue to the CDC. With more than a dozen birds dead and more than one hundred suspected cases of SLE reported, she also called the U.S. Army Medical Research Institute of Infectious Diseases, looking for a favor from an army pathologist she knew. Explaining her theory that the human and bird deaths were linked, she asked him to examine her bird samples.

The army lab at Fort Detrick, Maryland, where Bill Patrick had once turned germs into weapons and which was now devoted to biodefense research for the military, did not normally take on civilian assignments. But the pathologist agreed, and two days after receiving McNamara's samples, the lab ruled out SLE as a source of the illness. An isolate of the virus that had been made by the USDA, along with more Bronx Zoo samples, enabled the lab to run other tests. On September 24, Fort Detrick and the CDC confirmed that a "West Nile–like virus" had been found in the bird samples.

Tipped off by the bird findings, the CDC lab at Fort Collins used similar genetic fingerprinting techniques to analyze the human samples. At about the same time, tests were also being conducted by Ian Lipkin, the director of a University of California–Irvine laboratory devoted to studying emerging diseases; he was in New York as a visiting academic researcher whom state officials had asked for help. Lipkin told the New York State and New York City health departments that the West Nile encephalitis virus had also infected the humans. On

September 27, the CDC officially confirmed that a "West Nile–like virus" was responsible for the infections in humans as well as in the birds. The finding was astonishing: West Nile had never caused an epidemic in the Western Hemisphere.

The discovery unnerved many biodefense scientists and government officials who saw the New York outbreak as a dress rehearsal for a terrorist attack, a test of how well public-health officials would cope with the sudden spread of a disease not typically found in the United States. Many felt that the summer's events highlighted several weaknesses in the nation's network for detecting emerging diseases.

The public-health system had performed well in many respects. An alert physician had reported the cluster of mysterious illness to the authorities, who, in turn, had quickly identified the virus's likely vector as mosquitoes. Marci Layton noted that had Asnis not done so, it was not clear "when or if this outbreak would have been detected."

The city had also taken swift action to stop the virus's spread, despite the initial misdiagnosis. The spraying of insecticide had contained the outbreak of the globe-hopping virus among humans. Because of the city's investment in bioterrorism defense, local public-health and emergency-preparedness officials knew one another and had worked together fairly well. As Frank Young, the former HHS official, had observed, the site of a disaster was "not the time or place for Government officials to be exchanging business cards."

But the incident revealed some disturbing weaknesses. The daily conference calls involving as many as one hundred officials from eighteen different local, state, and federal agencies—often lasting up to two hours—left officials less time to deal with the crisis. In jurisdictional disputes, it was sometimes unclear which agency in which city, state, or county was in charge.

Although the outbreak was relatively small, a study by the General Accounting Office, Congress's independent auditing arm, concluded that it had severely strained the resources of one of the country's largest, most experienced urban public-health departments. If a relatively modest, naturally occurring epidemic had done that to the public-health system of New York City, what would happen if a broader epidemic erupted elsewhere in the nation?

The medical inquiry had also strained the federal government's limited lab resources. The CDC's lab at Fort Collins had analyzed more than two thousand specimens in three months, using almost all of the lab's capacity. And in 1999, Fort Collins was the only public-health lab in the country able to diagnose the West Nile virus. In its report on the episode, the GAO was unable to determine the number of American labs that could safely handle such dangerous pathogens. But it concluded that the nation desperately needed more of these labs, with even greater capabilities.

The GAO and a Senate committee inquiry also concluded that the CDC had failed to abide by one of its own mottos in disease surveillance and control: "Expect the unexpected." In testing samples from the initial victims, the CDC scientists had screened for only six viruses common in this country; they had not tested for several that have been linked to foreign epidemics or germ warfare, among them West Nile.

American intelligence officials briefly considered the possibility that the West Nile virus had been unleashed against America as part of a biological attack. In April 1999 a man claiming to be Saddam Hussein's former personal driver published a book in Britain that quoted the Iraqi leader as saying he one day might use the virus against his enemies.

Experts eventually discounted the idea that the West Nile outbreak was a germ assault. But scientists were unable to trace precisely how the virus had made its way from the Middle East to North America. The New York variant of West Nile was almost indistinguishable from a virulent strain found on an Israeli goose farm in 1998 and similar to a strain in Russia's collection. Wherever it came from, the virus survived the American winter of 1999 to re-emerge in insects and animals the following summer in a wide swath along the eastern seaboard.

In the months that followed, public-health and national-security officials learned more not only about missteps in detecting and handling the outbreak but also about the mysterious virus itself. Stephen M. Ostroff, the CDC's associate director for epidemiological science, said scientists discovered that some seventy-eight different bird species had died from the lethal virus and that between one and two million birds in the northeast were believed to have been infected. West Nile had also been discovered in rabbits, squirrels, chipmunks, raccoons, bats, and even horses.

While scientists had initially believed that the disease would not spread without mosquitoes, they now understood that birds could transmit it to one another without an intermediary. They also had learned that dead birds were the most reliable predictor of a human outbreak. New York had developed a "dead bird density index" to calibrate responses to the virus. The more birds were found, the more steps public-health officials would take.

In a subsequent review, city health authorities were chagrined to discover that the human outbreak had begun much earlier than they had initially thought. Looking back over hospital records, it became clear that the first case had been seen as early as August 8, more than two weeks before Asnis called in her report of a mysterious illness. Had the first case been recognized, some of the sixty-two people who contracted the disease, including the seven who died, might not have fallen ill. Ostroff said the West Nile epidemic underscored the need for cities to create new systems to monitor symptoms associated with serious infectious diseases. Ostroff and others agreed that the CDC's reluctance to take Tracey McNamara's claims seriously—and the lack of regular, respectful communication between the animal and human public-health communities—had impeded a correct diagnosis. Scientists from both worlds subsequently met for the first time at a conference in Chicago to discuss improved cooperation. McNamara, the animal doc, felt vindicated by the session, relishing the CDC's tribute to what it only half-jokingly called her "dogged" pursuit of the disease.

Most important, the outbreak showed the importance of investing in public health. Much was now being learned and done thanks to a relatively modest increase in public spending. After the disease appeared, Washington allocated $25 million on West Nile–related research and treatment in 2000. Scientists at the University of Texas in Galveston were making progress on adapting a vaccine for West Nile from existing protections against Japanese encephalitis and yellow fever. Researchers at the army lab at Fort Detrick and the CDC were working together on a vaccine for West Nile. Tom Monath, the scientist who had attended the meeting with President Clinton, was helping develop a vaccine at his company, Ora Vax, which was bought by Acambis. "It's remarkable what you can do if you have the resources," Ostroff said.

To White House officials, however, the missteps in tracking the West Nile virus were but one example of the practical difficulties they faced throughout the government as they struggled to turn President Clinton's broad commitment to germ defense into practical reality. Clinton's enthusiasm for the issue had translated into growing budgets for biodefense, and federal agencies were eagerly competing for their share. But no one had overall authority over how the money was spent. Dick Clarke continued to wield considerable power from his post as counterterrorism czar at the White House. But Clarke could not rewrite budgets or bring cabinet officers to heel. Biodefense crossed every imaginable line in the government, sprawling across the Defense Department, Justice Department, Central Intelligence Agency, Department of Health and Human Services, and Federal Emergency Management Agency. It was what one veteran of the Clinton administration would later call an "orphan mission," managed by everyone and no one.

THE Pentagon's effort to vaccinate American troops against anthrax ran into new difficulties in 1999. Mark Zaid, the lawyer who represented some of the first soldiers to refuse the shots, had obtained thousand of pages of documents through his Freedom of Information Act lawsuit. Some undercut the government's categorical assertions about the vaccine's safety and efficacy. Zaid issued press releases about the documents and provided copies to reporters.

At military bases around the country, several hundred soldiers had refused orders to take the vaccine, a tiny percentage of the two hundred thousand servicemen inoculated in the initial phase of the program. The cases were nonetheless high profile, and the military moved to court-martial soldiers who resisted.

In early 1998, Pentagon officials inadvertently broadened the rebellion with a decision they made in the name of fairness. The first round of vaccinations had been given to soldiers permanently stationed in regions that faced the greatest danger of biological attack, such as the Middle East. Reservists serving less than thirty days in such areas were not included. This made no sense to Pentagon officials who had been working for several years to integrate the reserves into military operations. Why should reservists be at greater risk in an anthrax attack? And so the Pentagon issued a new policy. Anyone serving in the Middle East or South Korea, even for a few days, would have to be immunized. The directive angered reservists, particularly pilots in the air national guard units. Many earned a comfortable living flying for commercial airlines. Unlike their active duty colleagues, who faced court-martial for refusing the shots, they could easily quit. Some did. Others complained to their congressmen.

The issue caught the eye of Chris Shays, a Republican congressman from Connecticut. Shays was an odd duck among House Republicans, a former Peace Corps volunteer and New England moderate. He had used his post as chairman of a House subcommittee to hold hearings that pounded the Pentagon for its handling of Gulf War Syndrome. Shays had a good sense of how to turn complicated issues into public dramas; his Gulf War hearings had pitted ailing veterans against officials from the Pentagon and the Veterans Administration. He assigned a longtime aide, Larry Halloran, to investigate the case of an Air National Guardsman from Connecticut who had quit the reserves rather than be immunized. Military officials claimed that the resignation was an isolated incident, but Halloran found eight pilots from his state alone who had left the guard over the issue. His suspicions aroused, he asked the GAO to investigate the program. He also filed a request with the Pentagon for every scrap of paper in its files on the anthrax debate.

The committee chairman to whom Shays reported, Representative Dan Burton of Indiana, was one of President Clinton's most visceral opponents. A darling of the conservative movement and leader of the impeachment drive, Burton never missed a chance to embarrass the Democrats. He was also a persistent critic of childhood vaccinations, questioning whether scientists were overlooking long-term side effects. Burton was delighted to have Shays's subcommittee examine the Pentagon's anthrax vaccination program.

In April 1999, at a hearing before Shays's subcommittee, the GAO announced its preliminary findings, bringing to light some of the complexities that Pentagon experts had been quietly weighing for years. Kwai-Cheung Chan, a GAO auditor, told the congressmen that the long-term safety of the vaccine had not been studied. "Therefore one cannot conclude there are no long-term effects." Animal studies offered conflicting evidence of its efficacy against inhaled anthrax, Chan said, though it appeared the inoculations offered "some protection."

He told the subcommittee that the only human trials of efficacy had involved a similar but not identical version of the vaccine. Another GAO official, Sushil K. Sharma, contradicted the Defense Department's brochure on the vaccine, which said it had "been safely and routinely administered in the United States to veterinarians, laboratory workers, and livestock handlers." In fact, Sharma said, the anthrax vaccine had been given to, at most, a few thousand people before 1990.

General Eddie Cain, head of the Joint Program Office for Biological Defense, was sent to mount a defense. The vaccinations were going well, Cain said. Only 8 of 260,000 people receiving the vaccine had reported reactions that caused them to be hospitalized or miss more than a day of duty.

The GAO offered no evidence that the vaccine was dangerous or a threat to health. But its testimony had raised some questions that did not have simple answers. Pentagon officials were shaken by the news coverage. Cain acknowledged in an e-mail to a senior Pentagon official that the program had taken a hit. "Two key areas we came up flat were GAO's assertion that #1 the anthrax vaccine licensed was NOT the one tested and #2, how can DoD say that reported Desert Storm illnesses were not caused by anthrax vaccine when we have no record of who received shots? If we cannot answer these questions we (DoD & the Administration) are in big time trouble." In fact, the GAO was correct on both counts.

Shays held five hearings on the anthrax issue in 1999 alone, taking more than twenty hours of testimony. Mark Zaid, the lawyer representing some of the soldiers who had refused shots, stirred more doubts when he came across documents the army had drafted to indemnify companies making the anthrax vaccine. One document dated September 1991 said manufacturers were entitled to special protection because of the "unusually hazardous" legal risks in making such a product. Lawsuits could result from adverse reactions to the vaccine or if the immunizations proved ineffective, it said. The FDA's testing of the vaccine did not involve enough people to predict what would happen when it was given to large numbers of people. "Only widespread use can provide this assessment," the document said. In addition, there was "no way to predict" whether it would be effective against the pathogens used in a biological attack. A later version of the agreement, written specifically for the Michigan lab, made the same points. Zaid provided the memos to a reporter from the *San Diego Union-Tribune,* who wrote a page-one story.

The Pentagon played down the significance of the documents, saying they were legal precautions, comparable to a homeowner buying insurance against a fire. Soldiers reading them on the Internet drew different conclusions.

A month later, the Pentagon promised to conduct a long-term study of the vaccine's safety, more than a year after the first shots were administered. Pentagon officials believed they could manage the problems on Capitol Hill. But in the latter months of 1999 the program confronted a more formidable challenge. BioPort, the company that had bought the Michigan facility, was having trouble meeting the FDA's standards. If production did not resume soon, the military would not have enough vaccine to go ahead with the immunizations. The company was financially shaky. In August the Pentagon agreed to raise the price it paid per dose of anthrax vaccine from $4.36 to $10.64.

The rewritten contract gave the company an additional $24 million, $18.7 million of which would be paid up front to tide over BioPort. Company officials said they had underestimated production costs when they bought the factory and negotiated the contract. The state of Michigan, it turned out, had been providing a much larger subsidy to the lab than had been understood.

The bad news kept coming. The retesting after the 1996 FDA inspection cost BioPort more than 3 million doses. The reasons varied. Nearly 1.5 million doses did not pass potency tests. Others were rejected by the FDA because BioPort had failed to follow procedures assuring sterility. The shrinking reserves prompted the Pentagon to scale back the vaccine program. In November 1999, federal inspectors delivered another blow. The renovations of the lab had been finished in January of that year, but BioPort was still not ready to resume production. A new inspection report found more than thirty deficiencies. The most significant involved the company's failure to comply with federal rules that each batch of pharmaceuticals meets the same specifications and is made the same way.

Drug regulators had significantly toughened their standards since the 1970s, when the original license was granted to the Michigan Department of Health. They were now demanding much greater precision. If the original formula said the anthrax bugs should be allowed to grow for twenty-two to twenty-eight hours, the modern FDA wanted a much more specific recipe, say twenty-two to twenty-four hours. Each change in the formula required a raft of test runs to prove the product remained the same. It was very slow going. In December 1999, the Pentagon had further slowed the vaccine program to stretch the dwindling supplies.

Pentagon officials began looking at alternatives to Michigan. Some proposed building a government-owned facility, the idea that had been killed in the early years of the Clinton administration. But that would not bring any immediate relief. Officials estimated that it would take seven years to build and license such a facility. Pentagon officials said that they hoped that BioPort would win federal approval for its production lines by the beginning of 2002, four years after the plant was temporarily closed for renovation.

Robert Myers, BioPort's chief scientific officer, said he was not daunted by the repeated setbacks. The FDA had toughened standards and the Michigan facility was hardly the only company struggling to meet them. Still, he believed production of anthrax vaccine would finally resume. "In the end, rather than being the gang that couldn't shoot straight, we'll be known as the little engine that could," he said.

In April 2000, Burton's Committee on Government Reform released a meticulously footnoted critique entitled "The Department of Defense Anthrax Vaccine Immunization Program: Unproven Force Protection." The anthrax vaccinations, the committee concluded, were a "well-intentioned but overwrought response to the threat of anthrax as a biological weapon." Gene splicing could produce new, vaccine-resistant strains, making the program a "medical Maginot Line, a fixed fortification protecting against attack from only one direction."

By some measures, the anthrax program was a success. The military had inoculated more than half a million troops, with only a handful reporting significant side effects. A committee of outside experts set up by the Pentagon to study the reports of "adverse events" found only 16 instances of serious reactions out of 1,530 reports to the FDA by mid-2001. Eleven people were hospitalized for swelling of the arm near the injection site. Four servicemen suffered systemic reactions to the vaccine, but none went into shock. One temporarily experienced severe lung problems that ultimately healed. According to the committee of experts, none of the illnesses persisted. The rate of adverse reactions appeared to be low, comparable to or better than those for childhood vaccines. There were no reported deaths.

But the Pentagon was nonetheless unable to sell many servicemen or key members of Congress on the value of the anthrax shots. The distrust stirred by the handling of Gulf War Syndrome was deepened by the initial sales pitch for the anthrax program. And critics questioned whether the voluntary reports to the FDA captured the full picture. Many servicemen, they said, were afraid to disclose their illnesses, fearing that their military careers would be harmed.

The Pentagon's initial public-relations strategy proved deeply flawed in an era in which every relevant document—from the scathing FDA assessments of the Michigan facility to the reports about adverse reactions—was available through the Internet. Pentagon officials dismissed the opposition to the vaccinations as Internet paranoia. But in fact, as Mark Zaid often pointed out, the most devastating questions were raised by the Pentagon's own documents. These are not my words, he would say, brandishing a memo, they are yours.

Over time, the military gave soldiers more accurate information about the vaccine. The first fact sheet answered the question "What are the side effects?" by mentioning only temporary maladies, such as sore arms or redness at the site of injection. In the fall of 2000, the Pentagon sent around a new version that was much more specific. "Any vaccine can cause serious reactions, such as those requiring hospitalization," it said. "For anthrax vaccine, they happen less than once per 200,000 doses. Severe allergic reactions occur less than once per 100,000 doses."

Pentagon officials acknowledged privately that they had oversold the program, particularly the broad claims about the vaccine's safety. A study completed in 2000 by a committee set up by the prestigious Institute of Medicine at the National Academy of Sciences concluded there was "inadequate/insufficient evidence" to assess whether the anthrax vaccine was associated with "long-term adverse health outcomes." There was a "paucity of published, peer-reviewed literature," which described "a few short-term studies." The research most frequently cited by the Pentagon in support of the vaccine's safety—the monitoring of 1,590 employees at Fort Detrick who received immunizations—did not settle the issue, the panel said. Participants in the twenty-five-year Fort Detrick study had not been specifically questioned about chronic symptoms.

In the years after the anthrax vaccinations began, some servicemen who had received the shots complained of dizziness, severe lethargy, and exhaustion similar to Gulf War Syndrome. Government health experts believed the vaccine had not caused their illnesses. But as the Institute of Medicine made clear, more research was needed. The panel said its findings were an "early step in the complex process of understanding the vaccine's safety."

The effects of the controversy were clear. Hundreds of reserve pilots quit the military. By mid-2000, more than four hundred servicemen had been disciplined for refusing to take the shots and fifty-one had been court-martialed. A few served brief sentences in the brig.

Some of the problems the Cohen team confronted in administering vaccinations against biological attack were inevitable. Doctors making the case for vaccines against dread diseases could point to measurable benefits that balanced the risks. For instance, before a measles vaccine was introduced in 1963, five hundred children a year died from the illness. Deaths fell to one or two a year after children were immunized, a huge gain when compared with the tiny number of serious adverse reactions.

Those advocating immunizations against biological attack could not show such clear benefits. Only the risks were certain. An otherwise healthy group of men and women received shots to protect against a health threat that might well never materialize. Other nations had calculated the risks and benefits and come to different conclusions. Britain made anthrax vaccinations for its troops voluntary. France's soldiers did not get them at all.

Pentagon leaders had seriously underestimated the anthrax vaccine program's technical difficulties. Making vaccine in a tough new regulatory environment was a daunting business, as BioPort's continuing struggles showed. Stephen Joseph, the Pentagon health official, had hoped that the anthrax program would mark the beginning of a new era in which immunizations were an element of military doctrine. But the anthrax vaccinations had seemingly had the opposite effect. Many Pentagon officials found it hard to imagine attempting another broad immunization program against, say, botulinum toxin anytime soon. Looking back after he left the Pentagon, Richard Danzig could only shake his head. He had not understood the full dimensions of the problems at Michigan when he pressed the case for inoculations in 1995 and 1996. In hindsight, the decision to broaden the program and immunize all 2.4 million active duty and reserve troops was a much closer call. Danzig believed that his original proposal to immunize the soldiers assigned to high-threat areas would have been more popular with servicemen. And the dwindling supplies of vaccine would not have been as harmful to a less ambitious program.

IN the last year of the Clinton administration, federal officials geared up to assess the effectiveness of the new civil defenses. In August 2000, they staged an exercise that a press release announced would cost $3.5 million, making it the largest emergency-preparedness drill ever conducted. Operation TopOff, for Top Officials, would test more than the ability of a large American city to respond to a germ attack. It was to be a several-day "consequence-management" extravaganza that would involve thousands of government employees from twenty-eight agencies. The exercise would have been even more ambitious had it not collided with financial, logistical, and political realities.

The exercise was Congress's idea. In 1999, after it had appropriated hundreds of millions of dollars to train first responders in cities throughout the nation, Congress had asked the Justice Department and the Federal Emergency Management Agency, in coordination with the National Security Council, to see if its investment in training and new programs was paying off. But by the time the elaborate exercise was scheduled, the novelty of tabletops and mock attacks was gone. Two months before the drill, Attorney General Janet Reno and FEMA Director James Lee were the only two top officials committed to participating.

As initially conceived, the exercise was to involve at least three cities. At a conference in the spring of 1999, government officials and the contractors whom the Justice Department had chosen to handle the exercise—Science Applications International Corporation and Research Planning Inc.—devised an elaborate scenario involving chemical, nuclear, biological, and cyber attacks in at least three and as many as six cities. By the following May, the exercise had been scaled back to a covert biological attack on Denver and a chemical-weapons strike in Portsmouth, New Hampshire. Officials had also conceded that if top federal officials were to participate, TopOff would have to run fewer than the planned ten days. They would keep secret the designated agents to be used, as well as the specific times and places of the attacks, to ensure an element of surprise and to make the tests as real as possible.

The government's response to the hypothetical mustard-gas attacks on runners at a charity footrace in Portsmouth on Saturday, May 20, 2000, was fairly effective. Though it took an exceedingly long time for the 170 "victims" of the bogus chemical sprayings to receive treatment, there were no riots and only limited chaos. After an FBI SWAT team stormed the fake hideout of the fake terrorists—the Group for Liberation of Orangeland and Destruction of Others—two postal workers being held hostage were released. Within forty-eight hours, order was restored. This, however, was not the case in Denver.

In Portsmouth, some three thousand government officials had been mobilized for the exercise; in Denver, important parts of the exercise occurred only on paper. Although some twenty-five hundred officials participated in the Denver drill, the diagnostic testing of the mysterious agent, the delivery of real medicine, and other key events in the exercise were simulated.

The scenario opened on Wednesday night, May 17, with a hypothetical lone terrorist's release of *Yersinia pestis*—pneumonic plague—through the fresh-air ducts of the Denver Center for the Performing Arts. By Saturday, May 20, when role-playing began, virtual "victims" of the attack were filling the emergency rooms of Denver Medical Health Center and two other hospitals in the area, complaining of chills, cough, fever, and other undiagnosed flulike symptoms.

Later that morning, the exercise continued as a maid at a make-believe motel that the contractors had set up at an abandoned army facility discovered a dead man in the room she was cleaning. The twenty-five-year-old man (actually a plastic mannequin) had blood and vomit around his mouth, a toe with gangrene, and skin lesions all over his body. Local police also found a weapon, terrorist literature, and instructions for making a spraying device in the room. The FBI was called in to investigate, and a four-block area around the motel was cordoned off.

By Saturday afternoon, some five hundred people had checked into local hospitals for treatment; twenty-five of them had died. After the CDC's lab at Fort Collins confirmed that the cause of death was plague, thirty-one real staff members were sent to Denver to help treat the sick and dying and contain the epidemic's spread. A public-health emergency was declared.

By late afternoon, some hospital staff were calling in sick. Antibiotics and ventilators to help people breathe were becoming scarce. Some doctors and nurses had begun wearing protective gear. The governor's office issued an order restricting travel by air, bus, and rail to and from the Denver area. The public-health authorities told area residents to go to a medical facility if they felt ill or had been in contact with someone who was. Other citizens were told to stay home. News alerts and press briefings were covered by a team of pseudoreporters whose black satin jackets were appliquéd with the words VIRTUAL NEWS NETWORK, the phony news organization the contractors had added to the drill rather than risk interaction with real reporters, a prospect that one participant called as daunting as plague.

By Sunday, May 21, although the CDC had alerted neighboring

states to the outbreak and its cause, cases of plague were being reported in several states, in England, and as far away as Japan. "Push packs" of antibiotics from the national pharmaceutical stockpile that the Clinton administration had created the previous year arrived in Denver. Several hospitals reached full capacity. So did city morgues. Some thirteen hundred ventilators were being sent to the area, almost half the nation's supply. By the day's end, more than eighteen hundred cases of plague were reported; almost four hundred people had died.

By Monday, the area's health-care system was shutting down from the strain, but the city and state had still not approved a distribution plan for antibiotics. The public-health authorities encouraged Denverites to wear face masks or, since those, too, were in short supply, the "Bronco" bandanas so popular in the football-crazed, cowboy-loving state. The CDC urged the governor's office to quarantine Colorado, but state officials refused, complaining that no provisions for shipping in extra food and supplies had been made and that such an edict could not be enforced. The highway patrol, too, was short of staff. Police were worried about rioting and people stealing precious antibiotics. The medical system, government offices, and society itself were collapsing.

The exercise ended on Tuesday, May 23, as the epidemic spread out of control. Estimates of how many people got sick and died varied widely. Some reports registered an estimated 3,700 plague cases with 950 deaths. Other participants estimated that more than 4,000 were sick and that 2,000 had died, almost twice the official death toll.

Federal, state, and local officials lost little time in proclaiming the catastrophe a successful exercise. While the "ten days in May" scenario had lasted only four and a half days, the drill had, in government-preparedness parlance, "stressed" the nation's emergency-management system to its limits. Reflecting on the value of the drill, Richard E. Hoffman, the chief medical officer of Colorado's Department of Public Health and Environment, said that much had been learned thanks to the planning group's decision to select the weapon he had recommended—plague, a highly contagious agent. All diseases were not equal; nor did they affect society equally, he observed. An attack of anthrax, which was not contagious, would resemble a chemical attack in impact. Terrible as it was, anthrax could be contained. But plague, one of President Clinton's "gifts that keep on giving," would be harder to stop. The drill had ended after the fourth day, Hoffman said, because, at least in this scenario, it was clear by then that "you would never catch up."

A major lesson was that democracies were probably ill prepared to make such emotionally charged decisions as whether scarce resources should be devoted to treating the sick or to trying to contain the epidemic. TopOff showed that even well-prepared cities would be hard-pressed to do both. According to several participants, the officials managing the outbreak had bitter, protracted fights over who would receive the antibiotics. Initial supplies had been rapidly depleted, and the state, everyone agreed, had been late in taking control of them.

Most public-health and emergency-preparedness officials insisted that emergency-management personnel, police, hospital staff, and their families had to have priority in getting drugs to enable them to remain healthy and cope with the disaster, as well as to encourage them to report for work. But others, especially elected officials, argued that it was unacceptable to give the lifesaving drugs to civil servants and their families when sick people needed them and the population remained unprotected. The politicians were overruled, perhaps only because Governor Bill Owens, a Republican who was skeptical of the Clinton biodefense initiative, had refused to participate in TopOff. Instead, political advice was offered by a group of senior advisers in his office.

Not all participants were happy with the selection of plague. "Gut-wrenching doomsday scenarios" that overwhelm a city and its health-care system are "easy, but not really helpful," said Tommy F. Grier Jr., the director of the Colorado Office of Emergency Management's department of local affairs. He believed that TopOff would have been far more useful had a more "realistically scaled" and, hence, more likely scenario been chosen.

The Denver metropolitan area, in fact, had received more than its share of emergency-preparedness training. All three of the cities involved in the drill—Denver, Aurora, and Colorado Springs—were among the first 120 American cities to have been trained under the congressional program. In 1993, Denver had skillfully handled a papal visit, and in 1998 an international presidential summit and the Oklahoma City bombing trial.

Several participants felt that failing to include real reporters had been a mistake, since without them there was no way to gauge how the scenario's unfolding events would be conveyed to the public.

TopOff had been very expensive. Grier's Office of Emergency Management—which got a $200,000 grant to offset the drill's costs—wound up spending far more than that, and far more than the Justice Department officially acknowledged, many officials agreed. The federal government's own figures showed that TopOff's real cost was closer to $10 million, much of it spent on contractors.

While state and local officials had worked fairly well together, there was tension among public-health officials, hospital representatives, and the FBI. They not only inhabited different worlds but spoke different languages. Many public-health workers, for instance, were unfamiliar with and frustrated by the steady repetition of the alphabet soup of emergency preparedness. "We weren't all exactly singing 'Kumbaya,' " said Greg Moser, a counterterrorism specialist in Colorado's Office of Emergency Management.

Moser and most other experts said that TopOff, especially its failures, had taught them invaluable lessons. As in New York's real outbreak of West Nile virus, the endless conference calls involving up to a hundred officials had been time-consuming and cumbersome. But often there was no other way to communicate: phones in the overworked public-health offices and even police stations went unmanned.

Most of Denver's hospitals had no contingency plans for treating mass-casualty incidents and only limited stocks of equipment, vaccines, and other drugs. Only three of Denver's twenty-two hospitals had been able to spend the time and money required to participate in TopOff, and one of them, overwhelmed by real patient loads, had had to drop out prematurely. Few of the city's morticians had even thought about coping with such an incident. The city's decision to commandeer refrigerated trucks from the Coors Brewing Company for body storage and transport was an inspired stopgap measure, but hardly a viable solution to emotionally and politically sensitive decisions about how and where to dispose of victims' bodies. "Our public-health-care infrastructure was in some ways better prepared in the thirties to handle an epidemic like this than they are today," said Tara O'Toole, a public-health expert who evaluated the exercise.

The national antibiotics stockpile, which faced its first test in TopOff, passed, but barely. While the CDC sent drugs to Denver within forty-eight hours of the request, an impressive achievement, the plane landed at Denver International Airport, which the mayor's office was about to close. Furthermore, the drugs could not be distributed quickly. There were few workers on hand to divide the giant antibiotic "push packs" into manageable sizes. A public-health worker went to Safeway to buy plastic bags in which the antibiotics could be distributed. The trip, TopOff organizers estimated, would have taken six hours through streets clogged with panicked citizens.

Officials complained about too many competing operations centers and not knowing who was in charge. There was also tension over making the right public-health decisions and the need to make them quickly. Despite all of Denver's training, plans for isolating contagious people and getting drugs to ordinary citizens were not in place. Plans the contractors imposed, which had worked in New York, proved inappropriate for Denver. The designated "points of distribution" for antibiotics could serve only 140 people in an hour, utterly inadequate to the demand generated by a city of 2 million.

When the exercise ended, officials were still arguing about whether a quarantine should and could be imposed on Colorado. People who hypothetically had already been confined to their homes for seventy-two hours were running short of food and supplies. Civil unrest was brewing. Hospitals were about to be stormed. Despite all of its previous training and drills, Denver was doomed.

THE disarray of TopOff came as no surprise to Amy Smithson, an analyst at the Henry L. Stimson Center in Washington. After eighteen months of interviews with public-health and emergency-preparedness officials in more than thirty cities across the nation, she had concluded in the fall of 2000 that the administration's directives and Congress's new laws had produced a hodgepodge of programs that were often conceptually dubious, bureaucratically duplicative, poorly coordinated, and disastrously implemented.

A southerner who loved politics and had witnessed several national-security fads come and go in her fifteen years in Washington, Smithson recognized a hot new trend when she saw one. The president's personal engagement meant there would be billions of new dollars available for agencies desperate to reinvent themselves after the Cold War. Aum's 1995 sarin-gas attack in the Tokyo subway and the 1998 bombings of two U.S. embassies in Africa helped the administration push huge budget increases through Congress to prevent terrorism and deal with its consequences. The political imperatives were obvious. Who could be opposed to fighting crazed zealots who were targeting Americans? No lawmaker or administration official wanted to have to explain in the aftermath of a devastating attack that lives could have been saved if only the government had not been pinching pennies. It was far easier for Congress and the executive branch to create new programs. Overall, the funding for counterterrorism rose from $6.5 billion in 1998 to $9.3 billion in 2001, while the part of that devoted to managing the consequences rose from $645 million in 1998 to $1.6 billion in 2001.

The experts most involved in pushing for stronger defenses against biological attack knew there would be waste. The government, several said privately, was slow to move and certain to stumble as it took steps in a new direction.

Even so, Smithson and some of the government's auditors felt that some of the initiatives were notably wasteful. Some of the first, most trenchant criticism came from the General Accounting Office, the congressional watchdog. In a series of reports and statements in late 1997, the GAO warned that the administration's program was based not on a rigorous assessment of the terrorist threat facing the nation but rather on the *potentially infinite* vulnerability of American civilians at home and U.S. armed forces abroad. Without cogent risk and threat assessments, the GAO argued, there was no way of knowing whether the government was spending too little, too much, or just enough money fighting terrorism, much less whether the money was being spent effectively.

The GAO reports underscored the singular nature of the bioterrorist threat. Despite huge investments in intelligence and readiness, there was probably no way to predict with confidence the source of an attack, its target, the agent that would be used, or, therefore, how best to defend the nation.

Many analysts noted that attacks using germs and unconventional weapons had been extremely rare—what national-security experts termed a "low-probability event." When the culprits had used or tried to use such agents, they had usually not chosen substances that would inflict mass casualties. Data collected by the Center for Nonproliferation Studies at the Monterey Institute of International Studies showed that there had been 285 incidents throughout the world since 1976 in which terrorists had used chemical or biological weapons. In 44 percent of those cases, no one had been killed or seriously injured; in 76 percent of them, five or fewer people had been hurt. Examining these numbers, many analysts argued that there was little statistical support for the "not-if-but-when" school of thought.

Amy Smithson and other analysts, though critical of such exaggerations and hyperbole about biothreats, at the same time also believed that given the rapid advances in biotechnology and the collapse of the Soviet germ-warfare empire, as well as the rise of a smaller but sophisticated South African program, the past might not be prologue. Statistics showed that although the number of terrorist incidents had declined in the past decade, their casualty rates had increased sharply. And America was increasingly often the target of such attacks. However low the probability of such an unconventional strike, the consequences might be so grave that any responsible government would have taken steps to protect the American people. Smithson's quarrel with Congress and the Clinton administration focused on how they had gone about it.

To begin with, only $315 million of the $8.4 billion that the government spent on counterterrorism in the year 2000 was devoted to training people in cities and states to respond to a covert bioterrorism attack. A small proportion of money, less than 4 percent, was being spent beyond the Washington beltway. A disproportionate share of this spending went to what public-health advocates derided as the "whistles and sirens," the fancy cars, faulty detectors, and other marginal gear. Only 6 percent of the funds to combat unconventional terrorism was allocated to strengthening the public-health infrastructure, the heart of meaningful biodefense preparedness.

The 1996 law that called for training of emergency personnel in the nation's 120 largest cities had proposed a simple structure: The Defense Department was in charge. But within four years, Smithson discovered, that statute had spawned a cottage industry of competing programs. Local officials could choose from among some ninety different courses offered by competing federal bureaucrats. The courses, moreover, differed dramatically in quality and content, but no federal office was charged with evaluating them or helping cities decide which best suited their needs. Government contractors, the so-called beltway bandits, were often allocated the bulk of the training money, leaving cities and states scrambling to find funds to buy protective clothing and emergency-preparedness equipment.

Smithson argued that the administration had shortchanged hospitals and public health in the early years. Financing for a national stockpile of vaccine and antibiotics had not begun until 1998, two years before the end of President Clinton's second term in office. Efforts to revive the nation's ailing infectious-disease surveillance system, critical to identifying naturally occurring and deliberately caused epidemics, did not begin in earnest until a year later. The Department of Health and Human Services was not even included in discussions involving bioterrorism initiatives until late 1998, and it was another year before the agency had secure communications facilities in which classified information could be examined and discussed.

Although they were well intentioned, Smithson concluded, the administration's homeland defense initiatives had become a triumph of political "pork over preparedness."

In most federal programs devoted to countering terrorism threats, germ agents were lumped together with other weapons of mass destruction, such as chemical, radioactive, and nuclear arms. And biological weapons were usually little more than an afterthought—barely the caboose of the counterterrorism train.

THE administration's effort to reduce the threat posed by former Russian germ scientists received generally higher marks than the scattered, duplicative domestic programs. But even this effort had its problems.

While in 1991 Senators Nunn and Lugar had allocated money to preventing former Soviet unconventional-weapons scientists from selling their arms or lethal expertise to rogue states or terrorists, the Clinton administration did not provide serious money for biological defense until years after the program was begun.

At the State Department, funds for the biological nonproliferation programs—the Moscow-based International Science and Technology Center, and other scientific exchange and cooperation projects with former Soviet bioweaponeers—were reduced year after year, as a result of budget battles between and within the executive and legislative branches. By 2000, annual State Department support for such projects reached a total of $12 million—its highest level but still a pittance compared to major government efforts.

The same was true at the Energy Department. Most of its nonproliferation funds went to Russian nuclear cities and programs; very little was spent on biology.

Andy Weber's efforts to win the allegiance of former Soviet germ warriors had always been modestly funded. It began with $5 million in 1996 and grew to $12 million by 2001, just 3 percent of the cooperative programs aimed at Russia. Most of that money was spent on reducing the threat posed by nuclear or chemical weapons.

Although Russian scientists had repeatedly warned the administration that Iranian officials were hunting for experts, particularly those with recombinant skills involving plant and animal pathogens, the Department of Agriculture received funding for antibioterrrorism programs only in 1999. And for two years in a row, either the administration's Office of Management and Budget or Congress had rejected the Agriculture Department's request for $70 million to expand and upgrade its own animal pathogen research and diagnostic center on New York's Plum Island.

THERE was no shortage of funding or congressional support for the Pentagon's effort to create rapid-reaction teams within the National Guard. By January 2001, more than $143 million had been invested in the project, which was racing ahead on an accelerated schedule.

In their 1998 memo to Cohen outlining the concept, army experts had said it would take three years to train and equip the new units, an exceptionally quick turnaround for such a sophisticated mission. Army officials overseeing the guard program set an even shorter timetable and planned to have the first ten teams ready for action by the end of 2000.

In a bow to civil liberties concerns, the Pentagon had dropped the hard-charging acronym for the teams—RAID—in favor of a much blander moniker. The new name, Civil Support Teams, was chosen to convey the guard's subordinate role to civilians.

In January 2001 the Department of Defense's inspector general delivered an independent assessment of the project. It depicted an unmitigated catastrophe.

The government's inspector generals are assigned the thankless task of auditing and reviewing the efficacy of federal programs. They are independent, and President Ronald Reagan once said he wanted them to be as mean as "junkyard dogs." Most of the auditors are decorous in their criticism, their language restrained. The report on the

National Guard initiative was unusually scathing, as damning an internal assessment as the Pentagon has ever made public. The inspector general concluded that the nineteen-person office setting up the program had so mismanaged the effort that it should be disbanded. Despite the years of work and training, none of the ten teams was ready to be sent anywhere. Because of management blunders, defective and inappropriate equipment, and poor training, the units that were supposed to be deployable by January 2000 might themselves be at risk in an urban catastrophe, the auditors found.

Although the units were supposed to be within 250 miles of 90 percent of America's population, the locations of existing military bases sometimes made that unrealistic, the auditors said. The proposed stationing of the Florida team at Camp Blanding, Florida, for instance, placed the team within 250 miles of Atlanta, which meant that it would overlap with the Georgia-based team, but it was more than 350 miles from Miami.

Only five of the ten teams had recruited all of their twenty-two members, and many of the members had not been able to participate in training exercises. Each team still needed some protective clothing; nine of the ten could not talk with headquarters because of lack of communications equipment. Only two teams had been given hand-held equipment to detect biological agents. When they conducted their initial exercises, none had received the mobile labs containing the equipment needed to identify a potential germ agent. When the vans with this equipment—each unit costing $1.6 million, the "cornerstone" of the project—finally arrived, some of their air filters were installed backward. Each van was supposed to contain a glove box, a sealed compartment for examining a biological agent. But because the boxes were badly designed and too big, there was too little room for team members to work in the vans. Many team members, moreover, had no idea how to use or maintain the glove boxes, which arrived without manuals.

The gas masks provided to the teams had never been tested and were equipped with incompatible parts. While the masks and the accompanying blowers probably would have worked when assembled, one team commander told the auditors, "I'm just not willing to bet my life on it."

During training exercises, some teams were issued partially filled air tanks to supply their masks and had to get refills from local firemen, which made the guardsmen a burden to the people they were sent to assist. Microscopes turned out to be too large to be installed in the vans, and their attachments required five electrical outlets; the vans had only three. As a result, the auditors said, many team members and their commanders, whom the report described as "dedicated," "highly motivated" people "committed to their missions," were demoralized. The inspectors noted that the commanders and their team members lacked "confidence in the unknown, untested, and unsubstantiated reliability of the equipment that they were issued," not to mention the program's managers.

One of the main confusions was over how, exactly, the team would help local officials. The sophisticated detection equipment could be useful for testing biological samples. But the FBI, which had set up its own hazardous materials unit, told the army in the fall of 2000 that it would not allow anyone at the scene of an incident to collect evidence without its permission. The FBI reminded the guard that it was the lead agency for the "crisis management."

Army officials disputed some of the inspector general's findings. They said, for example, that they had addressed the problem with the microscopes by issuing each team tents that could be pitched next to the vans. The auditors questioned how this plan would work if a terrorist incident occurred, say, on a day of heavy rains or high winds.

The Pentagon did not dispute the report's findings and moved to shut down the office managing the project even before it was officially filed. Charles L. Cragin, an acting assistant secretary of defense for reserve affairs charged with overseeing the National Guard, said in early 2001 that his organization was working with "great perseverance" to resolve the problems identified by the auditors and that many had already been fixed. Cragin blamed the failures on the pressure the guard had been under to mount so many sophisticated teams for a complex mission so quickly. Guard officials said the technical problems had been largely resolved and that the civil support teams would soon be ready to carry out their mission.

ANOTHER weakness in America's biodefenses was the continuing failure to produce a cheap, reliable detector that would sound the alarm if a germ attack was under way. The military had been trying to build such devices since the early days of America's offensive program. What the Pentagon wanted was a simple device like a smoke detector that could sniff the air and quickly identify a bug in a biological attack. Some progress had been made since the Gulf War, when commanders were startled to learn that they had nothing effective to warn them. But the detectors were still rudimentary and the stakes were rising. In the late 1990s, officials planning America's defenses against germ attack envisioned a crucial role for biodetectors. The United States could never immunize every citizen against every possible disease available to a terrorist or rogue nation. Detectors could be a key to building a credible civil defense. If the public-health authorities knew immediately that their city had been attacked, they could begin evacuations, quarantines, and treatments. But for the moment, the American people were the nation's main means of detecting biological attack. The alert would come through panicked visits to emergency rooms and reports of mysterious symptoms. In the case of anthrax, as the military knew all too well, such reports would come several days into the illness, too late for treatment.

The Clinton effort initially focused on projects mainly for the military. What the commanders wanted was a detector that would give enough warning to allow soldiers to escape an attack unharmed or to don protective gear. A warning that was slower, but still timely, might give infected people enough time to take vaccines and antibiotics or other measures to save their lives. The Pentagon had tested a truck-mounted lab to sniff the air, an airplane to track particulate clouds, and an electronic network at military bases that tied together many sensors to double-check results and lower the number of false alerts. All had serious flaws.

In September 2000, the Pentagon's inspector general criticized one of the more costly efforts. The Joint Biological Point Detection System, which used some of the most advanced technology, was developed by Lockheed-Martin. By 2000, the company had begun to build the first installment of 970 detectors in a $1 billion program. But the inspector general concluded that the prototype machine achieved only one of its ten critical goals. It broke down repeatedly and often failed to identify lethal pathogens. Lockheed insisted it had done the best it could given the difficulty of the technical challenge.

Making any detector is daunting mainly because the targets are literally everywhere. Even clear air contains billions of germs, and benign ones are hard to distinguish from the killers. That leads to "false positives," or mistaken identifications. For the military in time of war, false warnings might be acceptable. But the last thing New York or Washington would need is a germ alert that turned out to be wrong.

Frustrated by the lack of progress, military planners eventually turned to the weapons laboratories that had invented the atomic and hydrogen bombs. In the 1990s, these labs had found new work by addressing threats of mass destruction, especially germ weapons.

Lawrence Livermore National Laboratory, a square mile of federal science in central California, threw itself into the detector push. Zeroing in on genes, it sought to identify the most unambiguous signatures of microbial individuality—just as Venter had suggested to President Clinton at the Renaissance Weekend dinner. First, the lab developed a means of mapping a microbe's genetic material with the same kind of equipment that was being used to decode the human genome. These signatures were then copied into artificially made snippets of DNA. When these snippets were mixed with a candidate germ that had been unraveled—say, anthrax—the snippet would bind to the germ's genetic material, causing a bit of fluorescent dye to float free. Under proper illumination, the sample glowed.

Livermore then miniaturized its invention, reducing a huge mass of tubes, bottles, and electronic parts to a device that could fit in a suitcase or a handheld detector. By 1999, the lab had perfected a small chamber with a silicon chip that, theoretically, could identify germs in seven minutes. By 2001, it was packaged as HANAA, for Handheld Advanced Nucleic Acid Analyzer. Weighing two pounds, powered by batteries, highly computerized, HANAA underwent field trials at several sites, including the CDC in Atlanta, the FDA, and Los Angeles County's emergency operations office.

Livermore announced that it would transfer its invention to a company that would make and sell the detectors. The small device was promising. But further testing was needed to determine if it was really a breakthrough.

THE election of George Bush in the fall of 2000 raised questions about whether the cooperative programs nurtured by Andy Weber could survive. A review by the new administration gave the effort a qualified blessing. But problems persisted on both sides of the partnership. Russian officials had fired Yuri Kalinin, Biopreparat's former chief. Still, there was no shortage in Russian political circles of critics who opposed closer dealings with America. And not just within Russia. Some officials in the Uzbek and Kazakh governments had also tried to undercut the cooperative threat-reduction program by being less than cooperative on requirements that Congress and the Clinton administration had imposed. Despite quiet American pleadings, the Kazakhs by mid-2001 had still not ratified the treaty banning biological weapons. Uzbekistan, until the spring of 2001, had delayed signing an umbrella agreement under which cooperative assistance projects operated. But compared to Russia, and despite occasional setbacks, the programs with the central Asian states had made significant progress.

In Washington the gnawing suspicion that American aid was enabling Russia to do research that might benefit a weapons program continued to hinder the program. Weber himself had experienced disappointments that had given ammunition to critics of the program, and both the former Soviet institutes and the United States were to blame. Military labs were still closed to Americans.

At the Stepnogorsk complex, which Weber had dreamed of transforming into the "poster child" for nonproliferation, the joint venture to manufacture pharmaceuticals collapsed before it got off the ground. A $3 million contract was canceled in 1997; only $2 million of it was recovered. A plan to make disposable syringes in the former germ plant also collapsed the following year, when production stopped because of unpaid utility bills. Another plan to sterilize up to 2 million syringes in yet another part of the complex failed because of inadequate financing and a shortage of cobalt 60, a radioactive isotope used to kill germs. However, the Pentagon was systematically dismantling the plant. By the summer of 2000, the massive fermenters had been removed and melted down for scrap metal. All equipment that could be used for making germ weapons had been destroyed. Filtration systems that once kept contaminated air inside the labs had been ripped out and were stacked high in piles of debris outside the production buildings. That year, Kazakhstan and the Defense Department had decided to raze the buildings. But negotiations had stalled over the price.

While the plant no longer posed a threat, the same could not be said of its scientists and engineers. The Pentagon had helped destroy their life's work but had given them almost nothing to take its place. There was no more work at the plant itself. The small, well-equipped toxicology lab that Washington had paid $750,000 to help open employed about sixty former germ warriors in downtown Stepnogorsk. Hundreds more needed jobs.

The town's population had shrunk by half and unemployment was still rampant. Weber and other American officials argued that the United States should do more to ensure that the weapons scientists who had made anthrax would be able to find peaceful work, hopefully in Kazakhstan. But the idea had little support in Congress, which preferred to invest money in dismantling dangerous facilities. In fact, the physical infrastructure of death was mostly gone. What remained unsolved was the more vexing problem of how to keep scientists from selling their expertise to Iran or elsewhere.

12

The Future

In the spring of 2000, George Tenet, the director of central intelligence, met with some of the country's leading experts on biodefense—scientists, federal officials, health experts, and drug company executives. The group, known as the Non-Proliferation Advisory Panel, had been set up to give Tenet and his senior aides strategic advice. A former Senate staffer known for his blunt manner and creative use of expletives in private conversations, Tenet had been pushing his agency to do more to counter the threat of germ weapons. Early in his career he had focused on nuclear-arms control and how to monitor an adversary's missiles and warheads. Now he was confronting a problem that was far more complex, and he wanted help. He told the panel that its suggestions were "mundane." The CIA, he said, was looking for bold, imaginative solutions—something that would "break the back" of biological terrorism.

The Soviet scientists who worried Andy Weber figured prominently in Tenet's view of the threat. A few weeks earlier, he had warned Congress that a growing number of nations were trying to import the talent needed to make "dramatic leaps" in producing biological agents and delivery systems. "They can buy the expertise that confers the advantage of technological surprise," Tenet said. Terrorists were also experimenting with germ weapons. Osama bin

Laden, the Saudi exile, had been training his operatives to use chemical and biological toxins, Tenet said.

Although he did not say so in open session, the CIA and the Pentagon had been working separately for nearly three years on several highly classified projects to develop a better understanding of germ weapons and delivery systems developed by foreign scientists. The programs were among the government's most closely kept secrets, their code names known to only a handful of officials. Participants were required to sign special security forms in which they waived their rights to resist an investigation of unauthorized disclosures.

There were several reasons for the extraordinary secrecy. Some of the research focused on American vulnerabilities, and officials did not want terrorists or other nations to know which agents or techniques would work best against soldiers or civilians. Officials privately acknowledged another reason for their sensitivity: the projects were bringing America much closer to the limits set by the 1972 treaty banning biological weapons. Several federal officials had serious qualms about the research and had warned colleagues that a diplomatic disaster might ensue if the secret programs were revealed.

The treaty was maddeningly vague, and government lawyers had spent years trying to translate its provisions into practical rules. Government experts agreed that the pact allowed a broad range of experiments with germs and toxins, as long as the aim was defensive and the quantities of agents small. Studies of weapons were more problematic. The treaty banned nations from developing, producing, acquiring, retaining, or stockpiling a weapon or other means of delivering germ agents "for hostile purposes or in armed conflict." The government's legal experts had never formally wrestled with whether a country could buy, steal, or manufacture a germ bomb and use it to establish standards for testing vaccines or other defenses. Some experts believed such experiments were acceptable, as long as they were not intended for war. Other government officials contended that a weapon was, by definition, meant to inflict harm and was therefore out of bounds, even for defensive studies. A bomb was a bomb was a bomb, they would say.

America's biodefenders were wary of any work that came close to violating the treaty. Many had been in government in the late 1980s when Senator John Glenn's investigation and hearing prompted the much tighter limits on germ research. In the years that followed, scientists at Fort Detrick scrupulously confined their work to germs or toxins on the military's "validated" list of biological agents, a formal document negotiated among intelligence officials. Army researchers grumbled that the list was years out of date and did not include many of Russia's most lethal, futuristic germs. The difficulties of intelligence made "validating" a threat extremely difficult, and Glenn's action was seen as imposing a stifling standard on defensive research. Even so, nobody wanted to risk more trouble with Congress.

By the mid-1990s, Glenn had retired and the restrictions loosened a bit. But the army lab at Fort Detrick was facing budget cuts and had little money to take advantage of the more flexible rules. It was a different story at the CIA, where agency officials were significantly expanding programs that tracked the spread of nuclear, chemical, and biological weapons.

Iraq's statement to UNSCOM in 1995 that it had produced thousands of gallons of germ weapons came as no surprise to the CIA's analysts. They knew that Iraq had repeatedly tried to buy cluster bombs that might be turned into germ weapons. But Baghdad's admissions heightened the interest of senior policy makers—the "customers" as the agency calls them. Suddenly, many in Washington wanted to know more about biological weapons. The agency asked its spies to collect more information, as did the National Security Agency, which eavesdrops on the world's phone, fax, and e-mail communications.

Bill Patrick, who sometimes worked as a consultant for the agency, also felt the CIA could do a better job. For years, he argued, the nation's intelligence services had focused their efforts too narrowly, tracking mainly research and development of pathogens. Obtaining and culturing virulent germs was an important first step for an aspiring bio-warrior. But as Patrick had learned in the 1950s, that was only the beginning.

Producing industrial quantities of dried pathogens of the appropriate size was an equally daunting challenge, as was figuring out how to make effective bombs, missiles, or aerosol devices. The CIA, Patrick complained in a speech to air force officers, had long focused on "agent, agent, agent." It was, he said, "all they looked at." Patrick argued that it was equally important to understand the inner workings of munitions and their delivery systems. If you can't deliver most of the germs to a target, Patrick told the officers, "you have not done your job."

Tenet, who had been the agency's number two official, was sworn in as director in July 1997, and he immediately told senior aides that he wanted to see even more attention paid to biodefense. Soon after, he appointed John A. Lauder, a veteran CIA official, as director of the agency's Non-Proliferation Panel, a group of analysts who tracked the spread of unconventional weapons. More specialists were hired, and the agency pulled together the panel of outside experts, persuading Tom Monath, the virologist who had been part of the experts' meeting with President Clinton, to serve as its chairman.

The new focus on biological weapons was a boon for analysts like Gene Johnson, a protégé of Patrick's who had been hired by the agency. A virologist who had worked for years at Fort Detrick, Johnson was a creative man, a doer, a scientist who had risked his life to hunt for Ebola's hiding place in Africa. He was one of the most vividly drawn characters in *The Hot Zone,* Richard Preston's book about the Ebola virus, though colleagues said he disliked the depiction. "Gene Johnson is a large man, not to say massive, with a broad, heavy face and loose-flying disheveled brown hair and a busy brown beard and a gut that hangs over his belt, and glaring, deep eyes," Preston wrote. "If Gene Johnson were to put on a black leather jacket, he could pass for a roadie with the Grateful Dead."

Johnson, who shared many of Patrick's views on the importance of understanding munitions, began urging his colleagues to study the weapons systems developed by the Soviet Union to mount a biological attack. How were such bombs made? How well did they work? Could they be copied? A project took shape. CIA officials named it Clear Vision—an attempt to see into the future of biological warfare by understanding its past.

Intelligence officials said the CIA's main goal was to improve the type and quality of the information it collected for the government. It also wanted to be able to better assess the intelligence it had previously gathered about foreign germ weapons and the credibility of the people and other sources that had provided it.

The principal focus was on foreign weapons and the analysis was to be hands-on. Johnson had lots of ideas about enemy weapons systems that should be tested. He showed colleagues a blueprint of a biological warhead the agency might study.

CIA officials believed that the work was well within the bounds of the germ treaty, allowed by the provisions that permit defensive research. Clear Vision was under orders to pursue projects only when American intelligence received specific, credible reports about a threat posed by a potential adversary. It was not a license to roam, to try out blue-sky theories about what some evil scientist *might* do.

Clear Vision zeroed in on the weapons systems the Soviets had developed. Unlike the Americans, the Russians had built long-range missiles to hurl germ warheads between continents and had begun to make cruise missiles that would fly below radar, operating at ideal speeds for dispensing lethal clouds. A few years earlier, Alibek had told Patrick and CIA debriefers about Moscow's success in designing missiles and bombs. The two men later collaborated on a secret paper comparing the programs of their two countries. There were many questions. Some of the most significant centered on the miniature bombs, or bomblets in military parlance, that the Soviets had developed to spread germs. How did the bomblets fit into warheads? How and when were they scattered during an attack? How well did they work?

The first-generation biobombs produced by the United States and Russia in the 1950s were crude devices: liquid or dried agent poured or packed into a metal bomb filled with explosives. The blast and heat from detonation destroyed much of the biological agent. In his presentation to the air force officers, Patrick said that only 1 or 2 percent of the anthrax in the early American bombs survived their detonation, a "travesty."

Moscow and Washington had solved the problem over several years, each in its own way. The Americans had created bomblets and filled them with freon for cooling. The Soviets had found a different way, Alibek revealed, packing their bomblets with plastic pellets to soak up the explosion's heat and reduce the pressure on the germs.

Johnson was eager to analyze the properties of the Soviet biobomblet, and CIA headquarters directed American intelligence officers overseas to obtain one, perhaps in one of the former Soviet republics. The American teams sent out to Vozrozhdeniye Island in 1997 and 1998 to sample the anthrax that the Soviets had buried there were also enlisted in the hunt. They were told that even a fragment of a bomblet would be useful, and eventually found one. Johnson was delighted.

The Clear Vision team's desire to analyze a Soviet-style bomblet was prompted by reports in the mid-1990s that more than one country was running tests on what appeared to be biological munitions. At about the same time, the agency also picked up indications that Iraq was trying to rebuild its germ program. Baghdad's attempts to make germ bombs in the late 1980s were no more efficient than comparable early efforts by the Soviets or Americans. But what if Iraq hired a Soviet scientist who knew how to make a bomblet? And if the Iraqis made such a biobomblet, how efficient would it be?

"We wanted to learn all we could about past weaponization and future weaponization that might be used against us," said a senior agency official who was briefed on the program. "We wanted to go back and get the materials, the systems that had been used in the past and use them to better understand what we were likely to face." Testing a bomb's dispersal characteristics would illuminate how it could be used against Americans in an attack.

CIA officials believed the treaty permitted the testing of a fully operational weapon. But they chose not to. Intelligence officials said the devices manufactured at the orders of the Clear Vision team were not functional arms that could have been used on a battlefield. They lacked the explosive fuse needed to detonate a bomblet dropped from an aircraft. And they were filled with a simulant, not live agent. There was some confusion, even in internal agency documents, about what was being done. Memos referred to "reverse engineering" of a bomblet, a reference that the officials said was imprecise because the weapon was never completed. CIA lawyers reviewed the project and said it was consistent with the germ treaty. The agency went ahead without asking the White House for approval.

In early 1999, Johnson embarked on a new attempt to obtain equipment used in the Soviet germ program. The mission began when American officials found what appeared to be a bomblet-filling machine outside a reinforced bunker that once held germs at Stepnogorsk in Kazakhstan. Intelligence analysts who looked at photos of the machine then showed it to Ken Alibek, who said that he, too, thought it had been used to fill bomblets with agent. Johnson, eager to see firsthand how the device functioned, assess its capacity, and perhaps at long last secure traces of the Soviet superanthrax that Alibek's plant had made, asked Pentagon officials to purchase the machine from the Kazakhs and ship it back to the United States. Although the CIA authorized up to $150,000 for the purchase, officials at Stepnogorsk, somewhat puzzled by the request and the secrecy surrounding it, agreed to sell the device for only $10,000. In the spring of 1999, according to informed foreign and former American officials, the machine was quietly shipped to the United States and subjected to numerous tests aimed at detecting residue of anthrax or any other pathogen. After weeks of testing, the results were in. The suspect bomb-filling machine had no traces of anthrax. Instead, analysis showed that the putative machine for filling germ bomblets was in fact designed to put caps on bottles.

Officials at Stepnogorsk eventually solved the puzzle. When refrigeration units at Stepnogorsk's dairy plant had broken down in the late 1990s, Gennady Lepyoshkin, the germ plant's director, had invited the city's sole dairy manufacturer to move the machines into his bunker to continue producing milk for the city. The machine, and a dozen others just like it now standing idle in the neighboring dairy plant, had never been used to fill weapons, Kazakh officials insisted. They had processed milk bottles. A subsequent visit to the dairy plant and discussions with plant executives seemed to confirm this account. The CIA had spent $10,000 on a milk-bottling machine.

The verdict was by no means unanimous. Despite the lab tests, the interviews of Kazakh officials, and the visits to the dairy plant, some analysts remained convinced that the machine was not innocent. Ken Alibek, who had told the CIA that the machine was a bomblet filler, insisted in a subsequent interview that the machine was one of many "dual use" items in the Soviet germ program. The machine could fill containers with both powdered milk or, as at Stepnogorsk, powdered anthrax. He knew the machine well, he said. Earlier in his biowarfare career, he had been the director of the Soviet plant in Moscow that had built items like this.

In mid-1999 the CIA described Clear Vision to the Non-Proliferation Advisory Panel, the agency's group of outside advisers. Gene Johnson passed around an aluminum replica of the bomblet. While the panel was generally supportive of the program, one adviser, Josh Lederberg, was uneasy. Building any type of weapon, he told the CIA officials, skirted close to the germ treaty's prohibitions. He suggested a legal review. Wasn't this the sort of project, another panel member asked, that should be brought to the president for approval?

Some of the experts questioned whether the intelligence gains from the program were worth the political risks. The agency's lawyers might well be right about the legal fine points. But the appearances might be devastating if the project were disclosed. The United States had championed the germ pact and signed dozens of other arms control treaties, albeit out of self-interest, not altruism. Such a controversy could undermine a host of global accords that, however imperfect, had probably constrained potential foes.

For every objection, the agency had a counterargument in favor of Clear Vision. For instance, the CIA noted that it was responsible for assuring the government that every joint research project with Russia, such as Andy Weber's at the Pentagon and those of the Energy and State departments, did no damage to national security. CIA analysts needed to know the exact state of Russian research into germ munitions to assess whether the American expertise might help Moscow make better weapons.

At about the same time, key administration officials learned about the Clear Vision work on Soviet bomblets. Lauder, the head of the agency's Non-Proliferation Center, mentioned the project to two National Security Council aides. Curious and somewhat concerned, they asked the agency for a more detailed briefing, colleagues recalled.

Gene Johnson went to the White House to explain, equipped with charts and graphs. In early fall, he described the program to several National Security Council staffers. He said the CIA's lawyers believed it was consistent with the germ treaty. The White House officials were not entirely persuaded about the project's value, asking colleagues what, exactly, would be gained from producing a bomblet. They saw risks if the project were publicly revealed. Imagine what the United States would conclude if it discovered that, say, Iran had built and tested its own Soviet-style biological bomb.

The White House asked the CIA to brief the State Department and the Pentagon about the program. The CIA said it was already doing so. James E. Baker, the White House lawyer who reviewed secret intelligence matters for the National Security Council, was asked by White House officials to conduct a separate "scrub" of Clear Vision to ensure that it complied with America's treaty commitments. "Jamie" Baker, who was on loan to the White House from the State Department, had a reputation for giving fast, meticulous, and fair-minded counsel. He concluded that Clear Vision, as described by the CIA, did not violate the treaty's ban on developing weapons since it was meant to replicate foreign munitions and improve American defenses against a biological attack.

A White House aide drafted a memo to Sandy Berger, the national security adviser, describing Clear Vision and outlining its goals. He was not asked to make a decision or give his approval. The CIA had not sought a White House blessing for Clear Vision, officials later stressed. The agency was simply informing the NSC of its plans. White House officials say that President Clinton was never told of the program.

With the tacit approval of a few White House officials, the project nonetheless moved ahead.

In the ensuing months, Battelle, a military contractor in Columbus, Ohio, with sophisticated laboratories, conducted at least two sets of tests on a model of the biobomblet that measured, among other things, its dissemination characteristics and how it would perform in different atmospheric conditions. A series of experiments in a wind tunnel revealed how such bomblets, after release from a warhead, would fall through the air over targets. By mid-2000, the tests were complete and the effort pronounced a success.

At about this time, a new round of questions from other parts of the government arose about whether Clear Vision complied with the germ treaty. The CIA had briefed the congressional intelligence committees as well as other branches of the government. The program had become controversial, one senior intelligence official acknowledged, because "it was pressing how far you go before you do something illegal or immoral."

The CIA's deputy director, John A. Gordon, called for an interagency review. A broader group of lawyers that included officials from the Pentagon and State and Justice departments was introduced to Clear Vision.

The meetings dragged on for months. The State Department representative argued that the treaty ruled out any tests involving weapons. The CIA did not back down. Projects like Clear Vision, the agency argued, were a response to specific intelligence about a possible adversary. Hence they were defensive and clearly permitted by the treaty. The debate ended without resolution. Agency officials continued to assert that the treaty gave them the right to build and test germ munitions for defensive purposes.

Nonetheless, the senior officials subsequently took a pass on some of Gene Johnson's more aggressive proposals. "Did we really need rockets?" one official recalled asking. In the wake of the controversy over Clear Vision, the agency's leadership also encouraged the CIA's biodefense specialists to expand their efforts beyond the Soviet legacy. The agency needed to understand what a low-tech terrorist could accomplish. "What can you do with a suitcase? A vial?" the official said, recalling his questions to the agency's germ experts.

Although Clear Vision had initially focused on Soviet weapons, intelligence officials said its planners recognized from the start that they would have to address the military implications of gene splicing—a view reinforced by reports of such research abroad. Scientists working at the agency's direction took the first steps to create a superbug. Intelligence officials said this work included a number of genetic manipulations to identify genes that could be moved into pathogens to make them more deadly. But that work was halted. Intelligence officials said that no target gene was ever moved into a pathogen. Even so, intelligence officials said, the agency's lawyers remained convinced that the germ treaty allowed such research.

Whether or not Clear Vision provided the invaluable insights that its champions asserted, the effort itself fizzled. By early 2001, it was evident that the CIA had lost its taste for research that edged close to the treaty's limits, although some at the agency continued to defend the project's value. Nevertheless, the program was out of money. And the agency's leadership did not press for new appropriations. Gene Johnson, who had been the deputy director of the CIA's biological group, left his management post to become a senior scientist for germ defense. The panel of outside experts set up to guide the agency's thinking on the biological threat quietly stopped meeting. Clear Vision's recombinant work was put on hold.

As word of the secret research spread within the close-knit community of scientists who advise the government on arms control issues, some raised strong objections to the CIA's rationale. Barbara Rosenberg, who had discussed the germ-warfare ban with President Clinton, insisted that the treaty did not let nations make bombs or any other kind of biological weapon. She was amazed that the government had even debated the point. Although scholars had exhaustively dissected the treaty, Rosenberg said, she knew of none who had ever advanced this interpretation of the global accord. Moreover, she said, the research was unnecessary. Whether a bomblet emitted 5 or 50 percent of its deadly cargo was ultimately irrelevant if the germs were sufficiently potent. In terms of threat assessment and defense planning, the American military had to assume the worst.

THE Pentagon had its own tangle of ambitious, sometimes overlapping secret programs to develop defenses against biological and other unconventional weapons. They had exotic code names—Bite Size, Back Star, and Druid-Tempest, to name but a few—and several aims. Some gathered intelligence to help analysts assess the emerging biological threat. Others were searching for new ways to destroy an adversary's biological program without unleashing an epidemic.

Since the mid-1990s, the Pentagon had been burying components of germ and chemical facilities in the hardened underground tunnels at the government's Nevada Test Site to see whether the army's munitions and bunker-busting bombs could destroy them. The research had its origins in the Gulf War, when Schwarzkopf and Powell agonized over whether an attack on Iraq might kill tens of thousands of civilians or allied troops.

In 1998, the Pentagon began an ambitious venture to see if it could actually build a factory capable of making germ weapons with commercially available materials. The work was commissioned by the Defense Threat Reduction Agency. DTRA, known as "ditra" by its two thousand employees, was a new fiefdom in the Defense Department, created in 1998 through the merger of several Pentagon offices that had worked separately on proliferation issues.

The project was code-named Bacchus, an inside joke, one that scientists would appreciate. Bacchus was the god of fermentation, often an essential step in making a biological weapon. DTRA officials planned to buy the equipment in the United States and overseas and then set up a working germ factory that would make harmless bacteria to simulate the manufacture of large quantities of anthrax. The agency's director, Jay C. Davis, said Project Bacchus was looking for what intelligence agencies generally call "signatures"—telltale clues of hidden things or activities. In this case, the hunt was for distinctive patterns of equipment buying. A signature of a nuclear bomb program, for example, might be widespread purchases of uranium. Davis, a physicist with Lawrence Livermore National Laboratory who was DTRA's director until June 2001, said Bacchus posed several questions: "Can I find a small biosystem being put together? Can you see the signatures? And can I characterize it well enough to attack it?" Pentagon lawyers reviewed the project and ruled that it was permitted by the treaty. No biological agent was being produced, and the purpose was clearly defensive.

In March 1999, armed with $1.6 million in funding and commercial catalogues of lab equipment, a small team of DTRA employees quietly began buying new and secondhand equipment to build a small-scale germ facility at the Nevada Test Site. Within weeks, officials had acquired almost everything it needed. From a Lowe's hardware store they had purchased pipes and filters to clean contaminated air. It took about two months more to find the most crucial piece of equipment: a fermenter in which to culture and grow the germs. A fifty-liter unit was acquired from a supplier in Europe, DTRA officials said. A company in the Midwest sold the team an off-the-shelf milling machine for grinding down clusters of dried germs to the best size for human infection.

By the summer of 2000, the team had built a functioning facility that had turned out two pounds of "product"—anthrax simulants—in test runs. The low-tech plant produced *Bacillus thuringiensis,* the biopesticide Iraq made at Al Hakam after the Gulf War, and *Bacillus globigii,* an innocuous bacterium long used to simulate anthrax. The dried particles, white and taupe colored, were one to five microns wide—the ideal size for germ weapons. If anthrax spores had been dropped into the fermenters, the United States could have made enough biological agent to mount a deadly attack.

The project had proven its point—a nation or bioterrorist with the requisite expertise could easily assemble an anthrax factory from off-the-shelf materials. DTRA officials say the overseas purchases were not detected. The results suggested that with precious little money and off-the-shelf equipment, a state, or even a group of terrorists, could build and operate a small-scale germ weapons plant, probably without the intelligence agencies' knowledge. "That's one of the reasons why it was a useful exercise," said Davis.

In late spring of 2001, DTRA found yet another use for the production plant when a squad of military commandos attacked it in a mock raid. The mission was to "neutralize," the plant, to disable the fermenter and milling machine without spreading any germs. The operation was code-named Divine Junker, and the raid was deemed a success.

Project Bacchus and the other, related research was approved by a Defense Department panel set up to make sure such operations complied with arms control treaties. But few officials outside the Pentagon—and apparently even within the department—knew much about the project. A spokesman for the defense agency said DTRA did not know whether Secretary Cohen or any other senior Pentagon official had blessed Bacchus or even been told of it. The only person in Congress apparently informed was not a senator or a congressman but a staff aide on the House Government Affairs Committee. Senior National Security Council staff aides and several White House officials claim never to have heard of it.

Moreover, there does not appear to have been anyone in government keeping an eye on the full range of secret projects under way to improve biodefenses. If the White House had known that the United States had both built a factory capable of making germ agent and tested a biobomblet, it would have demanded a much broader legal review, one official said. But senior Clinton officials had been briefed only on Clear Vision, what a top official called "one part of the iceberg" that threatened to collide with the germ treaty.

White House officials say that neither President Clinton nor his senior aides ever reviewed the full package of ongoing efforts. Was it really wise for the United States to go ahead with both weapons experiments and a factory capable of producing large quantities of biological agent? The issue was never joined at the highest levels.

SECRETIVE efforts such as Bacchus and Clear Vision were unfolding at a diplomatically delicate moment. After years of argument, the world's nations were finally moving closer to adding some means of enforcing the biological weapons convention. New language was being drafted in a "protocol" to the treaty that would require nations to list laboratories in which they were doing advanced research and submit to independent inspections. The effort to draft a protocol had the strong support of Elisa Harris, a mid-level NSC official, but it was opposed by virtually every other agency.

The Commerce Department disliked the draft, fearing it would expose America's pharmaceutical manufacturers to intrusive, unnecessary inspections in which valuable trade secrets could be lost. National security officials felt that the treaty would be impossible to verify, even with the new language. Rogue nations would still cheat. But the new treaty provisions might force the United States to list some of its secret defensive research on germ weapons. For instance, the draft protocol would have required Clear Vision's planned recombinant work to be openly declared. The text, written by a Hungarian diplomat, Tibor Toth, called for countries to disclose any "arranging and manipulating nucleic acids, organisms, and microorganisms, to produce novel molecules or to add to them new characteristics."

Officials felt there were serious questions about how an inspection regime proposed in the protocol would work. At Congress's insistence, the administration officials had established teams of mock inspectors who between 1994 and 1996 walked through several American commercial laboratories, vaccine production plants, and university and medical school labs whose research was undeniably defensive. The results were dismaying to supporters of a treaty protocol. Even innocuous experiments looked as if they were part of a germ program when seen with a jaundiced eye. The visits that were aimed at reducing doubts about the treaty had only served to deepen them.

Undaunted, Harris continued to fight for the new protocol. But the agencies resisted. The result was gridlock. In mid-2001, after a review that involved all the relevant government agencies, the Bush administration announced its opposition to the protocol. The proposed agreement, it said, was overly intrusive and unlikely to deter cheaters. Most every country, even Washington's close allies, Britain and Japan, immediately criticized the move, voicing support for the draft protocol.

As the world's nations quarreled over how to improve the treaty, the new biology was racing ahead. The audacious work done by the Soviets, the details of which continued to emerge, was seen as a portent. American officials knew that the Russians had pressed well beyond the Pentagon's experimentation in the 1980s with recombinant germs, producing lethal weapons from designer bugs. As American officials pondered what the future might look like, they often found themselves focusing on men like Sergei Popov, who had moved biological warfare to the edges of modern science and beyond.

SERGEI Popov had defected to Britain in 1992 after the collapse of the Soviet Union and had then moved to Dallas. Now, in early 2001, he was working for Advanced Biosystems Inc., a new company outside Washington. Alibek was its president, and the company was a sponge for former Soviet weapons scientists, as well as federal research dollars. The aim was to use weapon skills for peaceful ends by doing research on germ defenses. The work was helping Popov think and talk about what he had done before. "For eight years I tried to forget," he said.

"But it's very difficult to get rid of the past and become a completely new person."

Alibek had described Popov's work and Moscow's vast program to develop recombinant arms in intelligence debriefings in his 1999 book, *Biohazard*. But there were grumblings that Alibek, a manager, had moved beyond his personal knowledge. There were no such doubts about Popov. He had *lived* the future. Hardly anyone in the world had done more with the new biology to create weapons. It went far beyond the traditional ways of trying to hurt and kill. Popov and his colleagues built germs meant to seize control of the human metabolism, causing a body to self-destruct.

Popov carned his doctorate in biochemistry in 1976, just as the recombinant revolution was getting under way. He was snapped up by Lev Sandakhchiev, the Soviet biochemist then in the process of founding Vector—the secret center for research on viral warfare that Andy Weber had visited. Vector had a few dozen people when Popov arrived. It would eventually employ more than four thousand. Popov said that in the late 1970s he and his Vector colleagues began synthesizing whole genes that expressed such substances as endorphins, the pain-killing hormones that produce feelings of euphoria when lovers kiss or runners push themselves hard. The goal was bliss overload. By creating designed bugs that oozed endorphins the Soviet scientists hoped to trigger an overdose that resulted in a coma and perhaps even death. But the plan failed. The scientists spliced these synthesized genes into viruses and then infected lab animals with the modified bugs. But there were no deadly repercussions. Popov and his colleagues regrouped in the early 1980s, as the Reagan administration was entering office. One team continued on the same path and after much work found what Popov described as the best places to insert foreign genes into viral genomes, ensuring good expression of lethal effects.

Popov then tried something more insidious. Rather than combining pathogens into unnatural hybrids or inserting foreign genes into microbes so as to express new toxins, he looked for ways to trigger autoimmune responses. The aim was to trick the body's defenses into self-destruction, as happens slowly in such diseases as lupus and rheumatoid arthritis or more suddenly in cases of anaphylactic shock. In contrast, toxins seemed crude. "Sometimes just small, very tiny quantities of foreign substances are enough," he said, to make the immune response "quite devastating." He disclosed an example: his team had inserted into viruses genes that make protein fragments of myelin, the fatty material that forms sheaths around nerves. The Soviet researchers found that when an animal was infected with these hybrids, they could trigger an autoimmune response that resulted in brain damage. It was a potential breakthrough. In effect, the team had discovered a way to rapidly produce multiple sclerosis, the degenerative disease of the central nervous system in which the nerve tissue is destroyed throughout the body, resulting in gradual paralysis and death.

In the mid-1980s, Popov was sent across the country to help scientists at the Obolensk laboratories, outside Moscow, do similar pioneering work—not with viruses but with bacteria, that center's expertise. They succeeded. In one case, Popov and his Obolensk peers reengineered *Legionella,* a troublesome but usually nonlethal germ that causes the kind of pneumonia known as Legionnaires' disease. The bug had been of little interest to germ warriors, but Popov said the Soviet tinkering had turned it into a killer.

The trick again was the myelin. The scientists injected the modified *Legionella* carrying the myelin genes into guinea pigs, which at first came down with mild symptoms of pneumonia. After the *Legionella* microbe and its mild aftermath came and went, and all signs of infection were gone, the metabolic reactions to the myelin fragments began a second, more pronounced wave of sickness. The guinea pigs first suffered brain damage and paralysis, then died, with the mortality rate close to 100 percent.

"But the symptoms were really, really unusual," Popov said. "The initial paralysis was of the rear legs, so half the animal was paralyzed, and half still active," the front legs struggling to move the dead hindquarters. In an unusual step, the director of Obolensk came to visit the animal house to witness the grim results. It was easy for the scientists to conceive in rough outline how the designer germ would affect people: the infectious agent would be gone by the time it began its killing, leaving few or no traces for medical detectives trying to understand the cause of the havoc.

Popov and his colleagues had created not only a killer but a very infectious one. With normal *Legionella,* many thousands of bacteria are required to sicken lab animals. But the recombinant *Legionella* was active with only a few cells. Military bioengineers didn't need to start with a disease and make it worse. Instead, they could make a new one, outdoing what evolution and nature had produced over the ages.

Around 1989 at Obolensk, Popov presented his findings to a group of more than fifty of Biopreparat's leaders, senior scientists, and military officials. Without naming Popov, Alibek recalled in *Biohazard* what happened after the bioengineer made his presentation: "The room was absolutely silent. We all recognized the implications. . . . A new class of weapons had been found. For the first time, we would be capable of producing weapons based on chemical substances produced naturally by the human body. They could damage the nervous system, alter moods, trigger psychological changes, and even kill."

Popov said that Alibek's recounting of the experiment in *Biohazard* was imprecise. The test used guinea pigs, not rabbits, as Alibek had written, and involved the body's own reactions rather than poisonings by natural substances. Popov said his experiment was more subtle and menacing than what Alibek had described.

According to Popov, the official reaction to such research was less enthusiastic than Alibek suggested. The Soviet military officials who regularly visited Obolensk and Vector to hear about the latest advances tended to be skeptical of the futuristic research. Very little produced by the genetic engineering programs was turned into weapons before the Soviet Union collapsed, he said. His recombinant *Legionella* was never made into munitions or deployed.

An exception was plague, one of the oldest and most dangerous germ weapons. As Pasechnik had described to Lederberg and others, the Soviets had worked hard to make the epidemic disease more lethal. Popov said the scientists at Obolensk went further by splicing into plague bacteria the gene that makes diphtheria toxin, which causes fever, chills, headache, nausea, and, if not treated, heart failure. The superplague with diphtheria toxin, Popov said, was highly virulent. "I know they tested recombinant strains in chambers of animals, primates, and that the results were very promising."

Did the scientists of Obolensk and Vector wrestle with moral questions or ethical regrets? "We never doubted that we did the right thing," Popov said, echoing Bill Patrick. "We tried to defend our country." Did they ever wonder if an enemy would be deterred by programs designed to be completely secret? Popov replied, "The majority of the people around me, and I think myself, we always thought more about scientific issues than military applications." The work of designing new kinds of pathogens was a way "to survive in that system. I had no alternative."

Popov said he believes the most devastating discoveries lie in the future. "It's a very powerful technique," he said of biotechnology. "If somebody wants to employ genetic engineering in creating weapons, the most exciting discoveries are still ahead. I think it opens—every day it opens—more and more possibilities to create something dangerous, to create a new kind of weapon."

Popov said he still found the work of his former life attractive, a fascination that deeply troubled him. It was hard, he said, to stop imagining new ways to make designer pathogens. Reflection on his own impulses had led him to believe that biologists needed to be more vocal about the dangers of recombinant science. "It's a good time to express a certain concern," he said. "The whole technology becomes more and more available. It becomes easier and easier to create new biological entities, and they could be quite dangerous."

Days after the interview, Popov reported that American and Italian scientists had just published a paper on myelin peptides in the journal *Nature Immunology* that echoed the secret research he did for the Soviet Union. The scientists said they had accidentally triggered a mechanism that caused mice to self-destruct in a condition known as "horror autotoxicus," a severe allergic reaction in which the immune system tries to destroy its own tissues.

"The whole story is similar" to his Soviet research, Popov remarked. "Very similar."

THE futuristic military research by Popov, Alibek, and other scientists presented America's biodefenders with a grim challenge. The vulnerability of soldiers and civilians to attack was growing. In what appeared to be a quiet biological arms race, much of which took place behind closed lab doors, offense was outpacing defense. Even before gene splicing became routine, an agent could be perfected in a few years, while a vaccine often took a decade to make and win approval. More than a decade after the Gulf War, America's vaccine against botulinum remained experimental. Antibiotics were losing the war against pathogens, which were performing their own, natural genetic engineering, mutating beyond the reach of the most powerful drugs.

In the late 1990s, the Pentagon dramatically increased funding to find new ways of fighting infectious disease, pouring hundreds of millions of dollars for biodefense into the Defense Advanced Research Projects Agency, or DARPA, the little-known agency that had invented the Internet and stealth technology. DARPA, an arm of the Pentagon, had no laboratories or scientists of its own. Its managers wanted to underwrite the most audacious research they could find. The hope was to spur avenues of inquiry that industry had ignored or abandoned. It was understood from the beginning that the research was high risk, that many of the projects would fail.

The first director of DARPA's Unconventional Countermeasures Program, Shaun Jones, had his own clear vision of the future. A doctor and navy commander who had been a member of the navy's elite commando unit, the SEALs, Jones had traveled the world on secret missions. He believed that defense against germ weapons required radical new approaches.

The medical breakthroughs of the late twentieth century had often been driven by profit. Pharmaceutical companies had made billions targeting individual diseases or maladies. There were blockbuster drugs to fight allergies, to slow baldness, to restore sexual prowess. Jones wanted to go in the opposite direction, to search for breakthroughs that would provide widespread protection. One focus was multivalent vaccines that could prime the body's immune system to ward off a range of microbial threats. Someday, perhaps, researchers could come up with a single shot that conferred immunity against, say, plague, anthrax, and botulinum. Jones was also fascinated by the potential of antiviral drugs. The viruses, which infiltrate and hide in human cells, had largely escaped medicine's weapons. But Jones believed new research might yield new ways to attack viral enemies like smallpox.

A young colleague of Lederberg's, Jones used the clout of the eminent scientist to recruit talent for his projects. Among the first researchers he signed up was Stan Cohen, the Stanford pioneer who, with Herb Boyer, had made the first recombinant breakthrough in 1973. By 1998, the program had scientists working on forty-three different projects.

Stephen A. Johnston of the University of Texas's Southwestern Medical Center in Dallas was typical. He had long nurtured blue-sky ideas. And the National Institutes of Health, the main source of federal funding for biomedical researchers, had consistently rejected his proposals as unlikely to work and unsuitable for financial support. Backed by a DARPA grant, Johnston used the new biology to break a pathogen's genes into hundreds of different bits that he then injected into hundreds of mice. His next step was to infect the mice with the original pathogen. Typically, most fell to the onslaught, but a few exhibited resistance. In that way, Johnston discovered which DNA parts could be used to bolster the immune response and fight disease. He called the innovative method Expression Library Immunization, or ELI. "The basic idea is to let the immune system tell you what works," he said. Peers hailed his research as surprising and elegant.

Johnston's work represented a major advance for gene vaccination, a young field that promised to revolutionize the science of immunization. Traditional vaccines use weakened or killed versions of disease organisms, or inactivated toxins or proteins from pathogens, to give the body's immune system advance warning of infection and time to build up defenses. Gene vaccines were just bare DNA—often plasmids. When injected into the body and incorporated into cells, the genes expressed a limited set of the pathogen parts that were nonetheless sufficient to trigger the immune response. The approach was like a scalpel. Patients would be injected with precisely what was needed to inoculate them against a disease. Though experimental, the method showed promise. It could eliminate the risk of infection associated with some live and weakened vaccines. It also could ease production and compliance with federal regulations, the complexities of which had stymied anthrax vaccines for so long. Gene vaccines were chemical, not biological. That cut the chance of contamination and spoilage. Finally, they were also highly stable. Unlike conventional vaccines that needed refrigeration, gene vaccines could be stored dry or in solution under many conditions and temperatures, making their distribution easier.

Johnston's research was important because it sped up the identification of suitable DNA snippets, reducing the search time from a year or more down to months. And he proceeded to accelerate the process further, finding ways to mechanize it with tiny robots. His goal, which he called instant immunization, was to make a new vaccine in a day. If successful, this promised to help scientists react very quickly to attacks with designer bugs that no one had ever encountered before.

Some of DARPA's most futuristic work was done by Maxygen, a small company in Redwood City, California, that Jones visited just after its founding. It had pioneered a type of artificial evolution it called DNA shuffling. The concept was an elegant elaboration of Cohen and Boyer's discoveries. Where the pioneers of gene splicing took a gene from one organism and moved it to another, Maxygen mixed up hundreds, even thousands of genes to produce a single, new product. Nature works by a similar process of trial and error. Over millions of years, bugs mutate and a tiny number become better equipped to deal with their environment. Maxygen had found a way to fast-forward the evolutionary process by recombining genes in hundreds, even thousands of new ways. One early project involved an enzyme used in detergents like Tide to dissolve grass stains, which researchers had been trying for years to improve. Maxygen's scientists shuffled thousands of genes in different combinations until they created a new genetic blueprint, one that had never before existed in nature which involved twenty-six genes, each sliced from a different kind of bacteria. The result was a much more powerful enzyme that could be used in detergents.

Jones immediately saw the applications to biodefense. In 1998, DARPA gave Maxygen a $3.8 million contract to refine the enzyme further, making it strong enough to dissolve not only grass stains but anthrax bacteria and other germs that form hardened spores. Perhaps someday the military would have a detergent that could be sprayed over people and neutralize an anthrax attack. Another Maxygen contract, for $7.7 million in 1999, focused on developing unusually strong gene vaccines that would stimulate the human body into superimmunity against viral and bacterial invaders. The military also asked the company to develop aerosol-based vaccines that could be inhaled to safeguard people against a broad range of pathogens. A cloud of vaccine, sprayed over many square miles, was seen as potentially the simplest way to protect people and animals from epidemics.

A main goal of the research was to shuffle the genetic material that made pathogen proteins and antigens, which spur the body to make protective antibodies. By tweaking the naturally occurring antigens of, say, anthrax, the company hoped to produce a more powerful immune response. Russell J. Howard, the president of Maxygen, said the initial results were encouraging and that it appeared possible to make vaccines that were not only more powerful but, perhaps, effective against several diseases at once.

At first, many of the DARPA projects were criticized because they tended to be so radical. Work on modifying red blood cells to knock out toxins and microbes was ridiculed because no one had ever tried it before, and it was judged, for the near future at least, as merely intriguing. But other projects showed quick promise, often raising commercial interest. Shapiro at Stanford, who discovered an enzyme common to many bacteria, was widely praised for advances that promised antibiotics of broad effectiveness. And Maxygen, whose claims seemed extravagant at first, was quickly proven right as rival companies rushed to exploit the shuffling technique.

A measure of the program's success, and Pentagon approval, was DARPA's expanding budget for defense against biological weapons, which included not only work on medical treatments but research on such devices as advanced germ detectors. The annual budget went from $59 million in 1998 to $162 million in 2001 and was projected to hit $205 million by 2005. Over that time, the agency was to spend $1.2 billion, making it a new power in the world of biomedical research. However, the benefits of the DARPA-funded research for biological defense, if any, would be unclear for years, even decades. And Jones acknowledged that some projects would surely be "extraordinary failures" and that the value of others would not be known until they were rigorously tested on people. After all, a promising response in a petri dish or a mouse was no guarantee of human benefit. That kind of evaluation required the slow, painstaking, carefully regulated process of clinical trials in which doctors and volunteers took on the responsibility of searching for unexpected side effects as well as proving safety and effectiveness.

But Jones and company saw the initiative as a good insurance policy, a cheap one given the stakes.

In the last days of the Clinton administration, the Pentagon gingerly moved toward doing its own recombinant work on pathogens. The question that officials weighed for several years was whether to reproduce the vaccine-resistant strain of anthrax made by the Russian scientists. The goal was eminently practical. Pentagon officials needed to know whether the Michigan vaccine being administered to millions of American soldiers was effective against the genetically modified pathogens.

The United States had been trying to obtain a sample of the strain since Russian scientists disclosed its existence in 1995 at a conference in Great Britain. American officials arranged generous grants for the institute at Oblensk, where the strain had been made. The Russians repeatedly promised to provide the germ to their American counterparts. But the arrangements always fell apart. Perhaps the problem was Russia's domestic politics; relations between Moscow and Washington were rocky for much of the 1990s. Perhaps it was the legendary

Russian bureaucracy. Or perhaps hard-liners in the former Soviet research establishments simply did not want to relinquish something the Americans wanted so badly. Whatever the reason, the superbug never made its way to an American lab, and so, in the late 1990s, intelligence officials began pushing for the United States to make its own version.

The CIA believed the germ treaty authorized such research. But agency officials were nonetheless reluctant to make the strain without specific intelligence that somebody, somewhere was trying to turn such a pathogen into a biological weapon. The mere existence of the strain, CIA officials argued, would not be sufficient justification.

The political implications of such recombinant research worried the Clinton administration. Among scientists, genetic manipulation had long since become routine. Still, intelligence officials hesitated. The project could produce tabloid headlines: U.S. Makes Killer Superbug.

In early 2001, in the first months of the Bush administration, Pentagon officials decided that they would wait no longer for Russia to deliver the anthrax strain. The job of making the bug was turned over to a secretive DIA program known as Project Jefferson. Established in early 1998 with a modest $2 million annual budget, the project drew its name from Thomas Jefferson's injunction "Eternal vigilance is the price of liberty." Initially, the Jefferson project studied the classical germ agents, the "oldie moldies" as Patrick called them. One early effort assessed whether 8,000 to 10,000 spores were really needed to infect a person with anthrax. Much of the research was done by reviewing classified and openly published scientific literature. Other work was accomplished using computers that allowed scientists to make virtual pathogens and weapons.

To make the genetically modified anthrax, the DIA turned to Battelle, its contractor in Columbus which had also worked on Clear Vision, the CIA project. U.S. intelligence officials said that they understood the significance of the step they were contemplating, and that this was the only genetic manipulation the Pentagon had even considered. To ensure compliance with the germ treaty's restrictions, the officials said, the DIA would make only small quantities of the superbug—a gram or less. Officials said that with such amounts, the intent could only be viewed as defensive.

Pentagon lawyers had reviewed and approved the secret project, which was to be done as part of Project Jefferson. Officials said they hoped the military contractor would be able to make the novel form of anthrax and test it in animals by the end of the year. Other military scientists doubted that such an ambitious schedule could be met. While the preparatory work was underway in mid-2001, the actual creation of the recombinant anthrax had not yet begun. Experts said it would probably require a year or more to make the bug and determine whether it could defeat America's vaccine.

National Security Council officials in the Bush administration were told about the project soon after taking office and viewed it sympathetically, a senior White House official said. No NSC approval was required, he added. The secretary of defense had authority to press ahead on his own. The official said that the new administration saw the value of such work and planned to encourage more of it to help build germ defenses and protect American lives. It would be "criminal," he said, to forgo such research.

Ken Alibek said he had no quarrel with the justifications advanced for secretly studying recombinant pathogens. But the niceties would not matter, he warned. If the secret research was eventually disclosed, many critics would have a different view. The United States would be accused of cheating on the germ treaty. The former physician turned biological warrior, who had spent much of his life working in a hidden weapons program, said that the work had to be done openly if done at all.

"It can't be classified," he said. "If done secretly, you can imagine the problems."

THE government's considered decision to take a first step into recombinant pathogens stood in sharp contrast to the frenetic, unregulated chaos of modern biology. Gene splicing in commercial labs was racing ahead, and at times stumbling into dangerous new territory. As Popov had suggested, scientists in commercial labs were beginning to do openly, and increasingly by happenstance, what the Soviet military had once done deliberately in secret.

In February 2001, Australian scientists announced that they had inadvertently killed dozens of mice by making a virus that had crippled their immune systems. At first the scientists informed only the Australian government and military about the experiments, which were done in 1998 and 1999. But after intense debate, they decided to publicize their findings. Others needed to know of the potential danger in case such discoveries were turned against people. They said they had produced their inadvertent killer while trying to make the mice infertile as part of a pest-control project. The agent they were experimenting with was a virus widely used in lab studies—mousepox, a cousin of the human smallpox virus. The scientists had inserted into the virus a mouse gene that controls the production of interleukin-4, a signaling molecule that helps the immune system fight off invaders. The goal was to make the immune system of a pregnant mouse so hypersensitive that she would reject her own eggs as foreign. The scientists did this by infecting the mice with mousepox viruses that had been modified to carry the interleukin-4 genes. The result was a surplus production of this powerful immune system controller.

But when the scientists infected the mice with their designer virus, the mice not only became infertile, they died. The scientists were surprised to find that even mice vaccinated against mousepox died. Birth control had become life control.

Because people also have the interleukin-4 immune system gene, the team realized that creating a surplus of that substance in humans might kill them too. Worse, the Australians realized, a rogue state or terrorist group might be able to achieve this result with smallpox, the devastating human virus. Just as the super-mousepox had defeated the mousepox vaccine, so might genetically enhanced smallpox defeat the human vaccine, the only existing protection should the disease reemerge.

"We thought it was better that the information came out in case somebody constructed something more sinister," said Ronald J. Jackson, the head of the research team. "We felt we had a moral obligation."

The news from Australia ricocheted through Washington, whose national security community had become increasingly concerned about the nation's vulnerability to the possible re-emergence of smallpox as a terrorist threat. The Soviet Union had made smallpox by the ton. Moscow's alleged successes in manipulating pathogens—and smallpox, in particular—had underscored a new dimension in the danger and helped prompt the Clinton administration to stockpile vaccine—not just for the military but also for civilians. The effort was not going smoothly. The administration had been struggling for almost three years to produce new batches. By 2001, there were two separate vaccine programs. The military contract with one company called for 300,000 doses at a cost of $22 million; the civilian contract placed with another firm would provide 168 million doses for $343 million. While the White House had wanted the Pentagon and the Department of Health and Human Services to combine their orders to secure a lower price, the military had refused to go along. The cost difference was substantial, with the military initially paying about $70 per dose versus $2 per dose for the much larger civilian order. The Pentagon said its costs would fall as it placed more orders. To produce the civilian stockpile, HHS hired the company run by Tom Monath, the scientist who headed the CIA's panel of outside experts. But neither the civilian nor military vaccine would be ready until 2005 or 2006 at the earliest.

Now the White House had something new to worry about. If a state or terrorist group could develop a designer smallpox laced, say, with the interleukin-4 gene, then, in theory, neither the military nor the civilian vaccine might protect Americans in an epidemic. Perhaps it would take a futuristic antiviral drug to defeat the new threat.

The interleukin-4 research in Australia, one government scientist said, prompted memos to officials on the National Security Council of the new Bush administration. The smallpox vaccine now being developed, the scientists warned, might be useless.

Discussion ensued among federal officials about whether Project Jefferson should replicate the Australian research so that the United States could better assess the potential threat. In the end, the idea was dropped. There was no clear defensive rationale, no credible way to argue that a foreign threat existed that demanded such a scientific investigation. Australia was an ally, not a foe. And mousepox was not on the DIA's validated threat list. The problem was simply that science was advancing rapidly.

It seemed possible that the troubling future envisioned by the scientists who invented the new biology more than a quarter century earlier was finally arriving. Deliberately or not, some researchers were coming up with new, more terrifying ways to kill.

Matt Meselson and Josh Lederberg, the two scientists who had done as much as anyone to shape this new science, continued to differ on how to respond to the threat of germ weapons. Lederberg saw the modern trends as worrisome and advocated as many prudent defensive steps as possible. Meselson took heart from the small number of germ attacks so far and believed that mankind's restraint could be reinforced with new global agreements. Lederberg tended to look for scientific solutions, assuming that the weapons would eventually be used. Meselson advocated political approaches.

But they agreed on much. Both men conceded that their hopes for the abolition of germ weapons had fallen short, and that a distressing number of their scientific colleagues, from those in the Soviet Union to Iraq to South Africa, had devoted their lives to producing pathogens that could kill in huge numbers. And they conceded that the defensive measures they separately championed had proven more complicated than expected. Efforts to strengthen the 1972 germ treaty were failing, despite Meselson's tireless advocacy. And the military program of anthrax vaccinations had proven far more complicated, practically and politically, than Lederberg could have imagined.

Both scientists worried that the Clinton administration had exaggerated the germ threat, doing more damage than good, at least in the short run. Lederberg faulted Bill Cohen's performance with the bag of sugar. Overstating the power of biological weapons could inspire terrorists to acquire them. Meselson wondered if America's new obsession with germ weapons might raise the political stock of scientists such as Wouter Basson, the leader of South Africa's apartheid germ program. People in other countries might see the sudden U.S. interest in germ weapons, the rumors of secret projects, and the backing away from negotiations to strengthen the germ treaty as a quiet return to an interest in offensive arms. As a hedge, they might embark on germ programs of their own.

Both men agreed that the germ threat was real. To critics who claimed it was exaggerated, Lederberg replied with a question. "Are there nuts in the world? Yes. So we could be in trouble."

Both scientists expressed doubts about the wisdom of wide vaccinations against future threats. For civilians, Lederberg said, his inclination was to put off the question of smallpox vaccinations, which had health risks and large financial costs. The focus should turn to developing antiviral drugs—few effective ones now existed—that in the future might make smallpox epidemics easier to control. His thinking was consistent with the federal plan, which was to stockpile new smallpox vaccine for civilians but use it only to help contain epidemics if they broke out.

Meselson's hope for the future was a global accord that would make the production and use of germ weapons a crime. The existing treaty applied to states, not individuals. A new accord based in criminal law might deter national leaders from seeking to develop germ weapons, discourage businesses from assisting them, and keep leading nations from looking the other way in the face of violations. He noted that the

United States is already party to seven treaties of this nature, including ones against hijacking airplanes, harming diplomats, and taking hostages. The new pact would simply give national courts jurisdiction over germ criminals from other countries.

Lederberg said that making superbugs had turned out to be more difficult than first thought. Perhaps only 1 percent of the efforts were destined to succeed. But with enough time and investment in scientific resources, he noted, that rate of success, though low, was enough to produce new dangers. After all, Popov had created new superbugs, and even civilian labs such as the one in Australia were doing work that might be misused.

Similarly, Lederberg was of two minds about ethnic weapons. Many experts dismissed them as science fiction, though Russia and South Africa had studied them. Lederberg noted that racial mixing and inherent ambiguity over definitions of ethnicity meant that no single genetic marker would give an attacker sufficient leverage. But in a half century or so, he said, science might become powerful enough so that it could find a subtle combination of markers common to particular races. That might work as an ethnic weapon, Lederberg said, but why bother? There were much easier ways to use germ weapons against particular ethnic groups. Anthrax or other traditional killers would be quite effective in spreading death.

Meselson also saw reassuring and alarming signs for the future. Foreign research on designer pathogens was often inept and illogical and done simply because ambitious scientists sought to please their military bosses. But while the topic of biological weapons had suffered much exaggeration, he said, the danger was serious for the long run.

In the future, Meselson said, germs might be designed not only to kill but to manipulate all the life processes—cognition, development, reproduction, everything. They would, in short, bestow the power to change what it means to be human. He posed a troubling question: might some group in the distant future use such powers to try to enslave others?

Hitler, he noted, wanted to enslave the Poles and keep them as a nation of workers for Germany. "Are we really so sure that we're completely enlightened after ten thousand years of recorded history, even though Hitler was not that long ago?" he asked. "Are we now cured of such things? I don't know."

Conclusions

Is the threat of germ weapons real or exaggerated? Our answer is both. At key points in recent years, senior officials overstated the danger of biological attack, harming their cause with hyperbole. In most conditions, a five-pound bag of anthrax could kill many people, but not as many as half the inhabitants of Washington, D.C. Similarly, political leaders undermined their credibility by asserting that a biological attack against the United States was inevitable in the next few years—a matter of "not if, but when."

No such certainty exists. Germ weapons played a minor role in twentieth-century warfare and terrorism, when scientists invented new means of slaughter on an industrial scale, from atomic bombs to gas chambers. Yet most nations chose not to use the germ weapons when they went to war. Terrorists have blown up buildings, hijacked scores of airplanes, and murdered indiscriminately in the past three decades. Only a handful of groups have attempted biological attacks and fewer still have succeeded. Thus far, the Rajneeshees' germ assault on the citizens of The Dalles, Oregon, in 1984 has proven to be an anomaly.

Nonetheless, we conclude that the threat of germ weapons is real and rising, driven by scientific discoveries and political upheavals around the world. As Aum Shinrikyo's failed efforts suggest, the crucial ingredient in a successful biological attack is not advanced laboratory equipment or virulent microbes alone, but knowledge. Such expertise is increasingly available. With the collapse of the Soviet Union, thousands of scientists skilled in biological warfare found themselves unemployed or penniless. The demise of South Africa's apartheid system and Iraq's defeat in the Gulf War have also added to the talent pool from which weapons scientists can be recruited.

Drugs and vaccines are now made throughout the world, giving many nations the ability to manufacture germ weapons. It took the United States and the Soviet Union more than a decade of trial and error to master the secrets in the 1950s. Three decades later, Iraqi scientists learned how to make thousands of gallons of anthrax and botulinum in just a few years.

The contrast to nuclear weapons illustrates why many call germ weapons the "poor man's atom bomb." A nation that obtains plans for a crude nuclear device is at the beginning of a complex technical challenge that requires staggering, easily detectable investments in mines, factories, and nuclear reactors. But scientists like Bill Patrick or Ken Alibek say they could teach a terrorist group how to make devastating germ weapons from a few handfuls of backyard dirt and some widely available lab equipment.

The emergence of the United States as the world's most powerful nation has made biological attack more likely. Adversaries that resent America's global dominance, envy its wealth, or fear its overwhelming military power can fight back most effectively with unconventional weapons. The attack on the U.S.S. *Cole,* in which a modern warship was crippled and nearly sunk in October 2000 by a dinghy packed with explosives and detonated by suicide bombers, showed how the seemingly powerless can strike a devastating military blow. In the coming years, those willing to die for their cause may well choose instead to become smallpox carriers or Marburg martyrs.

The pace of scientific advance has also intensified the germ threat. Cohen and Boyer's pioneering work a quarter century ago helped give scientists new tools to improve crops and cure disease. But genetic manipulation can also be exploited to disorient, maim, and kill. No one can predict the recombinant future. But for now it appears that the genetic revolution may produce few killers of the sort found in *The Cobra Event,* few agents such as brainpox. In real life, the most likely danger is that classic agents, the so-called oldie moldies, will be turned into custom pathogens that can defeat drugs, antidotes, and vaccines. The spread of recombinant knowledge through scientific exchanges and commerce has given even modestly skilled scientists the means to create havoc. In a decade or two, terrorists may no longer need to hire their own germ experts.

The threat is magnified by a unique feature of germ weapons—uncertainty. Explosive bombs leave few doubts about their toll. But in a biological attack, city officials would not immediately know the source and nature of the outbreak or the true number of victims. The authorities would be forced to assume the worst, Josh Lederberg has argued, and strive to protect as many people as possible, inevitably deepening panic.

The world's response to the growing dangers of germ weapons has fallen far short of what is needed. The 1972 treaty was a useful step, cementing an international consensus that this class of weapons is morally heinous. The subsequent conduct of the Soviet Union, Iraq, and South Africa, all of which signed the pact and then flagrantly violated it, underscores its inherent weaknesses. We believe that the recent effort to strengthen the treaty by providing some means of enforcement should be pursued. A balance can be struck between the need to protect legitimate research and the imperative of investigating suspected cheaters. The world's experience with Iraq clearly shows the limits of international inspections. But there is merit in reinforcing the moral taboos associated with germ weapons as well as in raising the practical and political costs of building biological bombs. Iraq kicked its germ efforts into high gear shortly after the world shrugged off Saddam Hussein's chemical attack on Kurdish civilians.

Leading scientists have quietly begun debating whether genetic engineers and other practitioners of the new biology should adopt their own version of the Hippocratic oath—do no harm. Such a step, while mostly symbolic, would help researchers become more sensitive to the ethical consequences of their work.

In recent years, the world's nations have created international tribunals to prosecute war crimes. We believe the weight of international law should be brought to bear on those who traffic in biological weapons, as Matthew Meselson and others have suggested. Closer international cooperation is also needed to limit the sales of deadly pathogens by germ banks. The United States wisely tightened its rules on such sales in 1996 after a right-wing survivalist attempted to buy plague germs from the American Type Culture Collection, the same company that had sold anthrax to Iraq in the late 1980s. We recognize that the global fight against infectious diseases requires the free exchange of dangerous germs among scientists. But once again, a balance must be struck.

Each nation should also enact tougher laws, as the United States has done. Ma Anand Sheela, a leader of the Rajneeshee cult convicted on charges of assault, attempted murder, arson, and poisoning hundreds of people in Oregon, was released from prison in 1988 after serving only twenty-nine months. (Sheela subsequently moved to Switzerland and began a new career as a nursing home operator.) Today, the punitive measures that Congress passed mean such a crime would likely earn a much longer sentence.

The Clinton administration made a real effort to confront the germ threat. Several of its initiatives deserve praise. After visiting facilities like Vector and Stepnogorsk, we conclude that the benefits of cooperative programs with the former Soviet Union far outweigh the potential dangers. Such programs, of course, have their risks. Some American officials believe that Russia's military is still making germ weapons at laboratories that remain closed to outsiders. And Moscow has continued to lie about its past biological efforts, even in the face of irrefutable evidence.

But the gains of cooperation have been substantial. Intelligence officials acknowledge that much of what they have learned about the former Soviet program came from the open scientific exchanges arranged by Andy Weber and other officials. In April 2001, Weber and his colleagues were the first Westerners to visit a scientific institute in Kazan, the capital of Tatarstan, a previously unknown corner of the Soviet germ empire where scientists developed biological weapons to infect livestock.

Such scientific exchanges build relationships that cross borders, creating constraints even for authoritarian governments. As Dave Franz, the former head of germ defense at Fort Detrick, points out, a Russia in which key scientists correspond regularly by e-mail with their Western counterparts will be less able to sustain a covert germ weapons program.

The Clinton administration's belated investments in public health, basic research, disease surveillance, and stockpiling of antidotes make sense. They are likely to save lives even if the United States never faces a germ attack. Even so, America in some ways is less prepared today for an outbreak of deadly disease than it was a half century ago. The largest hospital in The Dalles, Oregon, had 125 beds when the Rajneeshees struck in 1984. Today it has 49. Some public-health clinics lack microscopes, let alone computers or access to the Internet. New York's experience with West Nile virus demonstrated the crucial role of the public-health system in detecting and countering a germ outbreak, however it originates. A positive step would be the development of computerized systems that help epidemiologists track novel infectious diseases.

There was significant waste in the Clinton years. Biological defense turned into an entitlement program for federal agencies, private contractors, and government consultants. While many projects were well intentioned, the administration failed to give any single official the authority to eliminate duplications, enforce discipline, and direct the money to where it would do the most good. The National Guard's travails are but one example.

Although the president delivered two major addresses on the dangers of biological warfare, the administration never marshaled its resources to strengthen the germ treaty or to promote a coherent program of biodefense. Distracted by impeachment, the administration's attention to these issues was intense but episodic, a poor recipe for explaining a complex problem to the American people. It took Richard Danzig more than a year to persuade the Pentagon's commanders, an unusually well-informed group, to step up their responses to the germ threat. No one should expect Congress or the public to be more easily swayed. Programs like compulsory anthrax vaccinations must be explained in detail and candidly.

The same is true of the intelligence agencies' efforts to counter germ warfare. The 1972 treaty permits almost any kind of research in the name of defense. Some of this work is unquestionably justifiable. Other research edges closer to the blurriest of lines. We can imagine defensive research that a nation might legitimately keep secret—such as experiments exploring the vulnerabilities of existing vaccines. But American officials should disclose the existence of such research and its general outlines whenever possible. Few secrets are kept forever in a democracy, and the revelation of research into superbugs, or the building of germ bomblets, inevitably stirs deeper suspicions. The United States undermines the treaty it helped create and its own moral authority when it cloaks the most significant aspects of its defensive work in secrecy.

The experts who met with President Clinton in 1998 urged him to create a panel of experts who would provide an independent assessment of this secret work. While the CIA during the Clinton administration did create such a committee, it met infrequently and lacked the clout, access, and mandate to review programs that were spread throughout the government. We believe that the experts were right. Such a committee with real powers of oversight would build confidence at home and abroad in America's biological agenda.

A half century ago, a group of eminent citizens warned James Forrestal, the first secretary of defense, that the United States was defenseless against germ attacks. But its recommendations for better intelligence, more research, drug stockpiles, and medical surveillance systems were largely ignored. Over the next five decades, a series of American presidents confronted the problem, considered various remedies, and shuffled the issue into the "too hard" box. Such denial is understandable. Biodefense has no natural political constituency in Washington. The military-industrial complex that supports weapons systems has little interest in vaccines and public health.

"Plans should be prepared for the establishment of adequate laboratory and vaccine production facilities and stockpiles of essential basic medical supplies in the event the danger from enemy attack appears imminent," Forrestal's committee concluded in 1949. "Prompt action should be taken to establish a civil defense program."

Those words could have been written yesterday. The question is whether the United States will be able to wait another fifty years to act on them. If we as a nation believe that the germ threat is a hoax, we are spending too much money on it. But if the danger is real, as we conclude it is, then the investment is much too haphazard and diffuse. We remain woefully unprepared for a calamity that would be unlike any this country has ever experienced.

Notes

A Note on Sources

Parts of the narrative draw on archives of personal and federal documents available on the Internet. For the American program to make biological weapons, thousands of federal papers are accessible through the Defense Technical Information Center, Fort Belvoir, Virginia. Many of these documents are listed at http://www.dtic.mil. Copies can be ordered by mail or on the Internet. Reviewing a wider selection of the center's papers requires petitions under the Freedom of Information Act. Below, we refer to all papers from this archive as DTIC. Hundreds of Joshua Lederberg's papers, scientific reports, newspaper columns, and letters are available in the Joshua Lederberg Papers of the National Library of Medicine, Bethesda, Maryland, and at http://profiles.nlm.nih.gov/BB. We refer to these materials as Lederberg Papers. Finally, thousands of sensitive federal papers and declassified intelligence reports about Iraq, the 1991 Gulf War, and its aftermath are available at http://www.gulflink.osd.mil. We refer to these documents as Gulflink. Military officers are identified in the text by the ranks they held at the time of the events. In the notes, we refer to officers by the ranks they attained at retirement.

1. The Attack

This chapter is based on interviews that began in December 1999 and continued throughout 2000 with individuals in Oregon who were affected by the Rajneeshee salmonella attacks. They include Jeanie Senior, then a correspondent for the *Oregonian* newspaper, who covered the cult throughout its troubled tenure, including the prosecution of its members. Also interviewed were Michael Skeels, Carla Chamberlain, H. Robert Hamilton, Judge William Hulse, Dave Lutgens, Lynn Enyart, Dan

Ericksen, Arthur Van Eaton, Patricia Johnson, Karen Le Breton, and Dave Frohnmayer. Outside Oregon, interviews were conducted with scientists and experts on germ weapons and public health, including Thomas J. Török, Robert V. Tauxe, William C. Patrick III, and W. Seth Carus.

15 *a religious cult known as the Rajneeshees:* Hugh Milne, *Bhagwan: The God That Failed* (New York: St. Martin's Press, 1986); Lewis F. Carter, *Charisma and Control in Rajneeshpuram: The Role of Shared Values in the Creation of a Community* (New York: Cambridge University Press, 1990); "For Love and Money," a twenty-part series by Scotta Callister, James Long, and Leslie L. Zaitz in the *Oregonian,* June 30, 1985–July 19, 1985; "On the Road Again," a special report by Scotta Callister, James Long, and Leslie L. Zaitz in the *Oregonian,* Dec. 30, 1985; Frances FitzGerald, "A Reporter at Large: Rajneeshpuram—I," and "Rajneeshpuram—II," *New Yorker,* Sept. 22, 1986 and Sept. 29, 1986; Frances FitzGerald, *Cities on a Hill: A Journey Through Contemporary American Cultures* (New York: Simon & Schuster, 1986).

15 *The group had left Poona:* "For Love and Money," *Oregonian,* June 30, 1985. Satya Bharti Franklin, *The Promise of Paradise: A Woman's Intimate Story of the Perils of Life with Rajneesh* (Barrytown, N.Y.: Station Hill Press, 1992), chapters 1,2,3.

17 *Incorporation gave the Rajneeshee police:* Interview, W. Seth Carus, and W. Seth Carus, "The Rajneeshees (1984)," in *Toxic Terror: Assessing Terrorist Use of Chemical and Biological Weapons,* edited by Jonathan B. Tucker (Cambridge, Mass.: MIT Press, 2000), p. 119. Interviews, law-enforcement officials.

17 *Once the Rajneeshees controlled Wasco:* Interview, Dan Ericksen.

18 *identified the bacteria as:* The bacteria's formal name is *Salmonella enterica.* The specific serotype is Typhimuriam, a common cause of food poisoning. Though the bacteria is written as *Salmonella* Typhimuriam, we have chosen to use *Salmonella typhimuriam,* the style that is favored by *The Merck Manual,* a widely known medical reference book. Interviews with Carus, Chamberlain, and Skeels.

20 *the largest outbreak in Oregon's history:* Thomas J. Török, Robert V. Tauxe, Robert P. Wise, John R. Livergood, Robert Sokolow, Steven Mauvais, Kristin A. Birkness, Michael R. Skeels, James M. Horan, and Laurence R. Foster, "A Large Community Outbreak of Salmonellosis Caused by Intentional Contamination of Restaurant Salad Bars," *Journal of the American Medical Association* (Aug. 6, 1997), pp. 389–95.

20 *an exhaustive epidemiological investigation:* Interviews, Thomas Török, Robert Tauxe; Török et al., "A Large Community Outbreak."

21 *They had not tested him for salmonella:* Interview, Carla Chamberlain; The Dalles had not reported a single case since 1978, and before 1978, only one or two cases a year were reported. "The Town That Was Poisoned," remarks of Rep. James Weaver, Democrat of Oregon, *Congressional Record,* 99th Cong., 1st sess., Feb. 28, 1985, 131, no. 23.

22 *Hulse and the other commissioner suspected:* Interview, Hulse.

23 *In a preliminary report:* "Preliminary Report—Salmonellosis Outbreak, The

Dalles, Oregon, September 1984"; interoffice memo, State of Oregon, Nov. 7, 1984.

23 *report issued in January 1985:* Weaver, "The Town That Was Poisoned."

24 *formed a joint task force:* Interviews, H. Robert Hamilton, Dave Frohnmayer.

25 *a smoking gun:* Interview, W. Seth Carus, and Carus, "The Rajneeshees," pp. 227–28; interviews, Michael Skeels.

26 *that Puja had given Haldol:* Oregon Department of Justice and Oregon State Police, report of interview, Krishna Diva, Nov. 25, 1985, p. 42.

26 *ordered her assistants to fabricate records:* Ava Kay Avalos interrogation, transcribed Oct. 22, 1985, attorney general's files, pp. 9–10; interviews, law-enforcement officials.

27 *An invoice from the ATCC:* In a telephone interview, a spokeswoman for the ATCC initially denied that the germ bank had supplied any germs to the cult. But after being provided with a copy of the slide showing the ATCC invoice, the company said that it was "not at liberty to discuss the issue" since it was "under continuing litigation."

27 *the cult ordered and received:* Michael Skeels provided a copy of the invoice, dated Sept. 25, 1984, listing several dangerous agents that the cult had ordered between the two waves of the salmonella outbreak. It was apparently found by law-enforcement officials during their search of the ranch in 1985, but Dr. Skeels said it was not shared with him or with other public-health officials until 1998. Drs. Török and Tauxe also confirmed that they had not seen the slide or been told about the other agents that Puja had ordered.

27 *Francisella tularensis:* Interviews, Michael Skeels, William C. Patrick III.

27 *Puja had obtained orders:* Interview, Michael Skeels.

27 *Puja was also particularly fascinated:* Interviews with W. Seth Carus and Carus, "The Rajneeshees," pp. 126–27; interviews with law-enforcement officials and William C. Patrick III.

28 *Puja did some experiments:* Carus, "The Rajneeshees," p. 125.

29 *Giardia lamblia:* Carus, "The Rajneeshees," p. 126; interviews with law-enforcement officials, especially H. Robert Hamilton.

29 *finally settled on:* Carus, "The Rajneeshees," pp. 127–28; interviews with law-enforcement officials, Michael Skeels.

29 *to poison Judge Hulse:* Ava Kay Avalos interrogation, p. 16.

30 *attempts were never proven in court:* Interviews, law-enforcement officials.

31 *contingency plans to "snatch" the Bhagwan:* Interviews, law-enforcement officials.

2. Warrior

This chapter is based on interviews with Bill Patrick conducted between 1998 and 2001, as well as personal papers and biographical materials that he shared with us. Its journalistic predecessor is a profile of Patrick by William J. Broad and Judith Miller, "Scientist at Work: Bill Patrick; Once He Devised Germ Weapons; Now He Defends Against Them," *New York Times,* November 3, 1998, p. F1. Other interviewees included Herbert F. York, Robert S. McNamara, Philip D. Zelikow, Leonard A. Cole, Susan Wright, Riley D. Housewright, and Matthew S. Meselson. A number of current and former federal officials and germ experts, most speaking on the condition of anonymity, confirmed aspects of this secretive history or in some cases went beyond what Patrick was willing to say. One such individual was Housewright, who was scientific director of Fort Detrick between 1956 and 1969. We have endeavored to make such sourcing distinctions clear in the text and notes. In contrast to interviews and private papers, the references below tie the story of Bill Patrick and the American germ-warfare program to histories, science reports, and declassified federal documents. At times these documents, like some federal officials, go further in their disclosures than Patrick would.

34 *one worker falling ill:* "Detrick Infections: 1943–1969," viewgraph, U.S. Army, Fort Detrick, Maryland, undated.

35 *small size of the particles:* Frederick R. Sidell et al., editors, *Medical Aspects of Chemical and Biological Warfare* (Washington, D.C.: U.S. Army Surgeon General, 1997), p. 440. This textbook, 721 pages long, is the best single federal reference on the history and science of biological warfare.

35 *eight billion lethal doses:* This figure comes from comparisons of U.S. and potential Iraqi biological weapons made on the eve of the 1991 Gulf War. "Iraq Biological Warfare Threat," Armed Forces Medical Intelligence Center, Oct. 22, 1990. Gulflink.

35 *nearly a thousand American soldiers:* Robert L. Mole and Dale M. Mole, *For God and Country: Operation Whitecoat, 1954–1973* (Brushton, N.Y.: Teach Services, 1998), p. 35. Patrick said fewer than half of the 2,200 volunteers took part in the experiment.

35 *Ohio State Penitentiary:* Samuel Saslaw et al., "Tularemia Vaccine Study," *Archives of Internal Medicine,* May 1961, pp. 121–33.

35 *sprayed American cities:* Leonard A. Cole, *Clouds of Secrecy: The Army's Germ Warfare Tests over Populated Areas* (Totowa, N.J.: Rowman & Littlefield, 1988).

35 *456 times:* "Detrick Infections: 1945–1969."

35 *All but three survived:* Norman M. Covert, *Cutting Edge: A History of Fort Detrick, Maryland* (Fort Detrick: U.S. Army, 1997), p. 41.

37 *Germs and warfare are old allies:* Erhard Geissler and John Ellis van Cortland Moon, editors, *Biological and Toxin Weapons: Research, Development and Use from the Middle Ages to 1945* (New York: Oxford University Press and Stockholm International Peace Research Institute, 1999), pp. 8–34; Sidell et al., *Medical Aspects,* pp. 416–17.

38 *many countries that investigated:* Sidell et al., *Medical Aspects,* p. 32.

38 *Roosevelt publicly denounced:* Cole, *Clouds of Secrecy,* p. 12.

39 *degree of secrecy:* Covert, *Cutting Edge,* p. 17; for the army's view on early germ developments, see Sidell et al., *Medical Aspects,* pp. 427–28.

39 *metropolis of 250 buildings:* Richard M. Clendenin, "Science and Technology at Fort Detrick, 1943–1968" (Fort Detrick Technical Information Division, Frederick, Md., Apr. 1968), p. 19; for an overview of Detrick's early days, see Sidell et al., *Medical Aspects,* pp. 42–44.

39 *The post was ringed:* Covert, *Cutting Edge,* pp. xl, xli, 17–25, 37–40.

39 *a pound . . . could kill a billion people:* "Iraq Biological Warfare Threat." For an overview of botulinum, see Sidell et al., *Medical Aspects,* pp. 643–54.

40 *thousands of records from Japan:* Sheldon H. Harris, *Factories of Death: Japanese Biological Warfare, 1932–45, and the American Cover-up* (New York and London: Routledge, 1994); Ralph Blumenthal with Judith Miller, "Japan Rebuffs Requests for Information About Its Germ-Warfare Atrocities," *New York Times,* March 4, 1999, p. A12.

40 *In a secret report of July 1949:* Caryl P. Haskins et al., "Report of the Secretary of Defense's Ad Hoc Committee on Biological Warfare," July 11, 1949. The report was prepared at the request of James Forrestal, the first secretary of defense, but after his death in May 1949 was delivered to his successor, Louis Johnson.

41 *Theodor Rosebury . . . assailed germ weapons:* Theodor Rosebury, *Peace or Pestilence* (New York: Whittlesey House, 1949).

41 *Patrick signed a waiver:* Covert, *Cutting Edge,* p. 40, refers to the general waiver practice for all new employees.

41 *a hollow metal sphere:* Ed Regis, *The Biology of Doom: The History of America's Secret Germ Warfare Project* (New York: Holt, 1999), pp. 132–33.

42 *eleven patients:* Cole, *Clouds of Secrecy,* pp. 52–54, 75–104. The book devotes two chapters to the case of Edward J. Nevin, who died of the infection in the Stanford hospital, and portrays the trial judge in the 1981 suit against the government as sympathetic to federal defendants and hostile to the plaintiff, the grandson of the deceased man. The judge intervened at one point to bar a scientist from testifying for the plaintiff. Among the expert witnesses for the plaintiff was Matthew S. Meselson, a Harvard biologist. In the end, the judge found that the government could not be sued in the matter and that the death "was not the proximate or direct result" of the germ release.

42 *if left untreated:* Mark H. Beers and Robert Berkow, editors, *The Merck Manual of Diagnosis and Therapy,* 17th ed. (Whitehouse Station, N.J.: Merck Research Laboratories, 1999), pp. 1215–16.

42 *"many Negroes, whose incapacitation":* Charles Piller and Keith R. Yamamoto, *Gene Wars: Military Control over the New Genetic Technologies* (New York: William Morrow, Beech Tree Books, 1988), pp. 100, 272. See also Sidell et al., *Medical Aspects,* p. 429.

42 *code-named Project Saint Jo:* Matthew Meselson, "Averting the Hostile Exploitation of Biotechnology," *CBW Conventions Bulletin,* June 2000, pp. 16–19.

43 *The "dosage area," experimenters wrote:* K. L. Calder et al., "Preliminary Discussion of Methods for Calculating Munition Expenditures, with Special Reference to the St Jo Program" (Camp Detrick, Frederick, Md., Aug. 11, 1954), pp. 74, 76.

43 *skin lesions:* Sidell, et al., *Medical Aspects,* pp. 503–08.

44 *"may be particularly attractive": Ibid.,* p. 439.

44 *killed more than 99 percent:* Arnold J. Levine, *Viruses* (New York: Freeman, Scientific American Library, 1992), pp. 209–11.

44 *virus at Eure-et-Loire:* John Postgate, *Microbes and Man,* 4th ed. (New York: Cambridge University Press, 2000), p. 145.

45 *a part of the pox family:* Sidell, et al., *Medical Aspects,* pp. 541–42.

45 *studying how the smallpox virus:* F. Fenner et al., *Smallpox and Its Eradication* (Geneva: World Health Organization, 1988), pp. 102, 165, 195.

45 *fertilized chicken eggs were found ideal:* Sidell, et al., *Medical Aspects,* p. 44.

45 *injected viruses into eggs:* For a description of a parallel process of viral incubation in the germ-warfare program of the Soviet Union, see Ken Alibek with Stephen Handelman, *Biohazard: The Chilling True Story of the Largest Covert Biological Weapons Program in the World—Told from the Inside by the Man Who Ran It* (New York: Random House, 1999), pp. 111–12.

46 *yellow fever virus:* An article published in 1999 details several experiments conducted by the United States to perfect yellow fever as a weapon: Operations Bellwether, Big Buzz, Magic Sword, and Big Itch. Some were conducted by a military contractor, Lewis P. Gebhardt, then the head of the Department of Bacteriology at the University of Utah, and by the Chemical Corps. In two field trials conducted in 1959, known as Project Bellwether, ten men were "placed equidistantly around the perimeter of one circle" and "100 insect vectors were released." The study concluded that wind speed was the most important factor affecting biting activity. It also demonstrated that the "overall outdoor biting activity for the mosquito was estimated to be some 40 bites per hundred mosquitoes in the time period studied."

In Bellwether Two, with fourteen field trials between August and late October 1960, scientists learned that "Biting activity was lowest when the human volunteers were continually moving," and that "Individuals near buildings were bitten more often than those in open areas."

Operation Big Itch led the scientists to conclude that "fleas could be successfully transferred, reared to an appropriate stage of development, and delivered to a target with few if any of the fleas dying in the process."

Operation Big Buzz, conducted in Georgia and involving about a million mosquitoes, found that yellow fever was "the most likely antipersonnel BW system that could be used by the Warsaw Pact countries against the United States or the European NATO nations, since the mosquitoes could be stored for two weeks, placed in E-14 munitions, released at a height of 90 metres," and fly on the wind and their own steam as far as 610 miles downwind from the target release sight.

The authors of article had the following observation about the information they had collected thanks to America's Freedom of Information Act: "a political system confident enough to release this type of information has to be admired." Alastair Hay, "A Magic Sword or a Big Itch: An Historical Look at the United States Biological Weapons Programme," *Medicine, Conflict and Survival,* 15 (1999), pp. 215–34.

46 *fifty different viruses:* Today, Western counterproliferation experts recognize a core list of forty-three viruses, four rickettsiae, nineteen bacteria, and fourteen toxins. See Tom Mangold and Jeff Goldberg, *Plague Wars: The Terrifying Reality of Biological Warfare* (New York: St. Martin's Press, 1999), p. 379.

46 *Seventh-Day Adventists:* Mole and Mole, *For God and Country,* pp. 17–41.

47 *virus pilot plant:* Clendenin, "Science and Technology," p. 63, says that a virus and rickettsia laboratory was completed in 1955 at a cost of $4 million.

47 *Detrick's own eight ball:* Regis, *The Biology of Doom,* p. 171; Clendenin, "Science and Technology," pp. 54–55.

47 *Q fever can progress:* Sidell et al., *Medical Aspects,* pp. 528–30.

47 *About one in a hundred:* Beers and Berkow, *The Merck Manual,* p. 1233.

47 *a single invader:* Sidell et al., *Medical Aspects,* p. 526.

47 *first American field trial:* Regis, *The Biology of Doom,* pp. 3–6, 172–75.

48 *looked like Marilyn Monroe:* Videotape, United States Air Force Counterproliferation Center, Air War College, "Biological Warfare Briefing by Bill Patrick, Microbiologist," AUTV Video Production, Air University, Maxwell Air Force Base, Ala., F3569-99-0019, Feb. 23, 1999. Our thanks to the provider of this tape, who was not Bill Patrick and who shared it with us on the condition of anonymity.

48 *Georgi Zhukov, told:* Sidell et al., *Medical Aspects,* p. 54.

48 *U-2 spy plane:* Michael R. Beschloss, *Mayday: Eisenhower, Khrushchev and the U-2 Affair* (New York: Harper & Row, 1986), p. 121; the Federation of American Scientists in Washington, D.C., has assembled a fascinating gallery of declassified American spy photos of Soviet biological tests sites.

49 *glimpse of a large enterprise:* Valentine Bojtzov and Erhard Geissler, "Military Biology in the USSR, 1920–45," in Erhard Geissler and John Ellis van Cortland Moon, editors, *Biological and Toxin Weapons: Research, Development and Use from the Middle Ages to 1945* (New York: Oxford University Press and Stockholm International Peace Research Institute, 1999), pp. 153–67; Alibek, *Biohazard,* pp. 32–38, 111–12.

49 *make weapons from bacteria:* Sidell et al., *Medical Aspects,* pp. 50–51, 429.

50 *The disease strikes suddenly to cause: Ibid.,* pp. 572–73.

50 *Eisenhower was briefed:* Memorandum, "Discussion at the 435th Meeting of the National Security Council, Thursday, February 18, 1960," Ann Whitman file, Eisenhower papers, 1953–1961, Dwight D. Eisenhower Library.

51 *Eisenhower was trying to leave a legacy:* Stephen E. Ambrose, *Eisenhower: The President* (New York: Simon & Schuster, 1984), pp. 554–63.

52 *strongly endorsed a crash program:* Regis, *The Biology of Doom,* pp. 185–86.

52 *"line-source disseminator":* U.S. Army, Dugway Proving Grounds, "Field Evaluation of the General Mills Liquid Agent Line Source Disseminator, E42," Final Report, July 1963. DTIC, AD338223.

53 *720 bomblets:* Sidell et al., *Medical Aspects,* p. 59.

53 *undertaken in Okinawa: Ibid.,* p. 60.

53 *Expansion produced a new military bureaucracy:* Regis, *The Biology of Doom,* pp. 186, 198; William C. Patrick III, "A History of Biological and Toxin Warfare," in Kathleen C. Bailey, editor, *Director's Series on Proliferation,* vol. 4 (Livermore, Calif.: Lawrence Livermore National Laboratory, May 23, 1994), p. 17.

54 *Lemnitzer wrote McNamara in March 1962:* Memorandum, L. L. Lemnitzer, Chairman, Joint Chiefs of Staff, to the Secretary of Defense, "Subject: Justification for US Military Intervention in Cuba," Mar. 13, 1962. This formerly top-

secret letter, and an attached report on pretext strategies, can be found at the National Security Archive, www.gwu.edu/%7/Ensarchiv/news/20010430/. The proposal for pretext operations was disclosed in James Bamford, *Body of Secrets: Anatomy of the Ultra-Secret National Security Agency from the Cold War through the Dawn of a New Century* (New York: Doubleday, 2001), pp. 82–91.

54 *President Kennedy ordered:* Anatoli I. Gribkov and William Y. Smith, *Operation Anadyr: U.S. and Soviet Generals Recount the Cuban Missile Crisis* (Chicago: Edition Q, 1994), pp. 141–42; Dino A. Brugioni, *Eyeball to Eyeball* (New York: Random House, 1990), pp. 256, 371, 409–11; Aleksandr Fursenko and Timothy Naftali, *One Hell of a Gamble* (New York: Norton, 1997), p. 212; Seymour M. Hersh, *The Dark Side of Camelot* (Boston: Little Brown, 1997), pp. 341–42.

54 *Americans wounded or killed:* Hersh, *The Dark Side,* pp. 341, 355; Brugioni, *Eyeball to Eyeball,* pp. 297, 411.

55 *"The concept was that if we got into a shooting war":* Videotape, United States Air Force Counterproliferation Center, Air War College, "Biological Warfare Briefing by Bill Patrick, Microbiologist."

56 *The symptoms included chills:* Sidell et al., *Medical Aspects,* pp. 626–27.

56 *Its incubation period varied: Ibid.,* pp. 572–73.

56 *Its incubation period was ten: Ibid.,* pp. 524, 528–30.

57 *Marshall Plan was never implemented:* Over the decades, Cuba has charged that the United States repeatedly attacked it with germ weapons, even though no convincing evidence of such assaults has ever come to light. See Raymond A. Zilinskas, "Cuban Allegations of Biological Warfare by the United States: Assessing the Evidence," *Critical Reviews in Microbiology,* 24, no. 3 (1999), pp. 173–227.

57 *According to a once secret report of 1962:* U.S. Department of Defense, "United States and Allied Capabilities for Limited Military Operations to 1 July 1962," undated.

58 *half billion people:* Lawrence K. Altman, William J. Broad, and Judith Miller, "Smallpox: The Once and Future Scourge?" *New York Times,* June 15, 1999, p. F1; Fenner et al., *Smallpox and Its Eradication,* p. 1363, says that as recently as 1967 the number of smallpox victims ran as high as 15 million a year.

58 *its own supply:* U.S. Senate, "Unauthorized Storage of Toxic Agents," *Hearings of the Select Committee to Study Governmental Operations with Respect to Intelligence Activities,* 94th Cong., 1st sess., vol. 1, Sept. 16, 17, and 18, 1975, p. 190; also see Regis, *The Biology of Doom,* p. 214.

58 *official arsenal of weapons:* Lt. Col. George W. Christopher, Lt. Col. Theodore J. Cieslak, Maj. Julie A. Pavlin, and Col. Edward M. Eitzen Jr., "Biological Warfare: A Historical Perspective," *Journal of the American Medical Association,* Aug. 6, 1997, p. 414.

58 *it had been abandoned:* Regis, *The Biology of Doom,* pp. 10, 219.

59 *measuring four-tenths of a micron:* Thomas D. Brock et al., *Biology of Microorganisms,* 7th ed. (Englewood Cliffs, N.J.: Prentice Hall, 1994), pp. 185, 228–29.

59 *the virus can also be spread:* Sidell, et al., *Medical Aspects,* pp. 542–43.

59 *"Exposed monkey 3912":* Nicholas Hahon, "Screening Studies with Variola

Virus" (Biological Warfare Laboratories, Fort Detrick, Oct. 1958), p. 69. DTIC, AD322280.

59 *lyophilization, or freeze-drying:* Postgate, *Microbes and Man,* pp. 125–26.

59 *dilute the dried agent:* Robert S. Hutton et al., "Selection of Process for Freeze-Drying, Particle Size Reduction, and Filling of Selected BW Agents" (Army Biological Labs, Frederick, Md., October 1952). DTIC, AD325415.

59 *"no significant loss":* Hahon, "Screening Studies," pp. 15, 55.

59 *Virulence could thus be preserved for months and years:* Joshua Lederberg, editor, *Biological Weapons: Limiting the Threat* (Cambridge, Mass.: MIT Press, 1999), p. 56.

60 *a quarter of its strength . . . "easily disseminated":* "Miscellaneous Publication 7" (U.S. Army Biological Laboratories, Fort Detrick, July 1965), pp. 20, 32. In time, a black art emerged of protecting dry germs with coatings meant to make microbes less vulnerable to sunlight and other environmental hazards. See J. H. Nash et al., "Dissemination Properties of Encapsulated Particles" (Litton Systems, Inc., Aug. 1967). DTIC, AD826013.

60 *one in every twelve:* "Miscellaneous Publication 7," p. 34. The mock attack eventually made headlines. See Ken Ringle, "Army Sprayed Bacteria on Unsuspecting Travelers," *Washington Post,* Dec. 5, 1984, p. B1.

60 *"a long incubation period":* "Miscellaneous Publication 7," p. 20.

60 *its first symptoms:* Sidell et al., *Medical Aspects,* pp. 542–46.

61 *no smallpox outbreaks since 1959:* Fenner et al., *Smallpox and Its Eradication,* pp. 338, 345–46.

61 *American government took a leading role: Ibid,* pp. 338, 345–46.

61 *thousands of scientists signed a petition:* Susan Wright, "Evolution of Biological Warfare Policy: 1945–1990," in Susan Wright, editor, *Preventing a Biological Arms Race* (Cambridge, Mass.: MIT Press, 1990), pp. 38–39.

61 *included Meselson:* Seymour M. Hersh, *Chemical and Biological Warfare: America's Hidden Arsenal* (Garden City, N.Y.: Anchor, 1969), pp. 265–66.

61 *an exposé: Ibid.*

62 *A light aircraft could deliver enough to kill populations:* Matthew Meselson, "The United States and the Geneva Protocol of 1925," private paper, Biological Laboratories, Harvard University, Sept. 1969, 13 pp.

62 *projection of American military power:* For an account of the Pentagon's quiet role in promoting the Nixon ban, see Mangold and Goldberg, *Plague Wars,* pp. 54–56.

63 *"The U.S. . . . shall renounce the use of lethal biological agents":* Secondary sources differ widely on the exact language that Nixon used in his renunciation. This quote comes from the original White House news release, dated November 25, 1969. It can be found in U.S. Senate, "Unauthorized Storage of Toxic Agents," pp. 200–201.

63 *"The arms control benefits of our newly decided policy":* Matthew Meselson, "What Policy for Toxins?" private paper, Biological Laboratories, Harvard University, Jan. 22, 1970, 8 pp.

64 *entirely in America's self-interest:* Wright, *Preventing a Biological Arms Race,* p. 40.

64 *a point that Meselson, among others:* See, for instance, Matthew Meselson, review

of *Tomorrow's Weapons, Chemical and Biological,* by Jacquard Hirshon Rothschild, *Bulletin of the Atomic Scientists,* Oct. 1964, pp. 35–36. His argument in this review: "If the technical problems associated with the effective dissemination of biological agents could be solved, enormous destructive capability could become available at radically low cost, since the amounts of agent potentially sufficient to attack great areas are incredibly small. The introduction of radically cheap weapons of mass destruction into the arsenals of the world would not act as much to strengthen the big powers as it would to endow dozens of relatively weak countries with great destructive capability. Such weapons could even come within the reach of dissident private groups and individuals. It is obviously to the advantage of great powers to keep war very expensive."

3. Revelations

This chapter is based on interviews with Joshua Lederberg, Paul Berg, Douglas J. Feith, William Kucewicz, William C. Patrick III, John Weitz, Susan Wright, Richard Spertzel, Gary Crocker, Matthew S. Meselson, Douglas MacEachin, Bill Richardson, Col. Harry G. Dangerfield, Gen. Brent Scowcroft, Eileen R. Choffnes, Gen. Philip K. Russell, Philip Brachman, and a number of current and former scientists and intelligence analysts associated with Fort Detrick, including Col. David R. Franz, past head of the United States Army Medical Research Institute of Infectious Diseases. The literature cited below helped form the backdrop to many of the chapter's main settings.

66 *glowed in the dark:* Thomas F. Lee, *Gene Future: The Promise and Perils of the New Biology* (New York: Plenum Press, 1993), pp. 16–17.

67 *he won the Nobel Prize:* B. Lee Ligon, "Joshua Lederberg: PhD; Nobel Laureate, Geneticist, and President Emeritus of the Rockefeller University," *Seminars in Pediatric Infectious Diseases,* Oct. 1998, pp. 345–55. Lederberg Papers; see also Harriet Zuckerman, *Scientific Elite: Nobel Laureates in the United States* (New York: Free Press, 1977), pp. 166–67.

67 *a navy undergraduate program:* Joshua Lederberg, "Genetic Recombination in Bacteria: A Discovery Account," *Annual Review of Genetics,* 21 (1987), pp. 23–46. Lederberg Papers.

68 *played a major role in Fort Detrick's establishment:* Norman M. Covert, *Cutting Edge: A History of Fort Detrick, Maryland* (Fort Detrick: U.S. Army, 1997), pp. 17–20.

68 *Lederberg met Leo Szilard:* William Lanouette with Bela Szilard, *Genius in the Shadows: A Biography of Leo Szilard, the Man Behind the Bomb* (New York: Charles Scribner's Sons, 1993), pp. 317, 384, 390, 403, 457.

68 *began attempting to splice genes:* Susan Wright, *Molecular Politics: Developing American and British Regulatory Policy for Genetic Engineering, 1972–1982* (Chicago: University of Chicago Press, 1994), p. 71.

68 *"time is running out":* Joshua Lederberg, "Swift Biological Advance Can Be Bent to Genocide," *Washington Post,* Aug. 17, 1968, p. A14. Lederberg Papers. For six years, from 1966 through 1971, he wrote a weekly column for the *Washington Post* entitled "Science and Man."

68 *"Where will it all end?":* Letter, Joshua Lederberg to Riley D. Housewright, Mar. 11, 1969. Lederberg Papers.

69 *"hydrogen bombs available at the supermarket":* Wright, *Molecular Politics,* p. 119.

69 *In August 1970 he delivered a speech:* Clement J. Zablocki, Wisconsin Representative, "Dr. Lederberg Speaks Out on Biological Warfare Hazards," *Congressional Record,* 91st Cong., 2nd sess., Sept. 11, 1970, 112, part 23, pp. 31395–96. For an excerpt of the speech, see Joshua Lederberg, *Biological Weapons: Limiting the Threat* (Cambridge, Mass.: MIT Press, 1999), pp. 325–29.

69 *"repugnant to the conscience of mankind":* "Text of the 1972 Biological Weapons Convention," in *Preventing a Biological Arms Race,* edited by Susan Wright (Cambridge, Mass.: MIT Press, 1990), p. 371.

69 *they realized that by teaming up:* Stanley N. Cohen, "The Manipulation of Genes," *Scientific American,* July 1975, pp. 25–33; Bruce Wang, "Cohen: DNA Genius on the Farm," *Stanford Daily,* Nov. 10, 1999; James D. Watson et al., *Recombinant DNA,* 2nd ed. (New York: Freeman, Scientific American Books, 1998), pp. 73–74.

70 *genes from the South African clawed toad:* Nicholas Wade, *The Ultimate Experiment: Man-Made Evolution* (New York: Walker, 1977), pp. 26–27; Cohen, "The Manipulation of Genes," p. 31.

71 *called for an international treaty prohibiting:* Wright, *Molecular Politics,* p. 151.

71 *five scientists cast dissenting votes:* Wade, *The Ultimate Experiment,* p. 51.

72 *"promises some of the most pervasive benefits":* Joshua Lederberg, "DNA Splicing: Will Fear Rob Us of Its Benefits?" *Prism,* American Medical Association, Nov. 1975, pp. 33–37. Lederberg Papers.

72 *a new chapter of American history:* U.S. Senate, "Alleged Assassination Plots Involving Foreign Leaders: An Interim Report of the Select Committee to Study Government Operations with Respect to Intelligence Activities," 94th Cong., 1st sess., Nov. 20, 1975.

72 *planned to kill Patrice Lumumba:* Ed Regis, *The Biology of Doom: The History of America's Secret Germ Warfare Project* (New York: Holt, 1999), pp. 182–85.

72 *scheme to paralyze the dictator: Ibid.,* pp. 193–96.

73 *inventory surprisingly diverse:* U.S. Senate, "Unauthorized Storage of Toxic Agents," *Hearings of the Select Committee to Study Governmental Operations with Respect to Intelligence Activities,* 94th Cong., 1st sess., vol. 1, Sept. 16, 17, and 18, 1975, pp. 214–15.

73 *"We must insist": Ibid.,* p. 2.

73 *"suitable for dusting of clothes":* U.S. Senate, "Examination of Serious Deficiencies in the Defense Department's Efforts to Protect the Human Subjects of Drug Research," *Hearings Before the Subcommittee on Health and Scientific Research of the Committee on Human Resources,* 95th Cong., 1st sess., Mar. 8 and May 23, 1977, p. 248.

74 *"a successful operation":* U.S. Senate, "Unauthorized Storage of Toxic Agents," p. 6.

74 *"to prevent his appearance":* U.S. Senate, "Examination of Serious Deficiencies," p. 243.

74 *CIA document, buried in the back: Ibid.,* p. 246. The document is entitled "Summary Report on CIA Investigation of MKNAOMI," the agency's code name for its secret collaboration with the Fort Detrick biologists. We have found only one scholar, a German one, who reported the Schacht disclosure. See Erhard Geissler, *Biologische Waffen—Nicht in Hitler's Arsenalen: Biologische und Toxin-Kampfmittel in Deutschland, 1915–1945,* 2nd ed. (Muenster, Germany: LIT, 1999), pp. 336–37. John Weitz, an American biographer of Schacht, doubts the poisoning ever took place. His book, *Hitler's Banker: Hjalmar Horace Greeley Schacht* (New York: Little, Brown, 1997), makes no mention of it. In an interview, Weitz, a former OSS officer based in Europe, said that the CIA's congressional testimony about the poisoning may have been a lie meant to rationalize its own toxins program.

74 *financial brains of the Third Reich:* William L. Shirer, *The Rise and Fall of the Third Reich* (New York: Simon & Schuster, 1959), pp. 145–46, 189–90.

74 *The chosen toxin:* U.S. Senate, "Examination of Serious Deficiencies," p. 246. Though the CIA document states clearly that the toxin allegedly administered to Schacht was staphylococcal enterotoxin, it is interesting to note that the best histories of weapons development at Fort Detrick during World War II make no reference to the manufacture or use of that poison. For example, John Ellis van Courtland Moon, "US Biological Warfare Planning and Preparedness: The Dilemmas of Policy," in *Biological and Toxin Weapons: Research Development and Use from the Middle Ages to 1945,* edited by Erhard Geissler and John Ellis van Courtland Moon (New York: Oxford University Press and Stockholm International Peace Research Institute, 1999), pp. 239–42, 244–47. Still, the CIA in its document claiming responsibility for the Schacht attack implied that Ford Detrick was the source of the enterotoxin. "This agent was included in the materials maintained for the agency" by the Special Operations Division of Ford Detrick, the document said in describing the attack.

74 *an anti-Hitler conspirator:* Shirer, *The Rise and Fall,* pp. 411–13, 918, 1073–74, 1143.

75 *A senior Soviet diplomat:* Arkady N. Shevchenko, *Breaking with Moscow* (New York: Knopf, 1985), pp. 34, 172–74, 179, 202.

76 *ran a sketchy report:* Jeanne Guillemin, *Anthrax: The Investigation of a Deadly Outbreak* (Berkeley: University of California Press, 1999), pp. 7–9; Tom Mangold and Jeff Goldberg, *Plague Wars: The Terrifying Reality of Biological Warfare* (New York: St. Martin's Press, 1999), pp. 71–74.

77 *"disturbing indications":* Mangold and Goldberg, *Plague Wars,* p. 75.

77 *closed ranks around this scenario:* Leslie H. Gelb, "Keeping an Eye on Russia," *New York Times Magazine,* Nov. 29, 1981, p. 31; see also Guillemin, *Anthrax,* p. 9.

78 *lack of corresponding evidence on intestinal anthrax "cast doubt":* Interview, Matthew S. Meselson.

78 *unable to "reliably distinguish":* Letter to authors, Matthew S. Meselson, June 12, 2001.

78 *The refugees called it "yellow rain,":* Julian Robinson, Jeanne Guillemin, and Matthew Meselson, "Yellow Rain in Southeast Asia: The Story Collapses," in Wright, *Preventing a Biological Arms Race,* pp. 220–38.

78 *Alexander M. Haig Jr., pressed ahead:* Bernard Gwertzman, "U.S. Says Data Show Toxin Use in Asia Conflict," *New York Times,* September 14, 1981, p. A1; Barbara Crossette, "U.S. Presents an Analysis to Back its Charge of Toxin Weapons' Use," *New York Times,* Sept. 15, 1981, p. A1.

79 *Meselson and his allies developed an alternate theory:* Joan W. Nowicke and Matthew Meselson, "Yellow Rain—A Palynological Analysis," *Nature,* May 17, 1984, pp. 205–6; Thomas D. Seeley et al., "Yellow Rain," *Scientific American,* Sept. 1985, pp. 128–37.

79 *Meselson began to poke holes:* Gelb, "Keeping an Eye on Russia," p. 31; Guillemin, *Anthrax,* pp. 41–42.

81 *"Recombinant DNA and the Biological Warfare Threat":* British Medical Association, *Biotechnology, Weapons and Humanity* (Amsterdam: Harwood Academic Publishers, 1999), pp. 38, 130.

81 *could turn anthrax into: Ibid.,* p. 49.

81 *"could be applied to the creation":* U.S. Arms Control and Disarmament Agency, *Biological Weapons Convention: Confidence Building Measures,* Annex 6, "Declaration of Past Activities in Offensive and/or Defensive Biological Research and Development Programs," Apr. 1996, p. 23.

81 *Soviet Union was conducting:* Department of Defense, *Soviet Military Power* (Washington, D.C., 1984), p. 73.

81 *the* Wall Street Journal*'s editorial page:* William Kucewicz, "Soviets Search for Eerie New Weapons," *Wall Street Journal,* Apr. 23, 1984. Other stories in the series appeared on Apr. 25, Apr. 27, May 1, May 3, May 8, May 10, May 18, and Dec. 28, 1984.

82 *presidential commission charged:* Walter Stoessel et al., *Report of the Chemical Warfare Review Commission* (Washington, D.C.: U.S. Government Printing Office, June 1985), pp. 69–71; for a reprint of the chapter on biological warfare, see Wright, *Preventing a Biological Arms Race,* pp. 423–25.

82 *twenty-eight-page report:* Defense Intelligence Agency, "The Soviet Biological Warfare Threat," 1986, DST-1610F-057-86.

82 *"stunning advances":* Testimony of Douglas J. Feith, deputy assistant secretary of defense for negotiations policy, Subcommittee on Oversight and Evaluation of the House Permanent Select Committee on Intelligence, Aug. 8, 1986, reprinted in Wright, *Preventing a Biological Arms Race,* pp. 425–32; see also Charles Piller and Keith R. Yamamoto, *Gene Wars: Military Control over the New Genetic Technologies* (New York: William Morrow, Beech Tree Books, 1988), p. 16.

82 *The United States filed formal diplomatic protests:* U.S. Senate, "Global Spread of Chemical and Biological Weapons," *Hearings Before the Committee on Govern-*

mental Affairs and Its Permanent Subcommittee on Investigations, 101st Cong., 1st sess., May 17, 1989, p. 184.

83 *funds for scientific research on biodefense:* Seth Shulman, *Biohazard: How the Pentagon's Biological Warfare Research Program Defeats Its Own Goals* (Washington, D.C.: Center for Public Integrity, 1993), p. 33.

83 *"can be successfully mimicked":* Ralph R. Isberg and Stanley Falkow, "A Single Genetic Locus Encoded by *Yersinia pseudotuberculosis* Permits Invasion of Cultured Animal Cells by *Escherichia coli,*" *Nature,* Sept. 19, 1985, pp. 262–64.

83 *"level of pathogenicity":* Piller and Yamamoto, *Gene Wars,* p. 134.

83 *"We are doing research in genetic engineering":* William Kucewicz, "When Arms Control Falls Short," *Wall Street Journal,* May 18, 1984.

84 *An analysis of Pentagon research:* Piller and Yamamoto, *Gene Wars,* p. 133.

86 *was "highly reactogenic":* Department of the Army, U.S. Army Medical Research Acquisition Activity, Request for Proposals, May 16, 1985, DAMD 17-85-R-0078.

87 *Established in 1925:* The early history of the Michigan facility is described in a briefing for General Accounting Office officials of October 8, 1996. It was originally known as the Michigan Department of Health Biologic Products Division. Michigan officials were considering its sale to the private sector when the vaccine shortages hit in the mid-1980s, according to the document, which was provided to the House Committee on Government Reform.

87 *Between fifteen thousand and seventeen thousand doses:* U.S. Army Medical Materiel Development Activity, Memorandum for: Commander, U.S. Army Medical Research, Subject: Minutes of Source Selection Board for Request for Proposal, (RFP) DAMD17-88-R-0149, Sept. 8, 1988, p. 1. Written by Anna Johnson-Winegar, the memo acknowledges that the Michigan lab's "previous production experience has been somewhat limited."

87 *the army signed its first-ever contract:* U.S. Army Medical Research Acquisition Activity, Fort Detrick, Maryland, Contract with Michigan Department of Public Health, September 30, 1988, to September 29, 1993.

87 *"certainly not state-of-the-art":* U.S. Army, "Minutes of Source Selection Board," p. 4.

88 *"They are currently purchasing bacterial strains":* Defense Intelligence Agency, "Biological Warfare Production and Use," June 28, 1988. Gulflink.

88 *the company had sold an assortment:* Invoice, the American Type Culture Collection for University of Baghdad, Batch No. 010072; date of shipment: May 2, 1986.

89 *The order was placed by:* Invoice, the American Type Culture Collection for Iraqi Technical and Scientific Materials Import Division, Customer No. 022913; date of shipment: Sept. 29, 1988.

89 *The ban, said the announcement:* Department of Commerce, Bureau of Export Administration, "Removal of Unilateral National Security Controls; Additional Controls on Chemicals and Biological Agents and Precursors," *Federal Register,* 54 FR 8281, Feb. 28, 1989.

90 *"Nature isn't benign":* Laurie Garrett, *The Coming Plague: Newly Emerging Diseases in a World out of Balance* (New York: Penguin, 1995), p. 6.

92 *"Imaginations have run wild":* Piller and Yamamoto, *Gene Wars,* p. 116.

92 *Yamamoto testified:* U.S. Senate, "Global Spread of Chemical and Biological Weapons," pp. 204–5. Also see Shulman, *Biohazard,* pp. 25, 65.

92 *lacked a defensive rationale:* General Accounting Office, "Biological Warfare: Better Controls in DOD's Research Could Prevent Unneeded Expenditures," Dec. 1990, p. 2.

93 *one offered a side-by-side comparison:* Seeley, "Yellow Rain," pp. 132–33.

93 *laboratories in Britain and the United States had looked for mycotoxins:* Matthew Meselson, Julian Robinson, and Jeanne Guillemin, "Yellow Rain: The Story Collapses," *Foreign Policy,* fall 1987, pp. 100–17.

94 *Meselson said he found the Soviet Account "completely plausible":* R. Jeffrey Smith and Philip J. Hilts, "Soviets Deny Lab Caused Anthrax Cases: Tainted Meat Blamed for 1979 Deaths," *Washington Post,* Apr. 13, 1988, p. A1.

94 *could eliminate "any remaining reasonable doubt":* Matthew S. Meselson, "The Biological Weapons Convention and the Sverdlovsk Anthrax Outbreak of 1979," Federation of American Scientists, *FAS Public Interest Report,* Sept. 1988, p. 4.

94 *"The burden of the evidence available":* U.S. Senate, "Biological Weapons Proliferation and the New Genetics," p. 210.

94 *a top Soviet biologist defected:* Mangold and Goldberg, *Plague Wars,* pp. 83–85, 91–105; Ken Alibek with Stephen Handelman, *Biohazard: The Chilling True Story of the Largest Covert Biological Weapons Program in the World—Told from the Inside by the Man Who Ran It* (New York: Random House, 1999), pp. 137–45.

97 *Al Tuwaitha was doing "molecular biology and genetic engineering":* Defense Intelligence Agency, "The Nuclear Research Center," June 29, 1990. Gulflink.

4. Saddam

This chapter is based on interviews with Col. Arthur M. Friedlander, Margaret L. M. Pitt, Capt. Larry Seaquist, Col. Harry G. Dangerfield, Gen. Robert Belihar, Peter Collis, Gen. H. Norman Schwarzkopf, Col. James D. Bales Jr., Col. George E. Lewis, Col. Martin Crumrine, Adm. Edward Martin, Gen. John Jumper, Gen. Robert B. Johnston, Gen. Colin Powell, Gen. Brent Scowcroft, Anna Johnson-Winegar, and a number of other Bush administration officials.

98 *antibiotics alone, vaccines, or a combination?:* The study was described in interviews by two of its authors, Arthur M. Friedlander and Louise Pitt. It was first presented at the September 1991 Interscience Conference on Antimicrobial Agents and Chemotherapy in Chicago. Arthur M. Friedlander, Susan Welkos, Margaret L. M. Pitt, et al., "Postexposure Prophylaxis Against Experimental Inhalation Anthrax," *Journal of Infectious Diseases,* 167 (1993), pp. 1239–42.

100 *After a month, the results were in: Ibid.,* p. 124.

101 *In a war game in late July:* Michael R. Gordon and General Bernard E. Trainor, *The General's War* (Boston: Little, Brown, 1995), pp. 45–46; Anthony H. Cordes-

man and Abraham R. Wagner, *The Gulf War,* vol. 4 of *The Lessons of Modern War* (Boulder, Colo.: Westview Press, 1996), p. 43. Army officers disclosed to the authors that the exercise did not include biological warfare.

101 *"In all directives":* Interview, Gen. Robert B. Johnston.

102 *"would deploy these agents if needed":* Navy Operational Intelligence Center 0604327, "Iraqi Offensive Biological Warfare Capabilities," Aug. 6, 1990. Gulflink.

102 *"Iraq has a mature offensive BW program":* Armed Forces Medical Intelligence Center, "AFMIC Special Weekly Wire 32-90(C)(U)," Aug. 8, 1990. Gulflink.

102 *"available for weaponization":* The AFMIC report offers no supporting details. Iraq's use of biological weapons against Iran has never been confirmed.

102 *the CIA weighed in:* The Central Intelligence Agency, "Iraq's Biological Warfare Program: Saddam's Ace in the Hole," Aug. 8, 1990. Gulflink.

103 *More soldiers could be immunized:* This meeting was described in Department of the Army, Office of the Surgeon General, Memo for HQDA (DAMO-SWC), "Medical Biological Warfare (BW) Defense Plan, March 18, 1991." Gulflink.

103 *The answer was not reassuring:* Joint Chiefs of Staff, Logistics Directorate (J4), "Biological Defense Chronology," MPOD 3821, Feb. 12, 1992. This document, a 33-page chronology, is among the most detailed records of the handling of biological defense during the Gulf War. The officers working for the Joint Chiefs of Staff are divided by function. J4 handles logistics, including medical supplies. The chronology, which was declassified and posted on the Defense Department's Gulflink Web site, describes many of the key meetings on the issue. Whenever possible, we have verified the account it offers with other sources or documents. The author of the document did not record his or her name. The title "J4 BD Chronology, 12 Feb 92" is handwritten across the first page. We will refer to it in subsequent citations as Joint Staff Chronology.

103 *only 10,000 doses:* Central Command, "Memorandum for Chief, NBC Defense Division, CENTCOM, 25 August 1990, Subject: Meeting notes, 23 Aug 90." Gulflink.

103 *The laboratory had been pushing:* The status of vaccine production is described in an unsigned memorandum for the record entitled "Trip Report—27 Aug 90." The document comes from the files of the U.S. Army Medical Research Acquisition Activity.

104 *just 34,000 doses:* Central Command, "Memorandum for Chief, NBC Defense Division, CENTCOM, 25 Aug. 90, Subject: Meeting notes, 23 Aug. 90."

104 *Pentagon health officials pulled together a committee of experts:* The committee's meeting was described in "After Action Report: Inaugural Meeting of Biological Warfare Ad Hoc Working Group, 1100-1600, 20 August 1990." Gulflink.

105 *the rotting remains:* Daily Staff Journal, ARCENT, "Dead Livestock in Eastern Sector of AOR 29, Aug 90," Aug. 29–30. Gulflink.

105 *Mendez drafted a memo:* Joint Staff Chronology, p. 2.

105 *the threat was "tenuous":* Lt. Col. Sharon Falkenheimer, United States Air Force, "Background Paper on Medical Defense Against Chemical and Biological Warfare in Support of Operation Desert Shield," Sept. 24, 1990. Her memo

captures the conflicting impulses of senior officers. It begins by noting that Iraq was "likely to have weaponized anthrax and botulinum" and then says the chiefs made their September 21, 1990, decision to delay vaccination "until stock adequate to share with allies or threat changes." The chiefs based their conclusion on what they saw as the "tenuous nature of threat (no confirmation of weaponization) and inadequate number of doses." Gulflink.

105 *"The decisions were no longer medical":* Interview, Powell; Joint Staff Chronology, p. 4.

106 *"Jumper, I'm worried about BW defense":* Interview, Gen. John Jumper.

107 *a gigantic air-sniffing device:* Albert J. Mauroni, *America's Struggle with Chemical-Biological Warfare* (Westport, Conn.: Praeger, 2000) describes the 1960s history of biological detectors, pp. 99–101.

107 *could be mounted on an airplane:* Albert J. Mauroni, *Chemical-Biological Defense* (Westport, Conn.: Praeger, 1998) details the attempts to field a prototype detector and the plane crash that ended the program, pp. 34–35, 51.

107 *"I just did":* Interview, Collis.

107 *The conversations were frustrating:* Interview, Anna Johnson-Winegar.

108 *Of the 150 companies:* "Memorandum for the Record, 18 Oct. 1990, Third Tri-Service Task Force (Project Badger)." Gulflink.

108 *At a meeting in early October:* "Memorandum for Record, Tri-Service Task Force, Oct. 10, 1990." Gulflink. The meeting was also described in an interview with Col. Harry G. Dangerfield.

109 *I. Lewis Libby, a trim boyish lawyer:* Libby left office after President Bush's defeat to resume his law practice. He returned to government in 2001 as Vice President Dick Cheney's chief of staff.

110 *kill 90 percent:* Interview, Capt. Larry Seaquist.

112 *Condoleezza Rice, a Soviet specialist:* A decade later, Rice returned to the White House as national security adviser to President George W. Bush.

112 *General Carl Vuono:* Vuono's anger at the lack of biodefenses is described in U.S. Army, "Memorandum for Record, Army Surgeon General," Nov. 23, 1990. Gulflink.

112 *Schwarzkopf felt trapped:* Gen. H. Norman Schwarzkopf described his feelings about the lack of biological defense in a detailed interview in which he read from a day-to-day log of meetings and phone calls kept for him during the war.

112 *An October 22 report:* Armed Force Medical Intelligence Center, "Iraq Biological Warfare Threat," Oct. 22, 1990. Gulflink.

113 *"There is no reliable information":* Defense Intelligence Agency, "Chemical and Biological Warfare in the Kuwait Theater of Operations; Iraq's Capability and Posturing," undated. Gulflink.

113 *It could sow panic: Ibid.*, p. 7.

113 *"shock the coalition into a cease-fire":* Central Intelligence Agency, "Iraq as a Military Adversary," Nov. 1990. Gulflink.

114 *Some preliminary calculations:* Iraq Interagency Biological Warfare Working Group, Fusion Committee, "Estimate of Casualties Resulting from Coalition Air Strikes on Iraqi BW Related Facilities," Jan. 9, 1991. Gulflink.

114 *"There has got to be a penalty":* Gen. Horner described his experiences to an air force historian. They were quoted in: U.S. Air Force, "Oral History Interview of Lt. Gen. Charles A. Horner by: Perry Jamison, Rich Davis, and Barry Barlow." Mar. 4, 1992. Gulflink.

114 *the fusion group acknowledged:* "Estimate of Casualties Resulting from Coalition Air Strikes"; interview, William Patrick.

115 *Jumper, sent a message:* Gen. Jumper's cable was described by Gen. Schwarzkopf, his medical aide Col. James David Bales Jr., and Col. George E. Lewis, a germ weapons expert who worked for Gen. Jumper and helped draft the message. The proposal was not well known during the war and it contradicted the position of the army surgeon general which was, as one officer described it in an interview, to "vaccinate, vaccinate, vaccinate."

116 *"And how are you going to know that?":* Interview, Col. Bales.

116 *Schwarzkopf sent a private "eyes-only" message:* Joint Staff Chronology, p. 17. Schwarzkopf described the contents of his cable in an interview.

116 *"strenuous objections":* Schwarzkopf provided the date of the meeting from his log of Gulf War events.

117 *The British bought:* To boost the effectiveness of the anthrax shots, the British decided to give their soldiers an injection of the children's vaccine against pertussis, or whooping cough. A 1989 research paper suggested that the addition of a pertussis shot might prime the body for a quicker immune response. British experts decided to use the combination on their troops, even though it had never been tested on humans. In late December, the Department of Health in Britain told their military counterparts that preliminary tests on animals had found "severe loss of condition and weight loss." The fax transmitting that finding was misplaced, and British officials proceeded with the vaccine program. Subsequent research has found no evidence that the pertussis vaccine was harmful to British soldiers. Britain has declassified some of the documents relating to this incident, and they are described in: British Ministry of Defense, "Background to the Use of Medical Countermeasures to Protect British Forces during the Gulf War (Operation Granby)," Oct. 28, 1997. See also British Ministry of Defense, "Implementation of Immunisation Programme against Biological Warfare Agents for UK Forces During the Gulf Conflict 1990/91," Jan. 2000.

117 *The Saudis learned of the British plans:* Interviews, Schwarzkopf, Bales.

117 *severely punish Saddam Hussein:* Gen. H. Norman Schwarzkopf, *The Autobiography: It Doesn't Take a Hero,* written with Peter Petre (New York: Bantam Books, 1992) pp. 389–90.

117 *Think as Saddam Hussein would:* Efforts by Schwarzkopf to plan for how to distribute his limited supplies of vaccine were described by Belihar, Bales, and Schwarzkopf in interviews. Belihar also described his thinking in a January 12, 1996, appearance for the Presidential Advisory Committee on Gulf War Veterans Illnesses, which held a public hearing at the Westin Crown Center Hotel in Kansas City, Mo., pp. 95–96.

117 *The Americans arranged for blood samples:* Interview, Bales.

118 *hardly be any vaccine left:* Bales and Belihar described the dilemma in interviews. The amount of vaccine held in reserve is reported in a memo of January 15, 1991, written by Belihar for the deputy commander in chief, political adviser on the subject of "Friendly Forces and Anthrax." It said the United States planned to hold in reserve enough shots to vaccinate 20,000 soldiers. Belihar recommended that one half of the reserves be made available to "Friendly Forces" in the event of a suspected or proven anthrax attack.

118 *"infantryman's concerns":* The air force discussions about biodefense are described in a December 11, 1990, air force document, "Background Paper on Biological Warfare (BW) Inoculation Priority." Gulflink.

118 *"will temper media reaction":* Army surgeon general, "Iraq's Biological Capabilities, Update Assessment 11 December, 1990." Gulflink.

118 *On December 17, 1990, Powell:* Joint Staff Chronology, p. 21.

118 *Cheney and Powell took their plan:* Interview with Brent Scowcroft.

119 *received an encouraging bit of news:* Belihar testimony, p. 98; interview, Belihar.

119 *The FDA reluctantly agreed:* Letter, Enrique Mendez, assistant secretary of defense for health affairs, to Ronald Kessler, commissioner of food and drugs, December 28, 1990, and Kessler's reply, December 31, 1990, giving approval.

119 *overruling a proposal:* Described by Schwarzkopf in an interview and recorded in his personal log.

119 *Myatt was reluctant to play God:* Interview, Gen. Robert Johnston; e-mail interview, Gen. James M. Myatt.

119 *Schwarzkopf himself took an anthrax shot:* The scene was described by Bales in January 12, 1996, testimony before the Presidential Advisory Committee on Gulf War Veterans Illnesses in Kansas City, Mo., p. 124.

120 *General Cal Waller hated:* CENTCOM, "Personal for LTG Reimer from LTG Waller, 19 January, 1991," memo from Gen. Waller to Gen. Reimer. Gulflink.

120 *No one had much confidence:* Mauroni, *Chemical-Biological Defense,* pp. 71–72.

120 *Bush made his preferences clear:* Interview, Scowcroft.

120 *"will pay a terrible price":* Gordon and Trainor, *The General's War,* p. 197.

121 *"a gamble we had to take":* Colin L. Powell with Joseph E. Persico, *My American Journey* (New York: Random House, 1995), pp. 503–5.

122 *could not imagine the president:* Quayle's comments and their significance are recounted in William M. Arkin, "Calculated Ambiguity: Nuclear Weapons and the Gulf War," Sept. 22, 1996, *Washington Quarterly,* p. 3. Arkin's article, based on declassified documents and interviews, remains the definitive review of this issue.

122 *Replying to some skeptical questioning:* "Q & A's, Respond to Task 3864, 31 Jan., 1625 Hrs, Response to Question 3A." Gulflink. U.N. inspectors found no evidence that the "baby milk" factory had ever produced biological weapons. American officials continue to insist that it was a "backup" plant that could have been pressed into service.

122 *buried in temporary graves:* Memo to army surgeon general of January 18, 1991, from Ronald Williams, commander of the U.S. Army Medical Research and Development Command, Fort Detrick. Gulflink.

123 *"It has been three days since the attack":* Defense Intelligence Agency, assessment filed on January 23, 1991. Gulflink.

123 *reported flulike symptoms:* Interview, Bales.

5. Secrets and Lies

The chapter is based on interviews with Richard Spertzel, Rolf Ekeus, Theodore Prociv, Joshua Lederberg, Anthony Lake, Matthew S. Meselson, William C. Patrick III, William J. Clinton, Adm. Edward Martin, Col. David R. Franz, Robert Gates, Douglas MacEachin, Gen. Philip K. Russell, and various officials of UNSCOM, the Clinton administration, and the National Academy of Sciences.

125 *bombing had destroyed "all known":* Defense Intelligence Agency assessment, BDA-81-91, Mar. 22, 1991. Gulflink.

125 *most, if not all, of the Iraqi Scud attacks:* The army initially insisted that the Patriot had hit over 80 percent of the targets over Saudi Arabia and 50 percent over Israel. At a January 12, 2001, breakfast meeting with reporters, Cohen said the Patriot "didn't work" during the war. "There is no evidence on any destruction of Scud warheads" on videotapes taken during the fighting. Cohen's remarks were quoted in John Aloysius Farrell, "The Patriot Gulf Missile 'Didn't Work' Defense Secretary Speaks Out," *Boston Globe,* Jan. 13, 2001, p. 1.

125 *staying Saddam Hussein's hand:* The question of why the Iraqi leader did not use his germ weapons remains a mystery. Rolf Ekeus, executive chairman of UNSCOM, told the Senate Permanent Investigations Subcommittee in 1996 of a conversation he had with Iraq's foreign minister, Tariq Aziz. The Iraqi diplomat said Secretary of State James Baker's veiled comments in Geneva were taken as a threat of nuclear annihilation and were "decisive." "This is the story he, Aziz, tells," Ekeus said. "This is the story which they gladly tell everyone who talks to them. So I think one should be cautious at least about buying that story." Ekeus's testimony appears in U.S. Senate, *Hearing Before the Permanent Subcommittee of Investigations of the Committee on Governmental Affairs,* 104th Cong., 2nd sess., part 2, "Global Proliferation of Weapons of Mass Destruction," Mar. 13, 20, and 22, 1996, p. 92.

126 *"massive, offensive biological warfare programme":* Tom Mangold and Jeff Goldberg, *Plague Wars: The Terrifying Reality of Biological Warfare* (New York: St. Martin's Press, 1999), pp. 138–49.

127 *Tariq Aziz, ordered top officials:* The Iraqi concealment effort is described in a United Nations letter dated January 27, 1999, from the "Permanent Representatives of The Netherlands and Slovenia to the United Nations Addressed to the President of the Security Council," Jan. 29, 1999, S/1999/94. The role of Tariq Aziz is described in Scott Ritter, *Endgame: Solving the Iraq Problem—Once and for All* (New York: Simon & Schuster, 1999), p. 105, and by other UNSCOM inspectors.

127 *On April 18:* Peter James Spielmann, "Iraq Denies Biological Warfare Capabil-

ity—U.S. Objects," Associated Press, Apr.19, 1991. The 9-page document said "Iraq does not possess any biological weapons or related items."

128 *Amin handed the inspectors:* Tim Trevan, *Saddam's Secrets: The Hunt for Iraq's Hidden Weapons* (London: HarperCollins, 1999), pp. 28–29.

129 *UNSCOM could not support more than two missions:* Former inspector Ekeus related his view of Al Hakam in an interview.

129 *The source said Iraq had manufactured:* A 1991 summary of the intelligence report by the CIA's directorate of intelligence, "BW Missile Programs," does not describe the source. The summary is dated "early May, 1991." Gulflink. The report was described as "reliable source reporting" in the CIA's study "Intelligence Related to Possible Sources of Biological Agent Exposure During the Persian Gulf War," Aug. 2000, p. 4.

129 *a fuller account of what Iraq had produced:* Central Intelligence Agency, "Status of BW," Aug. 1991. Gulflink.

130 *reassess its view:* The CIA's September 1991 assessment of how many Iraqi facilities survived the war is described in "Intelligence Related to Possible Sources," p. 9.

130 *appeared to have made botulinum: Ibid.*, p. 9.

130 *none of the information:* Interviews with UNSCOM inspectors.

130 *"there is absolutely no evidence":* Trevan, *Saddam's Secrets,* pp. 117–18.

131 *filled with mustard gas:* Described by a UNSCOM inspector.

131 *biological agent, not sarin:* The experiences of UNSCOM 20 are described in "Intelligence Related to Possible Sources," p. 4; picture of bomb, p. 5.

131 *A was for botulinum:* An UNSCOM letter of January 25, 1999, from the executive chairman describes the Iraqi coding system.

132 *serious steps were needed:* Findings of Project Badger were described by team members.

132 *"minimal support":* U.S. Army, "Biological Defense Concept," Feb. 27, 1991, p. 11. Gulflink. The army proposals for peacetime vaccinations are described in attached documents. The officials who drew up the memo for the secretary of the army called for peacetime vaccinations in a separate document. See U.S. Army, "Medical Biological Warfare—BW-Defense Plan," Mar. 18, 1999, p. 2. Army officials recommended the immunization of "designated early deploying forces during peacetime against specific BW agents to maximize readiness and deployment posture."

132 *"sensitive BW topic in the public eye":* U.S. Army, Biodefense Concept Briefing, Apr. 8, 1991. Gulflink.

132 *those of "select allies":* Medical Biological Warfare Defense Plan, 8 Apr. 1991. Gulflink.

133 *termed it a "minimal goal":* Joint Staff Chronology, p. 37.

133 *cost more than $1 billion: Ibid.*, p. 32.

133 *asked whether the threat was "real": Ibid.*, pp. 31–32.

133 *"no vaccination policy": Ibid.*

134 *"I got your paper":* Mangold and Goldberg, *Plague Wars*, pp. 159–60; interviews with former Bush administration officials.

134 *"I smoked but I didn't inhale":* Interview with former Bush administration officials; Mangold and Goldberg, *Plague Wars,* p. 165.

134 *"our military development was the cause":* R. Jeffrey Smith, "Yeltsin Blames '79 Anthrax on Germ Warfare Efforts," *Washington Post,* June 16, 1992, p. A1.

134 *showed the visitors autopsy slides:* Jeanne Guillemin, *Anthrax: The Investigation of a Deadly Outbreak* (Berkeley: University of California Press, 1999), pp. 52–54.

135 *Meselson and his team presented their preliminary findings: Ibid.,* pp. 179–82.

135 *the* Washington Times: Bill Gertz, "Defecting Russian Scientist Revealed Biological Arms Efforts," *Washington Times,* July 4, 1992, p. A4; R. Jeffrey Smith, "Russia Fails to Detail Germ Arms; U.S. and Britain Fear Program Continues in Violation of Treaty," *Washington Post,* Aug. 31, 1992, p. A1.

135 *they accepted the deal:* Mangold and Goldberg, *Plague Wars,* pp. 174–76.

136 *"It wasn't until Bill Patrick walked through the door":* Ken Alibek with Stephen Handelman, *Biohazard: The Chilling True Story of the Largest Covert Biological Weapons Program in the World—Told from Inside by the Man Who Ran It* (New York: Random House, 1999), p. 262.

137 *more devasting assaults:* President Clinton described the significance of the attack on the Twin Towers to his thinking about germ terrorism in an interview with Judith Miller and William J. Broad on January 21, 1999. Asked how seriously he took the threat of biological attack, he said, "You asked me what keeps me awake at night, and that bothers me . . . this biological issue."

138 *Civex '93:* The scenario for Civex '93 was described by Frank E. Young and William Clark in interviews.

138 *would not deter dictators like Saddam Hussein:* Interview with former aide to Aspin.

139 *"no commercial interest":* The army's plans were outlined in a detailed memo prepared by the staff of the army surgeon general, "Biological Defense Vaccine Production Facility," Jan. 7, 1993. Gulflink.

139 *A fight soon erupted:* The dispute was described by several former government officials, including Gen. Philip K. Russell, commander of the army's medical research and development command from 1986 to 1990.

140 *Lederberg became chairman:* U.S.-Russian Collaborative Program for Research and Monitoring of Pathogens of Global Importance Committee, *Controlling Dangerous Pathogens: A Blueprint for U.S.-Russian Cooperation,* a report to the Cooperative Threat Reduction Program of the U.S. Department of Defense (Washington, D.C.: National Academy Press, 1997). The preface of this report gives some of the program's early history. It also notes the preliminary research under way, including a reference to the joint research being conducted with Vector. The U.S. scientists involved in the project were Peter B. Jarhling, of the army lab at Fort Detrick; Bernard Moss, of the National Institutes of Health; and Joseph Esposito, of the Centers for Disease Control and Prevention.

141 *"We urge DOD to take the initiative":* Defense Science Board, Summer Study, 1993. The document remains classified. The passage was quoted in Richard Danzig, "Biological Warfare: A Nation at Risk—a Time to Act," National De-

fense University Strategic Forum, Institute for National Strategic Studies, No. 58, Jan. 1996.

141 *the Pentagon committed:* Department of Defense, "DODD 6205.3 DoD Immunization Program for Biological Warfare Defense," Nov. 26, 1993.

141 *congressional opposition was growing:* U.S. Army, FY 95 PDB Executive Summary, "Vaccine Production Facility," Dec. 14, 1993. The document, a summary of budget plans for fiscal year 1995 to 1997, notes that $135 million has been deleted from the budget for construction of the Vaccine Production Facility "pending completion of a study to examine alternatives for acquiring necessary vaccines." The debate over the issue was described by former Pentagon officials.

141 *The issue came before John Deutch:* Deutch began at the Pentagon as undersecretary for acquisition, moving to deputy defense secretary in 1994 and director of central intelligence in 1995.

143 *"the world's juiciest targets":* Letter, Josh Lederberg to Anthony Lake, national security adviser, June 11, 1994.

143 *published their final conclusions:* Matthew Meselson et al., "The Sverdlovsk Anthrax Outbreak of 1979," *Science* (Nov. 18, 1994), pp. 1202–8. David Walker, the team's pathologist, who worked at the University of Texas and early on said publicly that the evidence suggested inhalation anthrax, was not among the paper's seven coauthors.

143 *"What we had thought and said to be plausible":* Letter, Matthew Meselson to authors, June 12, 2001.

143 *His position had evolved as the inquiry gathered new evidence:* Guillemin, *Anthrax,* pp. 188–91.

143 *The team obtained data on the direction of the winds: Ibid.,* pp. 190–91.

143 *size of the leak was: Ibid.,* pp. 192–93, 242.

144 *Alibek, too, had few doubts:* Alibek, *Biohazard,* pp. 70–86.

147 *an Israeli intelligence officer turned over documents:* The first contacts between Israeli intelligence were forged by Tim Trevan, a senior UNSCOM inspector, who describes the encounter in his *Saddam's Secrets,* pp. 267–69, 287–88.

147 *"TSMID's search for anthrax cultures":* The Armed Forces Medical Intelligence Center, "Iraq Biological Warfare Threat, Oct. 22, 1990." Gulflink.

147 *"tie these buildings to TSMID and BW":* Defense Intelligence Agency, "Respond to Task 3864, Pass to the ITF OPS Officer, 31 Jan. 91, 1625 Hrs"; U.S. Navy, "Iraqi Chemical and Biological Warfare Capabilities, 19 November, 1990" describes the purchases of microbial media. Gulflink.

150 *Chocolates were laced with anthrax:* John Murphy, "The Deadly Tools of 'Dr. Death,' " *Baltimore Sun,* May 26, 2000, p. 2A; Suzanne Daley, "In Support of Apartheid: Poison Whisky and Sterilization," *New York Times,* June 11, 1998, p. A3.

150 *Qadaffi was trying to hire South African scientists:* James Adams, "Gadaffi lures South Africa's Top Germ Warfare Scientists," *Sunday Times* of London, Feb. 26, 1995. See also James Adams, "The Dangerous New World of Chemical and Biological Weapons," in *Terrorism with Chemical and Biological Weapons,* edited by

Brad Roberts (Alexandria, Virginia: The Chemical and Biological Arms Control Institute, 1997), pp. 24–26.

150 *British and American representatives met with President Mandela:* William Finnegan, "The Poison Keeper," *New Yorker,* Jan. 15, 2001, p. 73.

6. The Cult

This chapter is based on interviews with Richard A. Clarke, Gen. Charles Krulak, Richard J. Danzig, John Sopko, Kyle B. Olson, Frank E. Young, Joshua Lederberg, Pamela Berkowsky, Adm. William Owens, Stephen C. Joseph, and various Clinton administration officials. Some of the literature it drew on is cited below.

152 *The next day, Clarke called a meeting:* The meeting was described by participants.

153 "*Eastern philosophy and religion":* The documents incorporating Aum as a New York company and registering it as a charity are reproduced in U.S. Senate, *Hearings Before the Permanent Subcommittee of Investigations of the Committee on Governmental Affairs,* 104th Cong., 1st sess., "Global Proliferation of Weapons of Mass Destruction," part I, Oct. 31 and Nov. 1, 1995, pp. 547–54.

153 *wide latitude in national security cases:* The government's handling of the Aum case was first detailed in Jim McGee and Brian Duffy, *Main Justice* (New York: Simon & Schuster, 1996), pp. 353–58. Several former government officials confirmed and elaborated on this account.

153 *It took American officials several years:* The best initial account of Aum's history can be found in David E. Kaplan and Andrew Marshall, *The Cult at the End of the World* (New York: Crown, 1996). Several authors have pointed out the significance of Aum's failure to produce workable biological weapons, including Sheryl Wu Dunn, Judith Miller, and William J. Broad, "Sowing Death: A Special Report; How Japan Germ Terror Alerted World" *New York Times,* May 26, 1998, p. A1; Milton Leitenberg, "The Experience of the Japanese Aum Shinrikyo Group and Biological Agents," in *Hype or Reality? The "New Terrorism" and Mass Casualty Attacks,* edited by Brad Roberts (Alexandria, Va.: Free Hand Press, 2000), pp. 159–70.

154 *It was called the Nuclear Emergency Search Team:* U.S. Senate, *Hearings Before the Permanent Subcommittee of Investigations of the Committee on Governmental Affairs,* 104th Cong., 2nd sess., "Global Proliferation of Weapons of Mass Destruction," part 3, Mar. 27, 1996, prepared testimony of Bill Richardson, p. 81.

159 *attack against visitors to Disneyland:* President Clinton referred to this incident on April 22, saying: "There was one recent incident with which I was intimately familiar, which involved a quick and secret deployment of a major United States effort of F.B.I." and emergency health forces. "We went to the place," he said, "and we were ready to respond. So we have been on top of this from the beginning." Stephen Labaton, "Threat to Disneyland, Mentioned by Clinton, Is Termed a Hoax," *New York Times,* Apr. 23, 1995, p. A36.

160 *Worldwide, Aum had fifty thousand members:* Sopko's discoveries about Aum are summarized in U.S. Senate, *Hearings Before the Permanent Subcommittee of Investi-*

gations of the Committee on Governmental Affairs, 104th Cong., 1st sess., "Global Proliferation of Weapons of Mass Destruction," part 1, staff study, pp. 47–102.

160 *An Aum member had already confessed:* Kaplan and Marshall, *The Cult at the End of the World,* pp. 95–96.

161 *"It is not far-fetched":* Olson's report is reprinted in U.S. Senate hearing, "Global Proliferation," part I, pp. 603–7.

163 *"now becomes more likely":* A transcript summarizing much of the closed-door sessions was printed as a paperback volume and distributed to a limited number of government officials. *Proceedings of the Seminar on Responding to the Consequences of Chemical and Biological Terrorism,* July 11–14, 1995, Sponsored by the U.S. Public Health Service, Office of Emergency Preparedness, Conducted at the Uniformed Services University of Health Sciences, 4301 Jones Bridge Road, Bethesda, Md., USA, section 1, pp. 1–7.

163 *"if an event occurs": Ibid.,* section 1, pp. 62-64.

164 *"I will help you do it": Ibid.,* section 3, p. 68.

7. Evil Empire

This chapter is based on interviews conducted in Washington, D.C., and during more than half a dozen trips to the former Soviet Union—Kazakhstan, Russia, Uzbekistan, Tajikistan—that Judith Miller made between 1998 and mid 2001. Some of those interviewed include: Andy Weber, Ken Alibek, Anne M. Harrington, Gennady L. Lepyoshkin, Lev S. Sandakhchiev, Igor V. Domaradskij, Valery M. Lipkin, Yevgeniy Severin, Peter and Marika Sveshnikov, Vladimir P. Zavyalov, Randall Beatty, David Kelly, Frank Miller, William Potter, Jonathan B. Tucker, Col. David R. Franz, former senator Sam Nunn, and Senator Richard Lugar.

165 *The plant was in Stepnogorsk . . . Soviet standby production facilities:* Visits, October 1998, May 2001; Ken Alibek with Stephen Handelman, *Biohazard: The Chilling True Story of the Largest Covert Biological Weapons Program in the World—Told from the Inside by the Man Who Ran It* (New York: Random House, 1999), pp. 88–106; Gulbarshyn Bozheyeva, Yerlan Kunakbayev, Dastan Yeleukenov, "Former Soviet Biological Weapons Facilities in Kazakhstan: Past, Present, and Future," Center for Nonproliferation Studies, Monterey Institute of International Studies, June 1999, pp. 8–15; Tom Mangold and Jeff Goldberg, *Plague Wars: A True Story of Biological Warfare* (London: Macmillan, 1999) pp. 186–89.

167 *"the Concern":* Biopreparat was also known as "the System," according to Alibek, *Biohazard,* p. 191; Mangold and Goldberg, *Plague Wars,* p. 65; and Anthony Rimmington, "From Military to Industrial Complex? The Conversion of Biological Weapons' Facilities in the Russian Federation," *Contemporary Security Policy,* 17 (Apr. 1996), p. 87.

167 *Created in 1973:* Alibek, *Biohazard,* p. 22; Mangold and Goldberg, *Plague Wars,* pp. 92, 95, 99–100; Rimmington, "From Military," pp. 80–112; M. Leitenberg, "The Conversion of Biological Warfare Research and Development Facilities to Peaceful Uses," in *Control of Dual-Threat Agents: The Vaccines for Peace Pro-*

gramme, edited by Erhard Geissler and John P. Woodall, Stockholm International Peace Research Institute Chemical and Biological Warfare Studies, no. 15 (Oxford: Oxford University Press, 1994), pp. 77–105.

171 *Weber then sent word:* The four American experts who accompanied Weber on this historic mission included: one intelligence analyst who wishes to remain anonymous, Ron Waggoner, Lester C. Caudle III, and Jeffrey Weiss.

171 *the flat steppe of Semipalatinsk:* Judith Miller, "One Last Explosion at Kazakh Test Site to Aid Arms Treaty," *New York Times,* Sept. 25, 1999, p. A1.

171 *thirty thousand Saiga antelope:* Judith Miller, "U.S. and Uzbeks Agree on Chemical Arms Plant Cleanup," *New York Times,* May 25, 1999, p. A3.

176 *Vozrozhdeniye, or Renaissance, Island:* The earliest unclassified reference to the true role of the island is in a recently declassified document entitled "The Soviet Biological Warfare Threat," by the Defense Intelligence Agency, DST-1610F-057-86, 1986. The report, which contains an assessment of the events at Sverdlovsk, does not mention Biopreparat. But it features the island on a map and the following identification: "Location of the candidate BW test and evaluation installation on Vozrozhdeniya Island in the Aral Sea." See also Rimmington, "From Military," p. 86, and Bozheyeva et al., "Former Soviet," pp. 5–8.

181 *Doctors Without Borders:* Judith Miller, "Poison Island: A Special Report; at Bleak Asian Site, Killer Germs Survive," *New York Times,* June 2, 1999, p. A1.

8. Breakthrough

This chapter is based on interviews with Richard Spertzel, Rolf Ekeus, Stephen Joseph, Adm. William Owens, Pamela Berkowsky, Gen. Richard Hearney, Gen. Dennis J. Reimer, Joshua Lederberg, Harold Smith, Billy Richardson, Gen. John M. Shalikashvili, Richard J. Danzig, Richard Preston, and various Clinton administration officials at the Pentagon and elsewhere.

184 *Iraq's admissions:* A copy of the document summarizing Taha's admissions entitled "Talking Points on the Biological Program" was provided to the authors by a former UNSCOM inspector.

185 *Kamel was confessing all:* Kamel provided only a sketchy outline of the Iraqi program to UNSCOM. On February 20, in a move that shocked Western officials, he returned to Iraq. Three days later, Iraq announced that Kamel and two of his brothers were gunned down by "relatives." Daniel Williams, "Iraqi Defectors Killed on Return to Baghdad; Relatives Reportedly Shot Brothers," *Washington Post,* Feb. 24, 1996, p. 1.

185 *A sign-out sheet for the album:* Interview, Spertzel. UNSCOM never directly confronted Taha with the inconsistency between her original statements on the program and the material she had signed out. The inspectors felt there was nothing to be gained from further embarrassing her.

185 *It bore the signature of Taha:* Interview, Spertzel.

185 *A flood of disclosures:* UNSCOM summarized the Iraqi admissions in "Letter Dated 27 January, 1991 from the Permanent Representatives of The Nether-

lands and Slovenia to the United Nations Addressed to the President of the Security Council," S/1999/94, pp. 84–157, Jan. 29, 1999.

187 *matched the information obtained by the CIA:* The agency offers this assessment in Central Intelligence Agency, "Intelligence Related to Possible Sources of Biological Agent Exposure During the Persian Gulf War," Aug. 2000, p. 13.

188 *"is quite convincing":* Stephen C. Joseph, assistant secretary of defense for health affairs, Memorandum for Vice Chairman, Joint Chiefs of Staff, Subject: Anthrax Vaccine, July 25, 1995. In this memo, Joseph writes that he is passing on an analysis prepared by his special assistant for chemical/biological matters, Gen. Russ Zajtchuk.

188 *questions about whether the Michigan vaccine:* The prepared testimony of Meryl Nass before the Committee on Government Reform, Subcommittee on National Security, Veteran's Affairs and International Relations, Apr. 29, 1999, details the wide range of results from the published studies of anthrax vaccine efficacy involving guinea pigs and mice.

188 *"may not provide universal protection":* U.S. Army, Medical Department Center and School, Fort Sam Houston, David C. Jackson, Colonel, director, Combat and Doctrine Development, Memorandum for See Distribution, Subject: Operational Requirements Document (ORD) for the Improved Anthrax Vaccine (IAV), Mar. 13, 1995, p. 2.

189 *might involve exposure to anthrax:* The early history of the vaccine is summarized in a November, 10, 1995, "information paper" prepared by Michael J. Gilbreath, medical project manager, Joint Program Office, Biological Defense. The document, stamped "for official use only," was circulated among Pentagon officials. It notes that early studies of the anthrax vaccine's effects on human subjects involved a product that was similar but not identical to the Michigan Department of Health product. That research, Gilbreath reported, found "insufficient data to show protection against inhalation (aerosol) disease." The vaccine's early use is detailed in testimony by Kwai-Cheung Chan, director special studies and evaluation, U.S. General Accounting Office, at a hearing before the Committee on Government Reform, subcommittee on National Security, Veterans Affairs, and International Relations, 106th Cong., 1st sess., Apr. 29, 1999, pp. 9–38.

189 *not licensed for the aerosol exposure:* SAIC Corporation plan, 29 Sept. 1995, enclosure to memorandum from Anna Johnson-Winegar (U.S. Army) to Dr. Robert Myers, U.S. Army Medical Research and Material Command, Fort Detrick, Md., 5 Oct. 1995.

190 *precise correlation:* Lt. Col. David L. Danley, deputy joint program manager for Medical Systems, Memorandum for See Distribution, Subject: Minutes of the Meeting on Changing the Food and Drug Administration License for the Michigan Department of Health (MDPH) Anthrax Vaccine to Meet Military Requirements, Nov. 13, 1995. The memo says the "lack of a suitable animal surrogate" presents serious complications for changing the license.

191 *"It is a reality":* U.S. Senate, *Hearings Before the Permanent Subcommittee of Investigations of the Committee on Government Affairs,* 104th Cong., 1st sess., "Global

Proliferation of Weapons of Mass Destruction," part 1, Oct. 31 and Nov. 1, 1995, p. 5.

191 *"We received no information": Ibid.*, p. 274.

191 *"we have not followed religious cults": Ibid.*, p. 276.

192 *"the report was ignored": Ibid.*, p. 280.

192 *"This is not routine": Ibid.*, pp. 279–80.

192 *The guru's chauffeur told a Japanese court:* Sheryl WuDunn, Judith Miller, and William J. Broad, "Sowing Death: A Special Report; How Japan Germ Terror Alerted World," *New York Times,* May 26, 1998, p. A1.

192 *These attempted strikes against American facilities: Ibid.* Administration officials acknowledged that they had been unaware of the court testimony about the attacks on the U.S. Navy bases in Japan.

193 *Sterne strain:* The finding was described in A. S. Kaufmann, P. Keim, K. Taniguchi, N. Okabe, and T. Kurata, "Kameido Incident: Documentation of a Failed Bioterrorist Attack," poster at the Fourth International Conference on Anthrax, Annapolis, Md., June 10–13, 2001.

193 *Pentagon officials were joined by Richard Preston:* Described by participants.

195 *The number of people working:* Estimate of staffing levels provided by USAMRIID in response to request by authors.

196 *a $1 billion cut:* The plans for the cut were described by former government officials and detailed in "Chemical and Biological Defense, Emphasis Remains Insufficient to Resolve Continuing Problems," General Accounting Office, Mar. 12, 1996, p. 7.

196 *"use of BW weapons":* Lt. Col. Glenn D. Baker, Joint Chiefs of Staff, J4, "Point Paper," 30 Jan. 1996. It says: "Chairman's Program Assessment recommends no immunizations, procure baseline stockpiles, and shift R & D efforts towards faster and quicker regimes and antidotes." The document also notes that the anthrax vaccine is seldom used, with 500 doses a year administered nationwide.

196 *A new GAO report:* General Accounting Office, "Emphasis Remains Insufficient."

197 *hearing before the Senate Judiciary Committee:* Senate Judiciary Committee, "Interstate Transportation of Human Pathogens," Serial No. J-104-67, Mar. 6, 1996.

197 *Congress passed a law:* WuDunn, Miller, and Broad, "Sowing Death."

198 *"It is not a matter of if":* Permanent Subcommittee on Investigations, Mar. 27, 1996, p. 22.

199 *Two teams of consultants:* Both inspections are described in e-mails from Michael Gilbreath, Joint Program Office for Biological Defense. The first inspection, by James Kenimer, a Virginia-based consultant, took place from February 12 to 16 and is detailed in an April, 8, 1996, e-mail from Gilbreath to Lt. Col. David Danley, Gen. John Doesburg, Winifrede Fanelli, and Lt. Col. Bob Ranhofer. The subject is Dr. Kenimer's report on MDPH. A second inspection, in May 1996, involved a team of inspectors assembled by SAIC. It is described in a May 20, 1996, e-mail from Gilbreath to Fanelli, Walter Busbee, Doesburg, Danley, and Richard Paul. The subject is: "Preliminary feed-back on MBPI facility."

199 *"seeds of doubt":* Walter Busbee to Finelli, Doesburg, Danley, Paul, May 20, 1996. The e-mail reads as follows: "I believe we should also be doing a parrallel [sic] analysis of impacts if we were to decide to cut our losses or FDA acts unilaterally. I planted the seeds of doubts at the last Danzig/JROC session that we really have no mobilization base for vaccines except MDPH/MBPI which is an antiquated facility by today's standards. No one really responded, but we owe it to DR's Prociv & Joseph to inform them if we have a major problem—prior to next session on this subj w/ all 'heavy hitters' on 31 May."

199 *would need six months to gear up:* Joint Program Office for Biological Defense, draft memorandum, May 29, 1996, from Michael Gilbreath to Anna Johnson-Winegar, subject: "Worldwide Anthrax Vaccine Manufacturers," purpose: "To provide information on current manufacturers of Anthrax vaccine."

201 *The* Washington Post *published a story:* Bradley Graham, "Military Chiefs Back Anthrax Inoculations," *Washington Post,* Oct. 2, 1996, p. A1.

9. Taking Charge

This chapter is based on interviews with a number of biological-warfare experts and government officials, including: Joshua Lederberg, John P. White, Andy Weber, Anne M. Harrington, John J. Hamre, Pamela Berkowsky, Col. David R. Franz, William C. Patrick III, Ken Alibek, Stephen C. Joseph, Richard A. Clarke, Richard J. Danzig, Philip C. Bobbitt, W. Seth Carus, Ashton B. Carter, Margaret A. Hamburg, Randall Beatty, Lev S. Sandakhchiev, Nikolai A. Staritsin, Alexei Stepanov, Igor V. Domaradskij, Nikolai N. Urakov, former and current civilian scientists and Clinton administration germ-warfare experts who asked not to be identified by name, and a number of current and former scientists and intelligence analysts associated with Fort Detrick. The literature and interviews cited below, in addition to providing specific information, helped form the backdrop to many of the chapter's main events.

202 *Are you guys out of your minds?:* Interview, John P. White, former deputy secretary of defense. Just a few weeks earlier, White testified before the presidential committee looking into Gulf War illnesses and said: "This issue is very important to us. It goes to the heart of our credibility as an institution." The transcript of his remarks "Statement of Deputy Secretary of Defense before the Presidential Advisory Committee on Gulf Veterans Illnesses," is available on Gulflink.

203 *to settle a crucial question:* Letter, Stephen C. Joseph, assistant secretary of defense for health affairs, to Michael A. Friedman, lead deputy commissioner of Food and Drugs, Mar. 4, 1997.

203 *"not inconsistent with the current label":* Letter, Michael A. Friedman, lead deputy commissioner of Food and Drugs, to Stephen C. Joseph, assistant secretary of defense for health affairs, Mar. 13, 1997.

203 *a litany of failings:* A redacted version of the report, with trade secrets removed, was released by the FDA. The 8-page report, dated Mar. 11, 1997, describes an

inspection conducted between Nov. 18 and Nov. 27, 1996. It is addressed to: "Robert Myers, DVM, Responsible Head, Michigan Biological Products Institute."

204 *"Are you sure you folks":* Memorandum, Michael Gilbreath, Joint Program Office for Biological Defense, to Gen. John Doesburg, Walter Busbee, Lt. Col. David Danley, Lt. Col. James Estep, Mar. 25, 1997. Gilbreath raised two questions: "1. By testing, what message is DoD imparting to the FDA? 2. Does this imply DoD does not think the vaccine's quality previously approved by FDA is good?"

204 *"the existing industrial base will be lost":* Memorandum, Stephen C. Joseph to Deputy Secretary of Defense Through USD (Personnel and Readiness), Subject: "Biological Warfare-Information Memorandum." The copy of this 14-page memo obtained by the authors is undated, but John White, the deputy secretary of defense, said it sums up the arguments that were made to him in the first months of 1997.

204 *The Soviet Union had paid a harrowing price:* Judith Miller, *One, by One, by One* (New York: Simon & Schuster, 1990).

206 *The Iranians had tried to recruit them:* Judith Miller and William J. Broad, "The Germ Warriors: A Special Report; Iranians, Bioweapons in Mind, Lure Needy Ex-Soviet Scientists," *New York Times,* Dec. 8, 1998, p. A1.

206 *Obolensk alone had lost 54 percent:* Judith Miller, "In a Gamble, U.S. Supports Russian Germ Warfare Scientists," *New York Times,* June 20, 2000, p. F1.

208 *poster display at a scientific conference:* Interview, Nikolai A. Staritsin, Obolensk, May 2000.

213 *a dramatic decision:* Eric Schmidt, "Cohen Details Faults of General in Bombing," *New York Times,* Aug. 1, 1997, p. 4. Cohen blocked Schwalier's scheduled promotion, and the general announced he would retire.

213 *"shown significant improvements":* The assessment is reported in Assistant Secretary of Defense for Health Affairs, *Chronology for Decision-Making in and Context for Developing Anthrax Protection Effort,* May 16, 2000. The internal document, prepared by Pentagon health officials, covers the key dates in the anthrax debate from 1970 to September 1998. The entry dated September 1997 states: "DoD team visits MBPI and determines that the facility has shown significant improvements and is moving forward to correct its FDA deficiencies."

214 *another fiasco:* Norm Brewer and John Hanchette, "FDA: Pentagon Ignored in Bosnia Experimental Vaccine Lessons Learned in Gulf," Gannett News Service, Oct. 8, 1997. The article quotes a July 22, 1996, letter from FDA commissioner Michael Friedman to the Pentagon which noted the omission of key warnings about the vaccine's possible side effects. The GAO looked at the records of 588 army soldiers known to have been vaccinated against tick-borne encephalitis and found that 24 percent lacked the proper notations. GAO, "Defense Health Care—Medical Surveillance Improved Since Gulf War, but Mixed Results on Bosnia," May 13, 1997, p. 13.

215 *Despite six years of the most aggressive inspection:* Interview, Charles Duelfer, deputy executive chairman, UNSCOM; see also United Nations, Report of

the Executive Chairman on the Activities of the Special Commission Established by the Secretary-General Pursuant to Paragraph 9(b)(I) of Resolution 687 (1991), S/1998/920, New York, Oct. 1998.

215 *ten thousand anthrax germs:* Interviews, William C. Patrick III; Ken Alibek.

215 *ricin, a protein toxin:* Jonathan B. Tucker and Jason Pate, "The Minnesota Patriots Council (1991)," in *Toxic Terror: Assessing Terrorist Use of Chemical and Biological Weapons,* edited by Jonathan B. Tucker (Cambridge, Mass.: MIT Press, 2000), p. 167.

216 *the germs might not harm anyone:* Interview, William C. Patrick III.

217 *$322 million, ten-year contract:* William J. Broad and Judith Miller, "Thwarting Terror: A Special Report; Germ Defense Plan in Peril as Its Flaws Are Revealed," *New York Times,* Aug. 7, 1998, p. A1.

217 *"We owe it to our people":* Susanne M. Schafer, "Military to Get Anthrax Inoculation," Associated Press, Dec. 16, 1997.

218 *"I know of no expert opinion":* Transcript, Federal News Service, White House Briefing, "Press Conference with President Bill Clinton," Dec. 16, 1997.

218 *December 1997 issue of* Vaccine: A. P. Pomerantsev, N. A. Staritsin, Y. V. Mockov, and L. I. Marinin, "Expression of Cereolysine AB Genes in *Bacillus anthracis* Vaccine Strain Ensures Protection Against Experimental Hemolytic Anthrax Infection," *Vaccine,* Dec. 1997, pp. 1846–50.

219 *In an interview:* Interview, Nikolai A. Staritsin, Obolensk, Russia.

219 *"out of the bag":* Interview, Joshua Lederberg.

221 *"illegal practices":* Judith Miller, "U.S. Aid Is Diverted to Germ Warfare, Russian Scientists Say," *New York Times,* Jan. 25, 2000, p. A6.

10. The President

This chapter is based on interviews with William J. Clinton, Joshua Lederberg, Pamela Berkowsky, Richard A. Clarke, Col. David R. Franz, Richard J. Danzig, Philip C. Bobbitt, William C. Patrick III, Thomas Schneider, Mark S. Zaid, Gregory T. Nojeim, J. Craig Venter, William A. Haseltine, Richard Preston, John J. Hamre, Ken Alibek, Lev S. Sandakhchiev, Andy Weber, Nikolai N. Urakov, John W. Huggins, Lucille Shapiro, Barbara H. Rosenberg, Thomas P. Monath, Jerome M. Hauer, Frank E. Young, Brad Roberts, Margaret A. Hamburg, William Clarke, Donald A. Henderson, Tara O'Toole, Yevgenia Ustinov, and other officials and experts who spoke on the condition of anonymity. The chapter also drew on the sources listed below.

223 *Clinton and Venter spent much of the evening:* Thomas Schneider, a friend of Clinton's who arranged the Renaissance dinners with Venter and joined his institute's board of trustees in 1997, recalled that the two men hit it off personally and intellectually. Schneider said Venter added a new dimension to the threat of terrorism—the danger of supergerms. The president's attention, he said, was immediately caught by Venter's account of "the dark side of biotechnology." Schneider said Clinton liked Venter's irreverence, openness, and creativity—

qualities the president saw himself as bringing to government. Clinton, he added, also liked the fact that Venter's wife, Claire M. Fraser, was a powerful woman in her own right, a world-class scientist. "There were a lot of parallels" between the two couples, Schneider recalled, "a lot of meeting of the minds."

226 *Alibek emerged from anonymity:* Tim Weiner, "Soviet Defector Warns of Biological Weapons," *New York Times,* Feb. 25, 1998, p. A1.

228 *analyzed soil samples:* Judith Miller, "Poison Island: A Special Report; at Bleak Asian Site, Killer Germs Survive," *New York Times,* June 2, 1999, p. A1.

228 *strain found on the island:* In 2001, subsequent tests performed at one of the U.S. national laboratories also showed that the anthrax buried on Vozrozhdeniye Island was virtually identical to anthrax samples from the site of the fatal 1979 accident at the Soviet germ-warfare plant at Sverdlovsk. As this book went to press, the results of that research, showing that the anthrax on Vozrozhdeniye Island had come from what the Americans alleged was a bioweapons plant at Sverdlovsk, had still not been made public.

229 *the "Vector model":* There are some indications that in the early days of the Clinton administration's effort to dismantle the Soviet germ-warfare empire by engaging Russian scientists, officials were not as rigorous as they should have been in reviewing the possible misuse of U.S. funds and potential application of expertise gleaned from these civilian projects to military ends. For example, a report conducted by NASA's inspector general and posted on the space agency's Web site on October 13, 2000, concluded that the agency had never reviewed grants it had made to Russian institutes between 1994 and 1997 through its "Russian Science Research Program" "for possible biological warfare connections." Moreover, the report concludes, "no site visits were scheduled to ensure that NASA funding was not supporting biological warfare research," although three of the grant recipients "had been part of the Soviet biological warfare program." The report, entitled "NASA Oversight of Russian Biotechnology Research, 1994–1997," urged NASA to "carefully coordinate" its activities in this area from now on with the State Department's Bureau of Nonproliferation (http://www.hq.nasa.gov/office/oig/hq/inspections/g-00-007.pdf).

The NASA inquiry was triggered by a *New York Times* article alleging that Biopreparat had diverted money from this and other grant programs intended for its institutes. Judith Miller, "U.S. Aid Is Diverted to Germ Warfare, Russian Scientists Say," *New York Times,* Jan. 25, 2000.

231 *Soviet military had tested smallpox:* Although Moscow has denied that it ever conducted open-air testing of smallpox, a detailed report prepared by the Monterey Institute of International Studies Center for Nonproliferation Studies also asserts that the Soviet Union did conduct such tests on Vozrozhdeniye Island. It cites no source for the assertion, but the report is widely considered one of the more reliable descriptions of the Soviet germ empire in Kazakhstan; see Gulbarshyn Bozheyeva, Yerlan Kunakbayev, and Dastan Yeleukenov, "Former Soviet Biological Facilities in Kazakhstan: Past, Present, and Future," Center for Nonproliferation Studies, Monterey Institute of International Studies, June 1999, p. 6.

232 *protect the civilian population:* Richard A. Falkenrath, Robert D. Newman, and Bradley A. Thayer, *America's Achilles' Heel: Nuclear, Biological, and Chemical Terrorism and Covert Attack,* BCSIA Studies in International Security (Cambridge, Mass.: MIT Press, 1998), pp. 1–27, 337–40.

232 *government's spending on biodefense:* Interviews, Michael T. Osterholm, former Minnesota State epidemiologist and chair and chief executive officer of Ican, Inc. (see also his *Living Terrors: What America Needs to Know to Survive the Coming Bioterrorist Castastrophe* [New York: Delacorte Press, 2000]; Donald A. Henderson, dean emeritus of the Johns Hopkins School of Public Health; Amy E. Smithson and Leslie-Anne Levy, *Ataxia: The Chemical and Biological Terrorism Threat and the U.S. Response* (Washington, D.C.: Henry L. Stimson Center, 2000).

232 *tabletop exercise:* Interviews with White House and National Security Council staff members. See also Judith Miller and William J. Broad, "Exercise Finds U.S. Unable to Handle Germ War Threat," *New York Times,* Apr. 26, 1998, p. A1.

233 *"You could make such a virus today":* Interviews, William A. Haseltine.

233 *The hybrid weapon:* Miller and Broad, "Exercise Finds U.S. Unable."

234 *"We can return to the Stone Age":* Federal Document Clearing House, "Secretary of Defense Delivers Remarks at the National Press Club," Mar. 17, 1998, p. 9.

234 *The story behind Cohen's announcement:* Former senior Pentagon officials described the origins of the National Guard program and the SAIC study of the issue.

235 *$7.5 million of that money:* The estimate comes from a portion of the SAIC report posted on the National Guard Web site (http://www.ngb.dtic.mil) and identified as "Section 6. Overview of the WMD Study Process." It said the study was to cover five congressionally mandated tasks and that the SAIC was "provided a total of $7,538,137." Three senior Pentagon officials asserted that the money was inserted into the defense spending measure by the guard's supporters on Capitol Hill and that the company had helped lobby for the appropriation.

235 *The study was quietly shelved:* Aides to Secretary Cohen said that the study had little or no effect on their planning for the guard's role in the fight against chemical, biological, or nuclear terrorism.

236 *invitations were being rescinded:* Interviews with experts.

236 *U.S. public-health system was a disaster:* Judith Miller, "U.S. Unprepared for Bioterrorism, Experts Say," *New York Times,* June 3, 1998, p. A14.

241 *Monath was widely admired:* Judith Miller and William J. Broad, "Thwarting Terror: A Special Report; Germ Defense Plan in Peril as Its Flaws Are Revealed," *New York Times,* Aug. 7, 1998, p. A1.

243 *The experts endorsed such "defensive research":* In an interview, Barbara Rosenberg said that she did not believe that such research should be done in secret, but some of the other panelists disagreed, highlighting the difficulty of the issue.

245 *Hamre told NATO officials:* Linda D. Kozaryn, "Hamre: Ancient Tactics, Modern Strategy," American Forces Press Service, July 2, 1998.

246 *The proposal alarmed civil liberties experts:* William J. Broad and Judith Miller, "Pentagon Seeks Command for Emergencies in the U.S.," *New York Times,* Jan. 28, 1999, p. A21.

246 New York Times *editorialized:* "Thwarting Tomorrow's Terrors," *New York Times,* Jan. 23, 1999, p. A18.

247 *President Clinton discussed the threat:* Judith Miller and William J. Broad, "Clinton Describes Terrorism Threat for 21st Century," *New York Times,* Jan. 22, 1999, p. A1.

252 *"we are all Indians":* Interview, Elizabeth A. Fenn, a smallpox historian at George Washington University, who said that the declared eradication of the disease and subsequent end of vaccinations meant that mankind was "approaching 100 percent susceptibility." Lawrence K. Altman, William J. Broad, and Judith Miller, "Smallpox: The Once and Future Scourge?" *New York Times,* June 15, 1999, p. F1.

253 *Bernard asked the National Academy of Sciences:* The report issued by the Academy's Institute of Medicine is entitled "Assessment of Future Scientific Needs for Live Variola Virus," National Academy Press, Washington, D.C., 1999.

254 *"Look at these killers":* Videotape, United States Air Force Counterproliferation Center, Air War College, "Biological Warfare Briefing by Bill Patrick, Microbiologist," AUTV Video Production, Air University, Maxwell Air Force Base, Ala., F3569-99-0019, Feb. 23, 1999. Our thanks to the provider of this tape, who was not Bill Patrick and who shared it with us on the condition of anonymity. Patrick in this presentation says Ken Alibek, the Soviet defector, told him of Moscow's production figures. In turn, Alibek echoes some of Patrick's chart data in Ken Alibek with Stephen Handelman, *Biohazard: The Chilling True Story of the Largest Covert Biological Weapons Program in the World—Told from Inside by the Man Who Ran It* (New York: Random House, 1999), pp. 99, 122, 166. He also says, on p. 230, that after his defection he collaborated with Patrick on "a secret history" of the Soviet and American germ programs. This collaboration was apparently done for American intelligence.

255 *to the town's cemetery:* Judith Miller accompanied Weber that day and on several of his tours to former Soviet weapons labs.

255 *"Variant U":* Alibek, *Biohazard,* p. 132.

11. Defenders

This chapter is based on interviews with Amy E. Smithson; Jerome M. Hauer; Tracey S. McNamara; Marcelle Layton; Mark S. Zaid; Robert Myers; Col. John Grabenstein; Tara O'Toole, deputy director, Johns Hopkins Center for Civilian Bio-Defense Studies, Baltimore; Michael L. Moodie, director, Chemical and Biological Arms Control Institute, Washington, D.C.; and several public-health and emergency-preparedness officials in Denver, among them David B. Sullivan, Greg Moser, Tommy F. Grier Jr., Richard Hoffman, Gregory M. Bogdan, and Peter T. Pons. In addition, it drew on the following studies, books, and literature.

256 *Asnis found their complaints:* U.S. General Accounting Office, "West Nile Virus Outbreak: Lessons for Public Health Preparedness," Sept. 2000, p. 9.

256 *unusual for that ailment:* Annie Fine and Marcelle Layton, "Lessons from the West Nile Viral Encephalitis Outbreak in New York City, 1999: Implications for Bioterrorism Preparedness," published electronically on Jan. 15, 2001, *Clinical Infectious Diseases,* 32 (2001), pp. 277–82, Infectious Diseases Society of America, p. 2: www.journals.uchicago.edu/CIK/journal/issues/v32n2/001285/001285.text.html.

258 *"confirmed to be associated:* Report of the Minority Staff, Senate Governmental Affairs Committee, "Expect the Unexpected: The West Nile Virus Wake Up Call," July 24, 2000, p. 11.

258 *the city decided to spray: Ibid.,* pp. 13, 15.

258 *Hauer's office quietly cornered:* Fine and Layton, "Lessons from the West Nile Viral Encephalitis Outbreak," p. 6; Jennifer Steinhauer and Judith Miller, "In New York Outbreak, Glimpse of Gaps in Biological Defenses," *New York Times,* Oct. 11, 1999, p. A1.

261 *West Nile had never caused an epidemic:* Report of the Minority Staff, "Expect the Unexpected," p. 17. The Senate minority report quotes Gubler as saying that the sudden appearance of the West Nile virus was "the most significant development in North American arbovirology in the past 50 years" (p. 20).

261 *"when or if this outbreak would have been detected":* Fine and Layton, "Lessons from the West Nile Virus Encephalitis Outbreak," p. 3.

261 *As Frank Young:* Interview, Frank Young; and see a similar quotation in Amy E. Smithson and Leslie-Anne Levy, *Ataxia: The Chemical and Biological Terrorism Threat and the U.S. Response* (Washington, D.C.: Henry L. Stimson Center, 2000), p. 124

263 *In a subsequent review:* Interview, Steven M. Ostroff; and Denis Nash, Neal Cohen, and Marcelle Layton, "West Nile Virus Infection in New York City: The Public Health Perspective," p. 15; manuscript provided by Layton.

263 *McNamara, the animal doc:* The conference, "West Nile Virus: Bridging the Gap Between the Zoo and Public Health Communities," was held at the Lincoln Park Zoo in Chicago, June 21–22, 2001. Its principal organizers were the CDC's Center for Disease Control; Dominic Travis, veterinary epidemiologist at Lincoln Park Zoo; and Tracey S. McNamara, Wildlife Conservation Society (Bronx Zoo).

264 *a decision they made in the name of fairness:* Prepared testimony of John J. Hamre, Gen. Anthony Zinni, Gen. John Keane, David Oliver, and Gen. Ronald R. Blanck, Subcommittee on Military Personnel, House Committee on Armed Services, Sept. 30, 1999. The "One Day Policy" of March 30, 1999, they said, required anthrax immunization for anyone serving more than a single day in the "high threat areas," p. 7.

265 *GAO announced its preliminary findings:* U.S. House, *DOD's Mandatory Anthrax Vaccine Immunization Program for Military Personnel,* Hearing before the Subcommittee on National Security, Veterans Affairs, and International Relations of the Committee on Government Reform, 106th Cong., 1st sess., Apr. 29, 1999, pp. 10–20.

265 *"routinely administered": Ibid.,* p. 31. The claim about routine shots for veterinar-

ians was made in the Pentagon fact sheet "What Every Service Member Needs to Know About the Anthrax Vaccine." It is undated but was released in 1998.

266 *Only 8 of 260,000 people: Ibid.,* p. 70. Gen. Cain's biography, reproduced on p. 65, notes that he was trained in the U.S. army's Chemical Corps.

266 *"big time trouble":* E-mail message from Gen. Eddie Cain to Col John V. Wade, Col. Bob Borowski. Subject: "Anthrax Vaccine." Wade replied that day that there was a "strong feeling" among Pentagon officials that "GAO carried the day during the questioning." Wade said that the secretary of defense would be calling FDA or the Department of Health and Human Services to complain that "we didn't do well on this one." The e-mail refers to a growing unease about the BioPort contract among Pentagon officials. Wade noted that David Oliver, the principal deputy undersecretary of defense for acquisition and technology, was demanding a "cradle to grave" review of how the Pentagon got into its deal with the Michigan company.

266 *One document dated September 1991:* U.S. Army, Secretary of the Army, Memorandum of Decision, September 3, 1991, Subject: Authority Under Public Law 85-804 to Include an Indemnification Clause in Contract DAMD17-91-C1086 with Program Resources, Inc.

266 *A later version:* U.S. Army, Secretary of the Army, Memorandum of Decision, September 3, 1998, Subject: Authority Under Public Law 85-804 to Include an Indemnification Clause in Contract DAMD17-91-C1086 with Michigan Biologic Products Institute.

266 *Zaid provided the memos:* Dwight Daniels, "Anthrax Shots Bad Medicine? Vaccine's Possible Dangers Admitted in Military Papers," *San Diego Union-Tribune,* June 29, 1999, p. 1.

266 *insurance against a fire:* Defense Department, DoD News Briefing, Kenneth H. Bacon, June 29, 1999, p. 1.

267 *raise the price it paid:* Steven Lee Myers, "U.S. Doubles Payment to Sole Source of Anthrax Vaccine," *New York Times,* Aug. 5, 1999, p. 16.

267 *cost BioPort more than 3 million doses:* Estimate of doses lost after retesting provided by Robert Myers, chief scientific officer of BioPort, in May 2001, interview.

268 *seven years to build:* Senate Armed Services Committee, Federal Document Clearing House transcript, July 12, 2000, p. 46. David Oliver, principal deputy undersecretary of defense for acquisition, said it would take seven years for a company that did not have BioPort's cooperation.

268 *"overwrought response":* Committee on Government Reform, "The Department of Defense Anthrax Vaccine Immunization Program: Unproven Force Protection," 106th Cong., 2d sess., House Report 106-556, Apr. 3, 2000, p. 2.

268 *"medical Maginot Line": Ibid.,* p. 2.

268 *anthrax program was a success:* Interview, Lt. Col. John Grabenstein. Figures were as of May 2001.

269 *"Any vaccine can cause serious reactions":* U.S. Department of Defense, "What Everyone Needs to Know About the Anthrax Vaccine," Nov. 10, 2000.

269 *"paucity of published, peer-reviewed literature":* Institute of Medicine, "An Assess-

ment of the Safety of the Anthrax Vaccine, A Letter Report," Washington, D.C., Mar. 30, 2000, pp. 5–6.

269 *"early step": Ibid.*, p. 6.

270 *tiny number of serious adverse reactions:* Centers for Disease Control, *Epidemiology and Prevention of Vaccine-Preventable Diseases,* 6th ed., Jan. 2001, pp. 123–24. In 1989 to 1991, the disease returned as the percentage of children vaccinated declined sharply and 123 people died, half of them children younger than 5 years old.

271 *by the time the elaborate exercise:* David S. Cloud, "U.S. Plans for Mock Terrorist Attacks Get Mixed Response from Top Officials," *Wall Street Journal,* Mar. 20, 2000, p. B14.

272 *the cause of death was plague:* Thomas Inglesby, Rita Grossman, and Tara O'Toole, "A Plague on Your City: Observations from TopOff," *Biodefense Quarterly* (Sept. 2000), p. 3; also at http://www.hopkins-biodefense.org/pages/news/quarter2_2.html.

273 *twice the official death toll: Ibid.*, p. 5.

273 *plague, a highly contagious agent:* Interview, Richard E. Hoffman.

273 *But plague, one of President Clinton's:* Even plague might not necessarily result in the doomsday scenario that some of the most pessimistic analysts in TopOff implied. Helen Epstein, a microbiologist who has battled AIDS in Uganda and studied infectious diseases in several poor countries, notes that a 1994 outbreak of plague in India killed only fifty people, despite India's badly dilapidated public-health services. Plagues themselves are often limited by environmental factors. Most Americans, for instance, are relatively well nourished and healthy and do not live in the overcrowded, unsanitary conditions that help spread disease.

274 *"Gut-wrenching doomsday scenarios":* Interview, Tommy F. Grier Jr.

275 *TopOff had been very expensive:* The uncertainty over cost is just one of the many unanswered questions about TopOff. Almost a year after the exercise, the Justice Department had yet to release the long-scheduled report evaluating the exercise's strengths and weaknesses. A draft of a 40-page evaluation obtained by the authors concludes, among other things, the following: Hospitals were "inadequate to meet demands during a large-scale WMD [weapons of mass destruction] incident." There was no "widely accepted concept and plan for management of a large-scale bioterrorism incident," and arrangements for handling the drill's hypothetical dead were "particularly confusing." As with the National Guard Civil Support Teams, "responders lacked the necessary personal protective equipment and detection equipment" and "procedures to restrict international travel of contagious infected persons" needed to be re-examined." Information about the epidemic was not "shared or coordinated in a timely manner," and it was "unclear" who was in charge of responding to the mock disaster. Getting drugs and treatment to infected people proved far harder than anyone had anticipated. Finally, the report concludes, budget shortages prevented some key government agencies and departments from participating." Office of Justice Programs and Federal Emergency Management

Agency, et al., "Top Officials (TopOff) 2000, Exercise Observation Report, Executive Summary," pp. 1–40.

275 *"We weren't all exactly singing 'Kumbaya' ":* Interview, Greg Moser.

275 *taught them invaluable lessons:* Interview, David B. Sullivan, deputy director, City and County of Denver Office of Emergency Management; see also David Sullivan, "Exercise TOPOFF 2000: Lessons Learned," *Bulletin of the International Association of Emergency Managers,* available at www.iaem.com.

275 *"Our public-health-care infrastructure":* interview, Tara O'Toole.

276 *to buy plastic bags:* Ingelsby, Grossman, and O'Toole,"A Plague on Your City," p. 8.

276 *utterly inadequate: Ibid.,* p. 8.

276 *Amy Smithson, an analyst:* Amy E. Smithson and Leslie-Anne Levy, *Ataxia: The Chemical and Biological Terrorism Threat and the U.S. Response* (Washington D.C.: Henry L. Stimson Center, 2000).

277 *funding for counterterrorism rose:* Executive Office of the President, Office of Management and Budget, "Annual Report to Congress on Combating Terrorism," May 18, 2000, pp. 47–51, 58–65.

277 *criticism came from the General Accounting Office:* See GAO reports, among them, "Combating Terrorism: Spending on Governmentwide Programs Requires Better Management and Coordination," GAO/NSIAD-98-39, Dec. 1, 1997; "Combating Terrorism: Threat and Risk Assessments Can Help Prioritize and Target Program Investments," GAO/NSIAD-98-74, Apr. 9, 1998; "Combating Terrorism: Observations on Federal Spending to Combat Terrorism," Statement of Henry L. Hinton Jr., Assistant Comptroller on National Security and International Affairs Division, Before the Subcommittee on National Security, Veterans Affairs and International Relations Committee on Government Reform, Mar. 11, 1999; "Combating Terrorism: Need for Comprehensive Threat and Risk Assessments of Chemicals and Biological Attacks," GAO/NSIAD-99-163; "Combating Terrorism: Linking Threats to Strategies and Resources," Statement of Norman J. Rabkin, Director, National Security Preparedness Issues, National Security and International Affairs Division, GAO/T-NSIAD-00-218, July 26, 2000; "Combating Terrorism: Need to Eliminate Duplicate Federal Weapons of Mass Destruction Training," GAO/NSIAD-00-64, Mar. 2000; "Chemical and Biological Defense: Units Better Equipped but Training and Readiness Reporting Problems Remain," GAO-01-27, Nov. 2000.

278 *the "not-if-but-when":* Smithson and Levy, *Ataxia,* p. 68. See also Jonathan B. Tucker and Amy Sands, "An Unlikely Threat," *Bulletin of the Atomic Scientists* (July–Aug. 1999), p. 47; W. Seth Carus, "Biowarfare Threats in Perspective," Press Briefing, "Biological Weapons and U.S. Security," Washington, D.C., Apr. 27, 1998: http://www.brook.edu/fp/events/19980427_carus.htm.

280 *their 1998 memo outlining the concept:* Dod Tiger Team, "Department of Defense Plan for Integrating National Guard and Reserve Component Support for Response to Attacks Using Weapons of Mass Destruction," Jan. 1998, chap. 4 (posted at www.defenselink.mil/pubs/wmdresponse/).

280 *unmitigated catastrophe:* Office of the Inspector General, Department of Defense, "Management of National Guard Weapons of Mass Destruction–Civil Support Teams," Report No. D-2001-043, Jan. 31, 2001. All quotations are from the 59-page report.

281 *Only five of the ten teams:* Associated Press, "National Guard Anti-Terrorism Teams at Risk; Faulty Equipment, Training Leave Units Unready After 3 Years, Review Finds," *Washington Post,* Feb. 26, 2001, p. A4.

283 *The Clinton effort initially focused:* Department of Defense, *Chemical and Biological Defense Program: Annual Report to Congress,* Mar. 2000, pp. A1–A25.

283 *The Pentagon's inspector general criticized:* Office of the Inspector General, Department of Defense, Audit Report, "The Low-Rate Initial Production Decision for the Joint Biological Point Detection System," Sept. 11, 2000, D-2000-187. For an overview of the field's problems, see David R. Walt and David R. Franz, "Biological Warfare Detection: A Host of Detection Strategies Have Been Developed, but Each Has Significant Limitations," *Analytical Chemistry* (Dec. 1, 2000), pp. 738a–46a.

284 *Zeroing in on genes:* Arnie Heller, "Uncovering Bioterrorism," *Science and Technology Review,* Lawrence Livermore National Laboratory (May 2000), pp. 4–12.

284 *signatures were then copied into artificially made snippets of DNA:* Katie Walter, "Reducing the Threat of Biological Weapons," *Science and Technology Review,* Lawrence Livermore National Laboratory (June 1998), pp. 4–9.

284 *a small chamber with a silicon chip:* Phillip Belgrader et al., "PCR Detection of Bacteria in Seven Minutes," *Science* (Apr. 16, 1999), pp. 449–50.

285 *At the Stepnogorsk complex:* Michael Dobbs, "Soviet-Era Work on Bioweapons Still Worrisome; Stall in U.S. Dismantling Effort Could Pose Proliferation Threat," *Washington Post,* Sept. 12, 2000, p. A1.

12. The Future

This chapter is based on interviews with Ken Alibek, Sergei Popov, Shaun Jones, Stephen S. Morse, Stephen A. Johnston, Russell J. Howard, Joshua Lederberg, Matthew S. Meselson, and many current and former government officials and scientists who spoke on the condition of anonymity. The chapter also drew on the literature cited below.

287 *Tenet had been pushing his agency to do more*: For an unclassified overview of the agency's initiatives, see John A. Lauder, special assistant for nonproliferation to the Director of Central Intelligence, statement to the Commission to Assess the Organization of the Federal Government to Combat the Proliferation of Weapons of Mass Destruction, Apr. 29, 1999.

287 *he had warned Congress:* Statement by Director of Central Intelligence, George J. Tenet, as prepared for delivery before the Senate Select Committee on Intelligence hearing on "The Worldwide Threat in 2000: Global Realities of Our National Security," Feb. 2, 2000.

288 *treaty banned nations:* Specifically, the treaty binds member states "never in any circumstances to develop, produce, stockpile or otherwise acquire or retain: (1) Microbial or other biological agents, or toxins whatever their origin or method of production, of types and in quantities that have no justification for prophylactic, protective or other peaceful purposes; (2) Weapons, equipment or means of delivery designed to use such agents or toxins for hostile purposes or in armed conflict."

289 *"agent, agent, agent":* Videotape, United States Air Force Counterproliferation Center, Air War College, "Biological Warfare Briefing by Bill Patrick, Microbiologist," AUTV Video Production, Air University, Maxwell Air Force Base, Ala., F3569-99-0019, Feb. 23, 1999. Our thanks to the provider of this tape, who was not Bill Patrick and who shared it with us on the condition of anonymity.

290 *"he could pass for a roadie with the Grateful Dead":* Richard Preston, *The Hot Zone* (New York: Random House, 1994), pp. 44–45.

294 *asked for a more detailed briefing:* In interviews, U.S. intelligence officials maintained that it was their usual practice to brief NSC officials and other officials who needed to know about such sensitive intelligence activities. They had already begun to brief senior officials about Clear Vision before the NSC asserts it asked them to do so, intelligence officials said.

295 *really need rockets?:* U.S. intelligence officials, in interviews, denied that the agency ever proposed to build a Soviet-style rocket to test.

297 *project was code-named Bacchus:* Although the authors have chosen to spell the project name as it is commonly written, the Pentagon refers to it in some secret documents as Bachus.

300 *The result was gridlock:* For how the Bush administration ended the bureaucratic impasse, see Michael R. Gordon and Judith Miller, "U.S. Germ Warfare Review Faults Plan on Enforcement," *New York Times,* May 20, 2001, p. Al.

300 *talk about what he had done:* Although Popov has been in the United States since 1992, he has only rarely been interviewed. In the best previously published interview, Popov said that neither the British nor American intelligence services had seemed interested in his germ warfare research in the Soviet Union, because "not a single person" came to discuss his work with him until "much later in Dallas." Even then, according to one well-informed U.S. source, the intelligence analyst who interviewed Popov asked only about the transfer of sensitive germ technology to Iran, Iraq, and other unfriendly states, and virtually nothing about his recombinant work. Popov discussed his gene modification research at length with retired air force Col. Randall Larsen, director of the ANSER Institute for Homeland Security. "Inside the Soviet Union's Biowarfare Program: Interviews with Dr. Ken Alibek and Dr. Sergeui Popov," ANSER Institute for Homeland Security, June 11, 2001; interview conducted on Nov. 1, 2000, p. 26. For an earlier version of this interview, see "Serguei Popov Interview," *Journal of Homeland Defense,* Nov. 13, 2000, available at www.homeland-defense.org.

303 *Alibek recalled in* Biohazard: Ken Alibek with Stephen Handelman, *Biohazard:*

The Chilling True Story of the Largest Covert Biological Weapons Program in the World—Told from Inside by the Man Who Ran It (New York: Random House, 1999), p. 164.

304 *American and Italian scientists had just published a paper:* Rosetta Pedotti et al., "An Unexpected Version of Horror Autotoxicus: Anaphylactic Shock to a Self-Peptide," *Nature Immunology* (Mar. 2001), pp. 217–22.

305 *his own clear vision of the future:* For an early description of the effort and the reaction of the scientific community, see Elliott Marshall, "Too Radical for NIH? Try DARPA," *Science* (Feb. 7, 1997), pp. 744–46. For an early report by DARPA's head, see Larry Lynn, director of the Defense Advanced Research Projects Agency, Statement to the Committee on Armed Services, Subcommittee on Acquisition and Technology, United States Senate, Mar. 11, 1997. Available at www.darpa.mil/body/NewsItems/lynn_03_11_97.html. See also Paul Jacobs, "Attack of the Killer Microbes," *Los Angeles Times,* Aug. 9, 1999, p. A1.

305 *working on forty-three different projects:* For a listing see Defense Advanced Research Projects Agency, "Unconventional Pathogen Countermeasures," undated program description, www.darpa.mil/dso/thrust/bwd/upc/navindex/index.html.

306 *he called the innovative method Expression Library Immunization:* Kathryn F. Sykes and Stephen Albert Johnston, "Linear Expression Elements: A Rapid, in vivo, Method to Screen for Gene Functions," *Nature Biotechnology* (Apr. 17, 1999), pp. 355–59; Michael A. Barry, Wayne C. Lal, and Stephen Albert Johnston, "Protection Against Mycoplasma Infection Using Expression-Library Immunization," *Nature* (Oct. 19, 1995), pp. 632–35.

307 *recombining genes in hundreds:* Andrew Pollack, "Selling Evolution in Ways Darwin Never Imagined; If You Can Build a Better Gene, Investors May Come," *New York Times,* Oct. 28, 2000, p. C1. Analysts have noted that the company's techniques, by enhancing such properties as antibiotic resistance, might also be used to make superbugs for war. See Steven M. Block, "Living Nightmares: Biological Threats Enabled by Molecular Biology," in *The New Terror: Facing the Threat of Biological and Chemical Weapons,* edited by Sidney D. Drell et al. (Stanford: Hoover Institution Press, 1999), pp. 57–58; and Carina Dennis, "The Bugs of War," *Nature* (May 17, 2001), pp. 232–35.

308 *The annual budget went from $59 million:* Defense Advanced Research Projects Agency, "RDT&E Budget Item Justification Sheet," Feb. 2000.

309 *DIA program known as Project Jefferson:* Defense Department officials maintain that much of the project's analysis and research is not classified and may even be published. But much of Jefferson is shrouded in secrecy. The project and its programs have been described in detail to congressional committees that oversee intelligence and biodefense research and discussed with other intelligence agencies. U.S. intelligence officials maintain that the White House under both the Clinton and Bush administrations were also briefed about Jefferson. But they acknowledge that the Clinton White House was probably not told about the new research involving genetic modification of anthrax because the request

to start work coincided with the transition between the Democratic and Republican administrations. Intelligence officials also say that the DIA inherited that project from the CIA, which decided against doing it under Clear Vision for reasons that neither the CIA nor DIA will discuss publicly. A rare recent reference to Jefferson's work appears in a study for the Pentagon which describes the project as an intelligence agency "tool" being developed "to scientifically validate threats." Brad Roberts, "Biological Weapons in Major Theater War," Institute for Defense Analyses, 1998, 1999, p. 6.

309 *reviewing classified and openly published scientific literature:* The information from Jefferson Project work is entered into a giant, largely secret database—BACHWORTH, the "biological and chemical warfare on line repository of technical holdings." U.S. intelligence officials say that civilian agencies, such as the departments of Agriculture and Health and Human Services, are increasingly turning to the giant DIA data bank for help in assessing threats to American crops, livestock, and public safety. Another group of government and outside advisers, known as the "2020" team, advises Jefferson in grappling with the military potential of advanced new research.

309 *DIA turned to Battelle:* Battelle, which permitted Judith Miller in April 2001, to visit its headquarters in Columbus, Ohio, and its facility in West Jefferson about twenty minutes away, has declined comment on any aspect of its work for the DIA's Jefferson Project or the CIA's Clear Vision, saying that the institute's "customers" do not want such work discussed. Battelle has withdrawn an undated brochure it published earlier about the project. The brochure states that Jefferson focuses on assessing foreign threats by "determining risks of current/new agents" through "literature studies" and "lab analysis." It also seeks to "evaluate weaponization and effectiveness of delivery capabilities" through means that U.S. intelligence officials have declined to discuss, and to "analyze biotechnological infrastructure and activities."

310 *Australian scientists announced:* Ronald J. Jackson et al., "Expression of Mouse Interleukin-4 by a Recombinant Ectromelia Virus Suppresses Cytolytic Lymphocyte Responses and Overcomes Genetic Resistance to Mousepox," *Journal of Virology* (Feb. 2001), pp. 1205–10.

311 *"We felt we had a moral obligation":* William J. Broad, "Australians Create a Deadly Mouse Virus," *New York Times,* Jan. 23, 2001, p. A6.

313 *Meselson's hope for the future:* Philip Heymann, Matthew Meselson, and Richard Zeckhauser, "Criminalize the Traffic in Terror Weapons," *Washington Post,* Apr. 15, 1998, p. A19.

314 *Russia and South Africa:* For a scholarly review of the issue of ethnic weapons, see British Medical Association, *Biotechnology, Weapons and Humanity* (Amsterdam: Harwood Academic Publishers, 1999), pp. 53–67.

Conclusions

315 *a five-pound bag of anthrax:* Interview, William C. Patrick III. An interesting demonstration of the resonance of Secretary Cohen's misstatement in his 1997

interview with ABC can be found in a *Wall Street Journal* editorial on Friday, July 27, 2001. The editorial asserted that "a quantity the size of a five-pound bag of sugar could wipe out New York City." *Wall Street Journal,* p. A8.

318 *Sheela subsequently moved to Switzerland:* First reported by Jeanie Senior, "Anand Sheela Tends Patients in Switzerland," *The Oregonian,* Dec. 26, 1999. And in a telephone interview, March 2000, Sheela Birnstiel, as she is now known, said she owns and operates two nursing homes near Basel which employ seventeen people and offer an atmosphere that is "friendly, warm, and nonthreatening." Sheela said that Alzheimer's patients were her favorite. Though they required some "medication to calm them down," she said, she never locked them up. She said she had put "all my bad memories behind me," and insisted that she was not guilty of the poisonings for which she was jailed. She had only followed the dictates "of the man I loved," she said, referring to the Bhagwan Shree Rajneesh. The guru died in 1990 in his native India, after having been convicted of immigration fraud in 1985 and deported from the United States. Sheela said she has no contacts with anyone from the cult and that her conscience is clear. She had pleaded guilty only to settle the charges and end the episode. "But I wasn't guilty; and I'm not guilty," she said. "Lovers don't commit crimes. I was a lover."

319 *largest hospital in The Dalles:* Interview, Gayle Jacobson, pathologist, Mid-Columbia Hospital.

320 *did create such a committee:* The committee was called the Non-Proliferation Advisory Panel, whose members included scientists, government officials, academics and public health specialists.

320 *warned James Forrestal:* Caryl P. Haskins et al., "Report of the Secretary of Defense's Ad Hoc Committee on Biological Warfare," July 11, 1949.

Select Bibliography

We found the books listed below particularly useful in our research and recommend them to readers interested in further information.

General

Cole, Leonard A. *The Eleventh Plague: The Politics of Biological and Chemical Warfare.* New York: W. H. Freeman, 1997.

Drell, Sidney D., Abraham B. Sofaer, and George D. Wilson, editors. *The New Terror: Facing the Threat of Biological and Chemical Weapons.* Stanford, Calif.: Hoover Institution Press, 1999.

Falkenrath, Richard A., Robert D. Newman, and Bradley A. Thayer. *America's Achilles' Heel: Nuclear, Biological, and Chemical Terrorism and Covert Attack.* Cambridge, Mass.: MIT Press, 1998.

Harris, Robert, and Jeremy Paxman. *A Higher Form of Killing: The Secret Story of Chemical and Biological Warfare.* New York: Hill & Wang, 1982.

Institute of Medicine. *Chemical and Biological Terrorism: Research and Development to Improve Civilian Medical Response.* Washington, D.C.: National Academy Press, 1999.

Lake, Anthony. *6 Nightmares: Real Threats in a Dangerous World and How America Can Meet Them.* New York: Little, Brown, 2000.

Lederberg, Joshua, editor. *Biological Weapons: Limiting the Threat.* Cambridge, Mass.: MIT Press, 1999.

Mangold, Tom, and Jeff Goldberg. *Plague Wars: The Terrifying Reality of Biological Warfare.* New York: St. Martin's Press, 1999. An insider account with many details on the Soviet and South African programs.

McDermott, Jeanne. *The Killing Winds: The Menace of Biological Warfare.* New York: Arbor House, 1987.

Osterholm, Michael T., and John Schwartz. *Living Terrors: What America Needs to Know to Survive the Coming Bioterrorist Catastrophe.* New York: Delacorte Press, 2000.

Roberts, Brad, editor. *Hype or Reality? The "New Terrorism" and Mass Casualty Attacks.* Alexandria, Va.: Free Hand Press, 2000.

Sidell, Frederick R., T. Takafuji, and David R. Franz, editors. *Medical Aspects of Chemical and Biological Warfare.* Washington, D.C.: Borden Institute, Walter Reed Army Medical Center; Falls Church, Va.: Office of the Surgeon General, United States Army, 1997. This textbook, though focused on diseases, also reviews the history of biological warfare, with emphasis on the American program.

Stern, Jessica. *The Ultimate Terrorists.* Cambridge, Mass.: Harvard University Press, 1999.

Stockholm International Peace Research Institute. *The Problem of Chemical and Biological Warfare.* Stockholm: Almqvist & Wiksell, 1971–75. This six-volume set, published between the United States' renunciation of germ weapons and the signing of the Biological Weapons Convention, is the best single overview of the field.

Tucker, Jonathan B., editor. *Toxic Terror: Assessing Terrorist Use of Chemical and Biological Weapons.* Cambridge, Mass.: MIT Press, 2000.

Wright, Susan, editor. *Preventing a Biological Arms Race.* Cambridge, Mass.: MIT Press, 1990. The appendixes reprint more than a dozen primary documents, including the National Security Decision Memorandum that spells out President Nixon's renunciation of germ weapons as well as the text of the 1972 Biological Weapons Convention.

Early History

Diamond, Jared M. *Guns, Germs, and Steel: The Fates of Human Societies.* New York: Norton, 1997. How the endemic diseases of invading peoples devastated the invaded.

Geissler, Erhard, and John Ellis van Courtland Moon, editors. *Biological and Toxin Weapons: Research, Development and Use from the Middle Ages to 1945.* New York: Oxford University Press and Stockholm International Peace Research Institute, 1999.

Harris, Sheldon H. *Factories of Death: Japanese Biological Warfare, 1932–45, and the American Cover-up.* New York: Routledge, 1994.

Kolata, Gina. *Flu: The Story of the Great Influenza Pandemic of 1918 and the Search for the Virus That Caused It.* New York: Farrar, Straus and Giroux, 1999.

The American Program

Cole, Leonard A. *Clouds of Secrecy: The Army's Germ Warfare Tests over Populated Areas.* Totowa, N.J.: Rowman & Littlefield, 1988.

Covert, Norman M. *Cutting Edge: A History of Fort Detrick, Maryland, 1943–1993.* Fort Detrick, Md.: U.S. Army, 1997. An institutional history with some interesting details.

Hersh, Seymour M. *Chemical and Biological Warfare: America's Hidden Arsenal.* Indianapolis: Bobbs-Merrill, 1968.

Mauroni, Albert J. *America's Struggle with Chemical-Biological Warfare.* Westport, Conn.: Praeger, 2000.

Mole, Robert L., and Dale M. Mole. *For God and Country: Operation Whitecoat, 1954–1973.* Brushton, N.Y.: Teach Services, 1998. A good source on the Seventh-Day Adventist volunteers.

Regis, Ed. *The Biology of Doom: The History of America's Secret Germ Warfare Project.* New York: Holt, 1999. The best popular overview of early U.S. efforts.

The Rajneeshees

Carter, Lewis F. *Charisma and Control in Rajneeshpuram: The Role of Shared Values in the Creation of a Community.* New York: Cambridge University Press, 1990.

Franklin, Satya Bharti. *The Promise of Paradise: A Woman's Intimate Story of the Perils of Life with Rajneesh.* Barrytown, NY: Station Hill Press, 1992.

Milne, Hugh. *Bhagwan: The God That Failed.* New York: St. Martin's Press, 1986.

Iraq and the Gulf War

Butler, Richard. *The Greatest Threat: Iraq, Weapons of Mass Destruction, and the Crisis of Global Security.* New York: Public Affairs, 2000.

Cordesman, Anthony M., and Ahmed S. Hashim. *Iraq: Sanctions and Beyond.* Boulder, Colo.: Westview Press, 1997.

Gordon, Michael R., and General Bernard E. Trainor. *The Generals' War: The Inside Story of the Conflict in the Gulf.* Boston: Little, Brown, 1995.

Hersh, Seymour M. *Against All Enemies: Gulf War Syndrome: The War Between America's Ailing Veterans and Their Government.* New York: Ballantine, 1998.

Mauroni, Albert J. *Chemical-Biological Defense: U.S. Military Policies and Decisions in the Gulf War.* Westport, Conn.: Praeger, 1998.

Powell, Colin L., with Joseph E. Persico. *My American Journey: An Autobiography.* New York: Random House, 1995.

Ritter, Scott. *Endgame: Solving the Iraq Problem—Once and for All.* New York: Simon & Schuster, 1999.

Schwarzkopf, General H. Norman, with Peter Petre. *The Autobiography: It Doesn't Take a Hero.* New York: Bantam Books, 1992.

Trevan, Tim. *Saddam's Secrets: The Hunt for Iraq's Hidden Weapons.* London: HarperCollins, 1999.

The Soviet Program

Alibek, Ken, with Stephen Handelman. *Biohazard: The Chilling True Story of the Largest Covert Biological Weapons Program in the World—Told from Inside by the Man Who Ran It.* New York: Random House, 1999.

Guillemin, Jeanne. *Anthrax: The Investigation of a Deadly Outbreak.* Berkeley: University of California Press, 1999.

Aum

Kaplan, David E., and Andrew Marshall. *The Cult at the End of the World.* New York: Crown Publishers, 1996.

Public Health

Garrett, Laurie. *Betrayal of Trust: The Collapse of Global Public Health.* New York: Hyperion, 2000.

———. *The Coming Plague: Newly Emerging Diseases in a World Out of Balance.* New York: Farrar, Straus and Giroux, 1994.

Lederberg, Joshua, Robert E. Shope, and Stanley C. Oaks, Jr., editors. *Emerging Infections: Microbial Threats to Health in the United States.* Washington, D.C.: National Academy Press, 1992.

Smallpox

Fenn, Elizabeth Anne. *Pox Americana: The Great Smallpox Epidemic of 1775–82.* New York: Hill & Wang, 2001.

Fenner, Frank, et al. *Smallpox and Its Eradication.* Geneva: World Health Organization, 1988.

Institute of Medicine. *Assessment of Future Scientific Needs for Variola Virus.* Washington, D.C.: National Academy Press, 1999.

The New Biology

Wade, Nicholas. *The Ultimate Experiment: Man-made Evolution.* New York: Walker, 1977.

Watson, James D. *A Passion for DNA: Genes, Genomes, and Society.* Cold Spring Harbor, N.Y.: Cold Spring Harbor Laboratory Press, 2000.

———, Michael Gilman, Jan Witkowski, and Mark Zoller. *Recombinant DNA,* second edition. New York: Scientific American Books, 1992.

Wright, Susan. *Molecular Politics: Developing American and British Regulatory Policy for Genetic Engineering, 1972–1982.* Chicago: University of Chicago Press, 1994.

Weapons and the New Biology

British Medical Association. *Biotechnology, Weapons and Humanity.* Amsterdam: Harwood Academic Publishers, 1999.

Piller, Charles, and Keith R. Yamamoto. *Gene Wars: Military Control over the New Genetic Technologies.* New York: Morrow, 1988.

Agriculture

Frazier, Thomas W., and Drew C. Richardson, editors. *Food and Agricultural Security: Guarding Against Natural Threats and Terrorist Attacks Affecting Health, National Food Supplies, and Agricultural Economics.* New York: New York Academy of Sciences, 1999.

Acknowledgments

We have accumulated many debts during more than three years of research and writing. Thanks go first of all to the experts, scientists, intelligence specialists, military officers, and federal officials who made this book possible. In addition to the individuals cited by name in the chapter notes, we would like to extend special thanks to Ken Alibek, Richard J. Danzig, Col. David R. Franz, Col. Arthur M. Friedlander, Stephen C. Joseph, John J. Hamre, Jerome M. Hauer, Joshua Lederberg, Matthew S. Meselson, William C. Patrick III, Barbara H. Rosenberg, Richard Spertzel, Frank E. Young, and Mark S. Zaid for their generous sharing of time, materials, and insights.

We are particularly grateful to former and current government officials whose aid came on the condition of anonymity and whose names go unmentioned as sources. Their help opened up a hidden world of reports, programs, and meetings. In our judgment, it will ultimately strengthen the roots of our democracy.

For assistance in many private and public matters, we thank Kelly D. Akers of the Defense Technical Information Center; Paul Berg, Christopher F. Chyba, and Sidney D. Drell of Stanford University; Robert Blitzer of Science Applications International Corp.; Lt. Cmdr. James E. Brooks of the public affairs office of the Defense Intelligence Agency; Queenie A. M. Byars of Defense Department public affairs; Leonard A. Cole of Rutgers University; Chuck Dacey of Fort Detrick; Col. Edward M. Eitzen Jr. and Caree Vander Linden of the United States Army Medical Research Institute of Infectious Diseases; Joseph J. Esposito, James M. Hughes, Scott R. Lillibridge, Brian Mahy, Joseph E. McDade, Stephen M. Ostroff, Barbara S. Reynolds, Robert V. Tauxe, and Thomas J. Török of the Centers for Disease Control and Prevention; Goutam Gupta and Paul J. Jackson of the Los Alamos National Laboratory; Bill Harlow of the Central Intelligence Agency; Artemis Housewright; Russell J. Howard of Maxygen; Stephen Aftergood and Kevin P. Kavanaugh of the Federation of American Scientists; Ronald P. Koopman and Susan Houghton of the Lawrence Livermore National Laboratory; Lt. Cmdr. Jeanette Lucas of the Armed Forces Radiobiology Institute; Stephen S. Morse of Columbia University; Janice O'Connell of the United States Senate; Col. Gerald W. Parker of the U.S. Army Medical Research & Materiel Command; Richard Preston; David Rigby of the public affairs office of the Defence Threat Reduction Agency; Gen. Philip K. Russell; Seth Shulman of the Center for Public Integrity; John F. Sopko; James Turner of Defense Department public affairs; Jan Walker of the Defense Advanced Research Projects Agency; Alan P. Zelicoff of the Sandia National Laboratories; Philip D. Zelikow of the University of Virginia; and Raymond A. Zilinskas of Johns Hopkins University and the Monterey Institute of International Studies.

For help with the Rajneeshee story, we would like to thank Jeanie Senior (and her husband and film journalist, Tom) who not only provided on-the-ground advice and support in Oregon but also shared with us her enormous knowledge and insight into the cult and the reactions of fellow Oregonians to it. We are also in debt to W. Seth Carus, whose meticulous reporting on the cult and other similar groups has been an invaluable source for our own work. Also, special thanks to Robert Hamilton and to Michael Skeels, whose continuing interest in the lessons of the attack have helped shape our conclusions.

For their assistance on the Gulf War, we thank Col. James D. Bales Jr., Col. Martin Crumrine, Col. Robert Eng, Anna Johnson-Winegar, Col. George E. Lewis, Adm. Edward Martin, Gen. H. Norman Schwarzkopf, Capt. Larry Seaquist, and Gen. Bernard E. Trainor. For their help in understanding the work of UNSCOM in Iraq, we would like to thank Rod Barton, Charles Duelfer, and David Kelly.

For help in Russia, the authors are indebted to the secretary of defense's offices of Cooperative Threat Reduction and Public Affairs, which permitted Judith Miller to accompany Andy Weber and other government officials on several trips to Russia and Kazakhstan and made Weber and other officials available for briefings on efforts to stop the spread of germ weapons. Thanks are due also to other American officials, among them Anne M. Harrington, Bill Richardson, and Floyd P. Horn, who helped explain the history, evolution, and accomplishments of the nonproliferation effort they helped shape. Much appreciated assistance came from the Moscow Bureau of the *New York Times*—specifically from Viktor Climenko, Nikolai Khalip, Natasha Bubenova, and Sophia Kishkovsky. Many thanks are also due to Russian scientists and institute directors, among them Lev S. Sandakhchiev, Nikolai N. Urakov, Vladimir P. Zav'yalov, Yevgeny Severin, Anatolij A. Makarov, Valery Lipkin, Igor V. Domaradskij, and Gregori Shcherbakov.

For help with Kazakhstan, many thanks to Vladimir Shkolnik; Erlan Idrissov; Olga Tyupkina; Bolat Nurgaliev; Gennady Lepyoshkin; Dastan Yeleukenov; Marina Voronova; the ambassador to Washington, Kanat Saudabayev; Roman Vassile; Sadigappar Mamadaliev; and Bakyt B. Atshabar. In Uzbekistan, we thank Sodyq Safaev, Isan M. Mustafoev, Djaloliddin A. Azimov, Islamov Abushair, Abdusattor Abdukarimov, and Ian Small.

Many experts, some in government, helped us grapple with the thorny public policy issues inherent in germ defense. Several preferred to remain anonymous. Others include Pamela B. Berkowsky, Philip C. Bobbitt, Frank Cilluffo, Richard A. Clarke, William Clark, Jay Davis, Margaret A. Hamburg, Frank Miller, Kyle B. Olson, Billy Richardson, and Brad Roberts.

We were also aided by several arms control and other nonprofit groups involved in the tracking of unconventional weapons. We relied heavily on their excellent guidance, information, and Web sites. Our special thanks go to the Monterey Institute of International Studies Center for Nonproliferation, and especially to William C. Potter, Jonathan B. Tucker, Amy Sands, and John V. Parachini; Michael L. Moodie of the Chemical and Biological Arms Control Institute; Ruth A. David and Randall J. Larsen of the ANSER Institute for Homeland Security; Donald A. Henderson, Tara O'Toole, and Thomas V. Inglesby of the Johns Hopkins Center for Civilian Bio-Defense Studies; Amy E. Smithson of the Henry L. Stimson Center; Joseph S. Nye, Graham T. Allison, Ashton B. Carter, Richard A. Falkenrath, and Paul M. Doty, of the John F. Kennedy School of Government; and John D. Steinbruner, director of the Center for International and Security Studies of Maryland University.

In New York City, we would like to thank Marcelle Layton and Tracey S. McNamara. In Washington, thanks for friendship and support of this book go to Marilyn Melkonian and Frances Cook.

We benefited enormously from authors who went before us, most especially Tom Mangold and Jeff Goldberg, *Plague Wars;* Seymour M. Hersh, *Against All Enemies;* Charles Piller and Keith R. Yamamoto, *Gene Wars;* and Ed Regis, *The Biology of Doom.*

This book grew out of a series of investigative reports for the *New York Times* that began in early 1998. We thank the editors and reporters who aided and encouraged that endeavor, including Joseph Lelyveld, Bill Keller, Andrew Rosenthal, Susan Chira, Michael Oreskes, Lawrence K. Altman, John Broder, and Sheryl WuDunn. We are also grateful to *Times* colleagues who gave us continuing support over the years, especially Raymond Bonner, Diane Ceribelli, Cornelia Dean, Michael R. Gordon, Gina Kolata, Jane Perlez, Eric Schmitt, and Nicholas Wade.

Many experts, some who wish to remain anonymous, have been kind enough to read parts of the manuscript and help us root out errors. We thank Rocky Collins, Edward Engelberg, Helen Epstein, Paul S. Fishleder, Robert J. Kanasola, Bill Keller, Joe Sexton, and Stuart Wachs. It goes without saying that any mistakes that may remain belong to us alone.

We are indebted to our agents, Andrew Wylie and Peter Matson, as well as the team at Simon & Schuster that remained a source of professional insight and calm amid our periodic storms. Many thanks to our editor, Alice Mayhew, as well as Roger Labrie, Martha Schwartz, Bonnie Thompson, Victoria Meyer, Aileen Boyle, and Leslie E. Jones.

For encouragement and support, we especially want to thank our families—in particular Tanya Mohn, Gabrielle Glaser, and Jason Epstein, as well as Max, Izzy, and Julie Broad, and Ilana and Moriah Engelberg. Their years of germ terror are over.

Finally, we would like to acknowledge one another. The decision of three people to collaborate on a book is surely the triumph of hope over logic. Few human endeavors are more individualistic than writing and we struggled, mightily at times, to mesh our contributions and styles. We considered spelling out who wrote each chapter. But in the end, there is hardly a paragraph or sentence that does not reflect the work and thought of us all. The synergy that developed has produced a book that, we are confident, no one of us could have written alone.

New York City
August 10, 2001

Judith Miller
Stephen Engelberg
William Broad

Index

About the Authors

JUDITH MILLER, a correspondent for *The New York Times* since 1977, has reported from throughout the world and concentrated on the Middle East and the former Soviet republics. Her most recent book is *God Has Ninety-nine Names.* STEPHEN ENGELBERG has reported on national security for over a decade and is now investigations editor for the *Times.* WILLIAM BROAD, a science writer for the *Times* since 1983, has twice shared the Pulitzer Prize. His most recent book is *The Universe Below.* All three live in the New York City area.